T0212308

Trends in Mathematics is a series devoted to the publication of volumes arising from conferences and lecture series focusing on a particular topic from any area of mathematics. Its aim is to make current developments available to the community as rapidly as possible without compromise to quality and to archive these for reference.

Proposals for volumes can be sent to the Mathematics Editor at either

Springer Basel AG
Birkhäuser
P.O. Box 133
CH-4010 Basel
Switzerland

or

Birkhauser Boston
233 Spring Street
New York, NY 10013
USA

Material submitted for publication must be screened and prepared as follows:

All contributions should undergo a reviewing process similar to that carried out by journals and be checked for correct use of language which, as a rule, is English. Articles without proofs, or which do not contain any significantly new results, should be rejected. High quality survey papers, however, are welcome.

We expect the organizers to deliver manuscripts in a form that is essentially ready for direct reproduction. Any version of TeX is acceptable, but the entire collection of files must be in one particular dialect of TeX and unified according to simple instructions available from Birkhäuser.

Furthermore, in order to guarantee the timely appearance of the proceedings it is essential that the final version of the entire material be submitted no later than one year after the conference. The total number of pages should not exceed 350. The first-mentioned author of each article will receive 25 free offprints. To the participants of the congress the book will be offered at a special rate.

Affine Flag Manifolds and Principal Bundles

Alexander Schmitt
Editor

Birkhäuser

Editor:

Alexander Schmitt
Freie Universität Berlin
Institut für Mathematik
Arnimallee 3
14195 Berlin
Germany
e-mail: alexander.schmitt@fu-berlin.de

2000 Mathematics Subject Classification: 14-06

Bibliographic information published by Die Deutsche Bibliothek. Die Deutsche Bibliothek lists
this publication in the Deutsche Nationalbibliografie; detailed bibliographic data is available in
the Internet at http://dnb.ddb.de

Cover illustration: with friendly permission by Freie Universität Berlin, Bernd Wannenmacher

© 2010 Springer Basel AG
Softcover reprint of the hardcover 1st edition 2010
P.O. Box 133, CH-4010 Basel, Switzerland
Part of Springer Science+Business Media
Printed on acid-free paper produced from chlorine-free pulp. TCF ∞
Cover Design: Alexander Faust, Basel, Switzerland

ISBN 978-3-0348-0309-0   ISBN 978-3-0346-0288-4 (eBook)

9 8 7 6 5 4 3 2 1   www.birkhauser.ch

Contents

Preface ix

U. Görtz

Affine Springer Fibers and Affine Deligne–Lusztig Varieties 1

  1 Introduction 1
  2 The affine Grassmannian and the affine flag manifold 3
  3 Affine Springer fibers 15
  4 Affine Deligne–Lusztig varieties 25
    References 47

T.L. Gómez

**Quantization of Hitchin's Integrable System and
the Geometric Langlands Conjecture** 51

  1 $\mathcal{D}$-modules on stacks 53
  2 Chiral algebras 55
  3 Geometry of the affine Grassmannian 59
  4 Hecke eigenproperty 62
  5 Opers 69
  6 Constructing $\mathcal{D}$-modules 73
  7 Hitchin integrable system I: definition 76
  8 Localization functor 78
  9 Quantum integrable system h 81
  10 Hitchin integrable system II: $\mathcal{D}$-algebras 83
  11 Quantization condition 85
  12 Proof of the Hecke eigenproperty 86
    References 88

G. Hein

**Faltings' Construction of the Moduli Space of Vector Bundles
on a Smooth Projective Curve** 91

  1 Outline of the construction 91
  2 Background and notation 92

3 A nice over-parameterizing family 98
4 The generalized Θ-divisor 100
5 Raynaud's vanishing result for rank two bundles 104
6 Semistable limits 109
7 Positivity 113
8 The construction 117
9 Prospect to higher dimension 119
   References 121

J. Heinloth

Lectures on the Moduli Stack of Vector Bundles on a Curve 123

   Introduction 123
   Lecture 1: Algebraic stacks 124
   Lecture 2: Geometric properties of algebraic stacks 131
   Lecture 3: Relation with coarse moduli spaces 136
   Lecture 4: Cohomology of Bun_n^d 141
   Lecture 5: The cohomology of the coarse moduli space
              (coprime case) 146
   References 152

N. Hoffmann

On Moduli Stacks of G-bundles over a Curve 155

1 Introduction 155
2 Algebraicity 156
3 Lifting principal bundles 157
4 Smoothness 158
5 Connected components 159
   Reference 163

H. Lange and P.E. Newstead

Clifford Indices for Vector Bundles on Curves 165

1 Introduction 165
2 Definition of γ_n and γ_n' 169
3 Mercat's conjecture 171
4 The invariants d_r 175
5 Rank two 186
6 Ranks three and four 188
7 Rank five 190
8 Plane curves 195
9 Problems 198
   References 200

F. Reede and U. Stuhler

Division Algebras and Unit Groups on Surfaces 203

    Introduction 203

  1 Classical finiteness results: The case of a curve 203

  2 Locally free sheaves of modules over Azumaya algebras:

     The case of a surface 208

  3 Elementary modifications and connectivity 211

    References 217

K.-G. Schlesinger

A Physics Perspective on Geometric Langlands Duality 219

  1 Introduction 219

  2 $N = 4$ supersymmetric gauge theory 220

  3 S-duality 221

  4 Topological twisting 222

  5 Dimensional reduction 223

  6 Wilson operators 224

  7 Mirror symmetry 226

  8 Higher-dimensional operators 228

  9 The six-dimensional view 229

  10 Conclusion 230

    References 231

M. Varagnolo and E. Vasserot

Double Affine Hecke Algebras and Affine Flag Manifolds, I 233

    Introduction 233

  1 Schemes and ind-schemes 235

  2 Affine flag manifolds 256

  3 Classification of the simple admissible modules

     of the double affine Hecke algebra 279

    References 287

Affine Flag Manifolds and Principal Bundles

Trends in Mathematics, ix–xii

© 2010 Springer Basel AG

Preface

The present volume contains material from the following events:

- German–Spanish Workshop on Moduli Spaces of Vector Bundles[1]
- Workshop on Affine Flag Manifolds and Principal Bundles[2]
- Vector Bundles on Algebraic Curves: Derived Categories and the Langlands Programme.[3]

These events were devoted to the theory of vector and principal bundles, especially on curves, and their applications in various fields such as representation theory and the Langlands program.

At the start of the theory of vector and principal bundles on a compact Riemann surface X, the connection to representations of the fundamental group $\pi_1(X)$ played an important role. André Weil characterized those vector bundles on X which arise from representations of $\pi_1(X)$ in a general linear group. The fundamental theorem of Narasimhan and Seshadri gives a homeomorphism between the moduli space of representations of $\pi_1(X)$ with values in the unitary group $U_n(\mathbb{C})$ and the moduli space of semistable vector bundles of rank n and degree zero. The representation space of $\pi_1(X)$ is easily seen to be a real analytic variety whereas the moduli space $\mathcal{M}$ of semistable vector bundles can be constructed as a projective variety. Now, $\mathcal{M}$ is an interesting projective variety and has been the subject of an intensive program of research. Since vector bundles are basically linear objects, one can easily write down large families of them, e.g., using extension techniques. In this way, one can show that $\mathcal{M}$ is irreducible and that the subvariety $\mathcal{M}_{\mathcal{L}}$, parameterizing vector bundles with fixed determinant $\mathcal{L}$, is unirational.

---

[1] Essen, February 12–16, 2007, and Madrid, February 19–23, 2007, sponsored by DAAD and Ministerio de Educacíon y Ciencia de España ("Acciones Integradas Hispano-Alemanas" – Ref. HA 2004-0083), and DFG (Schwerpunkt "Global Methods in Complex Geometry");

www.uni-due.de/~mat907/workshop.html;

www.mat.csic.es/webpages/conf/german-spanish-mod07.

[2] Berlin, September 8–12, 2008, sponsored by DFG (SFB/TR 45 "Periods, Moduli Spaces and Arithmetic of Algebraic Varieties");

userpage.fu-berlin.de/~aschmitt/workshopFU.html.

[3] Berlin, June 15–19, 2009, sponsored by DFG (SFB 647 "Space-Time-Matter");

userpage.fu-berlin.de/~aschmitt/VBAC2009.html.

For principal bundles, such families are harder to construct. Here, the affine Graßmannian which classifies principal G-bundles on X with a trivialization outside a point[4] is a valuable parameter space. It is an ind-scheme, i.e., a(n) (infinite-dimensional) limit of schemes, projective in this case. The affine Graßmannian was applied in the work of Kumar, Narasimhan and Ramanathan to prove that the moduli space of semistable principal G-bundles[5] on X is unirational and to determine its Picard group.

Last but not least, one may forget about the semistability condition for principal G-bundles and look at the parameter space of all principal G-bundles. This is an algebraic stack. Although it is neither of finite type nor separated, it is finite-dimensional and smooth and, therefore, still a nice object. Assuming G to be semisimple again, the stack of principal G-bundles admits a "uniformization", i.e., it can be written as a double quotient of the loop group of G. This description reflects the fact that a principal G-bundle can be obtained by gluing trivial bundles on $X \setminus \{pt\}$ and a formal disc around pt. It is then also obvious that the stack is a quotient of the affine Graßmannian. Now, the theory of stacks and ind-schemes provides an appropriate technical framework to explore fascinating relations between the theory of principal bundles and other areas. A famous example is the proof of the Verlinde formula by Beauville/Laszlo/Sorger, Faltings and Kumar/Narasimhan/Ramanathan. This formula is a formula of mathematical physics. Its relationship to principal bundles on curves comes from the interpretation of the "spaces of conformal blocks" of quantum field theory as spaces of sections of line bundles on the moduli spaces.

In the work of Drinfeld and Lafforgue on the Langlands program for GL_n over function fields, the moduli stack of shtukas plays a leading part. (This time it is a Deligne–Mumford stack, locally of finite type.) Shtukas with structure group G are principal G-bundles with certain extra structures.[6] Beilinson and Drinfeld proposed the geometric Langlands program over $\mathbb{C}$ as an analog to the arithmetic one. It is entirely formulated in terms of moduli stacks of principal bundles and principal Higgs bundles. Kapustin and Witten suggest a string theoretic reading of the expected duality.

Speaking of the Langlands program, we note that its "fundamental lemma" relates orbital integrals of different algebraic groups. These orbitals can be expressed in terms of affine Springer varieties. Likewise twisted orbital integrals are related to affine Deligne–Lusztig varieties.

Finally, there is another interesting link to representation theory: using affine Graßmannians one can construct the so-called Cherednik or double affine Hecke algebras. These were introduced some 15 years ago by Cherednik in order to prove certain combinatorial identities conjectured by Macdonald. In the meantime, these algebras have become an important topic in geometric representation theory. They

---

[4]The structure group G is assumed to be semisimple, here.

[5]In that work, G is simple and simply connected.

[6]A detailed account for an arbitrary reductive structure group is given in the work by Varshavsky.

relate to fields such as algebraic geometry, combinatorics, homological algebra, integrable systems and Lie theory.

It was the aim of the above meetings and is the aim of this volume to understand various topics alluded to in the above summary in greater depth with the help of expositions of renowned experts. The reader will find in this volume the latest developments in the theory of vector bundles on curves and surfaces (Hein, Lange/Newstead, Reede/Stuhler), the basic theory of moduli stacks of bundles on curves (Heinloth, Hoffmann), the theory of affine flag manifolds and related objects in algebraic geometry and representation theory (Görtz, Varagnolo/Vasserot), and an exposition of the geometric Langlands program, including its string theoretic formulation (Gómez, Schlesinger).

The contributions of Gómez, Hein and Heinloth are based on their lectures at the German–Spanish Workshop on Moduli Spaces of Vector Bundles. The article of Gómez gives an introduction to the geometric Langlands program as proposed by Beilinson and Drinfeld. It is based on lectures of Gaitsgory at the Tata Institute of Fundamental Research (Mumbai, India). Gómez discusses the technical framework ($\mathcal{D}$-modules on stacks, chiral algebras, Hecke eigensheaves, opers) and exposes the work of Beilinson and Drinfeld on the Langlands correspondence for opers.

Hein presents the alternative construction of Faltings of the moduli spaces of semistable vector bundles on curves. By focusing on the case of rank two vector bundles with canonical determinant, he makes the central ideas especially transparent. The article introduces the reader to recent work on ϑ-functions on moduli spaces and also gives some historical background.

In Heinloth's paper, the reader will learn about the theory of moduli stacks of vector bundles on curves and its relationship to the moduli spaces of semistable vector bundles. Heinloth describes the cohomology ring of the moduli stack and explains how one can use the result to compute the cohomology of the moduli spaces. He also shows how one can settle the existence or non-existence of Poincaré families on moduli spaces by an elegant argument on moduli stacks.

The contributions of Görtz, Hoffmann, Lange/Newstead, Reede/Stuhler and Varagnolo/Vasserot correspond to expositions delivered at the Workshop on Affine Flag Manifolds and Principal Bundles in Berlin. Görtz presents the theory of affine Graßmannians and flag manifolds and emphasizes their interpretation as parameter spaces of lattices and lattice chains respectively. After reviewing the case of "classical" Springer fibers and Deligne–Lusztig varieties, he introduces the corresponding subvarieties of affine flag manifolds. He gives a detailed account of known results on their basic geometric and cohomological properties (many of them due to him and his collaborators) and briefly discusses their role in the understanding of (twisted) orbital integrals and the proof of the "fundamental lemma". The extensive bibliography will help the reader to pursue the subject further.

Hoffmann gives a proof for the fact that the connected components of the moduli stack of principal bundles with reductive linear algebraic structure group G are in bijection to the elements of the fundamental group of G. Though this is well known, there seems to be no published reference for the result in its full generality.

The article by Lange and Newstead is a research paper on Clifford indices for vector bundles. The Clifford indices are invariants of a (smooth, projective) curve defined in terms of semistable vector bundles of a given rank. The rank 1 case is classical, but for higher rank these invariants are new. The authors relate the Clifford indices to the gonality sequence and therefore to the geometry of the underlying curve. They also discuss the link to Brill–Noether theory as well as open problems.

Reede and Stuhler study division algebras over function fields of curves and surfaces. In the case of curves, they reprove some classical finiteness results for the unit groups of such algebras, using their actions on an appropriate Bruhat–Tits building and the Riemann–Roch theorem. They go on to explore this approach for algebraic surfaces.

The paper by Varagnolo and Vasserot is the first of a series of articles devoted to the construction of the double affine Hecke algebra via affine flag manifolds. The aim of this series is to make the foundations of the theory accessible to a wider audience. The paper in this volume reviews sheaves of modules on non-noetherian schemes and ind-schemes, discusses affine flag manifolds and an affine version of the Steinberg variety, and defines the (equivariant) K-theory of the latter. The authors introduce a new *concentration map* which, among other things, leads to a simplified proof of the classification of simple modules in the category $\mathcal{O}$ of the double affine Hecke algebra. This is a major result due to Vasserot.

Schlesinger's article is an expanded version of his talk at the conference Vector Bundles on Algebraic Curves: Derived Categories and the Langlands Programme in Berlin. It explains the physics approach to geometric Langlands duality, due to Kapustin and Witten. Schlesinger emphasizes the role of gauge and string theory in the picture. The paper is a valuable companion to the notes by Gómez.

The editor would like to thank all speakers and participants of the aforementioned events for their input which helped make them stimulating and successful meetings. Special thanks go to the contributors of this volume for their great effort in putting their oral expositions into written form so that they can serve a wider audience as introductions to fundamental areas of current research in algebraic and arithmetic geometry and representation theory. Many thanks also to the people who served as referees and made numerous valuable comments. Last but not least, I am grateful to Thomas Hempfling of Birkhäuser for professionally handling the whole project.

Berlin, November 2009   Alexander Schmitt

Affine Flag Manifolds and Principal Bundles
Trends in Mathematics, 1–50
© 2010 Springer Basel AG

Affine Springer Fibers and Affine Deligne–Lusztig Varieties

Ulrich Görtz

Abstract. We give a survey on the notion of affine Grassmannian, on affine Springer fibers and the purity conjecture of Goresky, Kottwitz, and MacPherson, and on affine Deligne–Lusztig varieties and results about their dimensions in the hyperspecial and Iwahori cases.

Mathematics Subject Classification (2000). 22E67, 20G25, 14G35.

Keywords. Affine Grassmannian, affine Springer fibers, affine Deligne–Lusztig varieties.

1. Introduction

These notes are based on the lectures I gave at the Workshop on Affine Flag Manifolds and Principal Bundles which took place in Berlin in September 2008. There are three chapters, corresponding to the main topics of the course. The first one is the construction of the affine Grassmannian and the affine flag variety, which are the ambient spaces of the varieties considered afterwards. In the following chapter we look at affine Springer fibers. They were first investigated in 1988 by Kazhdan and Lusztig [41], and played a prominent role in the recent work about the "fundamental lemma", culminating in the proof of the latter by Ngô. See Section 3.8. Finally, we study affine Deligne–Lusztig varieties, a "σ-linear variant" of affine Springer fibers over fields of positive characteristic, σ denoting the Frobenius automorphism. The term "affine Deligne–Lusztig variety" was coined by Rapoport who first considered the variety structure on these sets. The sets themselves appear implicitly already much earlier in the study of twisted orbital integrals.

We remark that the term "affine" in both cases is not related to the varieties in question being affine, but rather refers to the fact that these are notions defined in the context of an affine root system. We include short reminders about the corresponding non-affine notions, i.e., Springer fibers and Deligne–Lusztig varieties.

No originality for any of the results in this article is claimed; this is especially true for Chapter 3 where I am really only reporting about the work of others: Goresky, Kottwitz, Laumon, MacPherson, Ngô, ...

1.1. Notation

We collect some standard, mostly group-theoretic, notation which is used throughout this article. Let k be a field (in large parts of Section 2 we can even work over an arbitrary commutative ring). In Section 3, k is assumed to be algebraically closed. In Section 4, k will be an algebraic closure of a finite field.

Let $\mathcal{O} = k[\![\epsilon]\!]$ be the ring of formal power series over k, and let $L = k(\!(\epsilon)\!) := k[\![\epsilon]\!][\frac{1}{\epsilon}]$ be the field of Laurent series over k. In Section 4, we let $F = \mathbb{F}_q(\!(\epsilon)\!)$.

Let G be a connected reductive group over k, or – in Section 4 – over a finite field $\mathbb{F}_q$. We will assume that G is split, i.e., that there exists a maximal torus $A \subseteq G$ which is isomorphic to a product of copies of the multiplicative group $\mathbb{G}_m$. We fix such a split torus A. We also fix a Borel subgroup B of G which contains A. We denote by W the Weyl group of A, i.e., the quotient $N_G A / A$ of the normalizer of A by A, and by $X_*(A) = \mathrm{Hom}(\mathbb{G}_m, A)$ the cocharacter lattice of A.

For notational convenience, we assume that the Dynkin diagram of G is connected. We denote by Φ the set of roots given by the choice of A, and by Φ^+ the set of positive roots distinguished by B. We let

$$X_*(A)_+ = \left\{ \lambda \in X_*(A); \ \langle \lambda, \alpha \rangle \geq 0 \text{ for all } \alpha \in \Phi^+ \right\}$$

denote the set of dominant cocharacters. An index $-_{\mathbb{Q}}$ denotes tensoring by $\mathbb{Q}$, e.g., we have $X_*(A)_{\mathbb{Q}}$, $X_*(A)_{\mathbb{Q},+}$. Similarly, we have $X_*(A)_{\mathbb{R}}$, etc. Let ρ denote half the sum of the positive roots. For $\lambda, \mu \in X_*(A)$, we write $\lambda \leq \mu$ if $\mu - \lambda$ is a linear combination of simple coroots with non-negative coefficients.

Now we come to the "affine" situation. We embed $X_*(A)$ into $A(L) \subset G(L)$ by $\lambda \mapsto \lambda(\epsilon) =: \epsilon^\lambda$, where by $\lambda(\epsilon)$ we denote the image of ϵ under the map

$$L^\times = \mathbb{G}_m(L) \to A(L) \subset G(L)$$

induced by λ. The extended affine Weyl group (or Iwahori–Weyl group) $\widetilde{W}$ is defined as the quotient $N_{G(L)}T(L)/T(\mathcal{O})$. It can also be identified with the semi-direct product $W \ltimes X_*(A)$. On $\widetilde{W}$, we have a length function $\ell \colon \widetilde{W} \to \mathbb{Z}_{\geq 0}$,

$$\ell(w\epsilon^\lambda) = \sum_{\substack{\alpha>0 \\ w(\alpha)<0}} |\langle \alpha, \lambda \rangle + 1| + \sum_{\substack{\alpha>0 \\ w(\alpha)>0}} |\langle \alpha, \lambda \rangle|, \qquad w \in W, \quad \lambda \in X_*(A).$$

Let $S \subset W$ denote the subset of simple reflections, and let $s_0 = \epsilon^{\tilde{\alpha}^\vee} s_{\tilde{\alpha}}$, where $\tilde{\alpha}$ is the unique highest root. The affine Weyl group W_a is the subgroup of $\widetilde{W}$ generated by $S \cup \{s_0\}$. Then $(W_a, S \cup \{s_0\})$ is a Coxeter system, and the restriction of the length function is the length function on W_a given by the fixed system of generators.

We can write the extended affine Weyl group as a semi-direct product $\widetilde{W} \cong W_a \rtimes \Omega$, where $\Omega \subset \widetilde{W}$ is the subgroup of length 0 elements. We extend the Bruhat order on W_a to a partial order on $\widetilde{W}$, again called the Bruhat order, by setting $w\tau \leq w'\tau'$ if and only if $\tau = \tau'$, $w \leq w'$, for $w, w' \in W_a$, $\tau, \tau' \in \Omega$.

The subgroup $G(\mathcal{O})$ is a "hyperspecial" subgroup of $G(L)$, so we sometimes refer to a case relating to $G(\mathcal{O})$ as the hyperspecial case. We also consider the Iwahori subgroup $I \subset G(\mathcal{O})$ which we define as the inverse image of the opposite Borel $B^-(k)$ under the projection $G(\mathcal{O}) \to G(k)$, $\epsilon \mapsto 0$.

In the case $G = \mathrm{GL}_n$, we always let A be the torus of diagonal matrices, and we choose the subgroup of upper triangular matrices B as Borel subgroup. We can then identify the Weyl group W with the subgroup of permutation matrices in GL_n, and we can identify the extended affine Weyl group $\widetilde{W}$ with the subgroup of matrices with exactly one non-zero entry in each row and column, which is of the form ϵ^i, $i \in \mathbb{Z}$. The subgroup $\Omega \subset \widetilde{W}$ is isomorphic to $\mathbb{Z}$. For $G = \mathrm{SL}_n$, we make analogous choices of the maximal torus and the Borel subgroup.

2. The affine Grassmannian and the affine flag manifold

We start by an introduction to the affine Grassmannian and affine flag variety of the group G. Both affine Springer fibers and affine Deligne–Lusztig varieties live inside one of these. As general references for the construction we name the papers by Beauville and Laszlo [4], Pappas and Rapoport [63] and Sorger [75]. Let k be a field (at least in 2.1–2.4 we could work over any base ring, though).

2.1. Ind-schemes

We first recall the notion of ind-scheme. Roughly speaking, an ind-scheme is just the union of an "ascending" system of schemes. To say precisely in which sense the union is taken, it is most appropriate to use the functorial point of view on schemes.

Definition 2.1. A *k-space* is a functor $F\colon (\mathrm{Sch})^o \to (\mathrm{Sets})$ which is a sheaf for the fpqc-topology, i.e., whenever $X = \bigcup_i U_i$ is a covering by (Zariski-) open subsets, then the sequence
$$F(X) \to \prod_i F(U_i) \rightrightarrows \prod_{i,j} F(U_i \cap U_j)$$
is exact, and whenever $R \to R'$ is a faithfully flat homomorphism of k-algebras, then the sequence
$$F(\mathrm{Spec}\, R) \to F(\mathrm{Spec}\, R') \rightrightarrows F(\mathrm{Spec}\, R' \times_{\mathrm{Spec}\, R} \mathrm{Spec}\, R')$$
is exact. A morphism of k-spaces is a morphism of functors.

Here by exactness we mean that the map on the left-hand side is injective, and that its image is equal to the subset of elements which have the same image under both maps on the right-hand side. The first axiom is called the Zariski-sheaf axiom, for obvious reasons. The second condition is called the fpqc sheaf axiom, where fpqc stands for fidèlement plat quasi-compact.

Remark 2.2.

1. Because of the Zariski-sheaf axiom, one can equivalently consider (contravariant) functors from the category of *affine* schemes to the category of sets which satisfy the fpqc condition, or similarly (covariant) functors G from the category of k-algebras to the category of sets for which $G(R) \to G(R') \rightrightarrows G(R' \otimes_R R')$ is exact for every faithfully flat homomorphism $R \to R'$.

2. Every k-scheme Z gives rise to a k-space, also denoted by Z, by setting $Z(S) = \operatorname{Hom}_k(S, Z)$, the S-valued points of Z. The sheaf axiom for Zariski coverings is clearly satisfied, and Grothendieck's theory of faithfully flat descent shows that the fpqc sheaf axiom is satisfied as well.

3. The Yoneda lemma says that the functor from the category of schemes to the category of k-spaces is fully faithful. Those k-spaces which are in the essential image of this functor are called representable.

There is a standard method of "sheafification" in this context (see Artin's notes [2]). Therefore, many constructions available for usual sheaves can be carried out for k-spaces as well. For instance, one shows similarly as for usual sheaves that inductive limits exist in the category of k-spaces, and that quotients of sheaves of abelian groups exist.

Definition 2.3. An ind-*scheme* is a k-space which is the inductive limit (in the category of k-spaces) of an inductive system of schemes, where the index set is the set $\mathbb{N}$ of natural numbers with its usual order, and where all the transition maps are closed immersions.

Of course, one could also allow more general index sets. Often one does not make the restriction that all transition maps must be closed immersions, and speaks of a *strict* ind-*scheme* if this is the case. Since in the sequel this additional condition will always be satisfied, we omit the *strict* from the notation. See Drinfeld's paper [20] for generalities on ind-schemes and remarks about the relation to the notion of formal scheme.

We call an ind-scheme X of ind-*finite type*, or ind-*projective*, etc., if we can write $X = \varinjlim X_n$ where each X_n is of finite type, projective, etc. We call X *reduced* if it can be written as the inductive limit of a system $(X_n)_n$ where each X_n is a reduced scheme. This is a somewhat subtle notion because usually there will be many ways to write a reduced ind-scheme as a limit of non-reduced schemes.

Lemma 2.4. *Let $X = \varinjlim_n X_n$ be an ind-scheme. Let $S \to X$ be a morphism from a quasi-compact scheme S to X. Then there exists n such that the morphism $S \to X$ factors through X_n.*

Note that this does not follow from the universal property of the inductive limit. Rather, the reason is that since S is quasi-compact, the S-valued points of the sheafification of the inductive limit $\varinjlim X_n$ are just the S-valued points of the presheaf inductive limit, i.e., $X(S) = \varinjlim X_n(S)$ for quasi-compact S.

We will see many examples of ind-schemes (which are not schemes) below. A simple example which is quite helpful is the following: Suppose we want to view the power series ring $k[\![\epsilon]\!]$ over the field k as a k-scheme. This is easy because a power series is just given by its coefficients, so $k[\![\epsilon]\!]$ is the set of k-valued points of the countably infinite product $\mathbb{A}^\infty = \prod_{i \geq 0} \mathbb{A}^1$, a perfectly reasonable scheme, even if it is infinite-dimensional. In fact, this scheme is just the spectrum of the polynomial ring in countably many variables. Similarly, we can express the set of doubly infinite power series $\sum_{i=-\infty}^\infty a_i \epsilon^i$ as an infinite product of affine lines. Now suppose we want to express the field $k(\!(\epsilon)\!)$ of Laurent series $\sum_i a_i \epsilon^i$ with $a_i = 0$ for all but finitely many $i < 0$ in a similar way. Clearly, it is contained in the product $\prod_{i=-\infty}^\infty \mathbb{A}^1$. But the condition that only finitely many coefficients with negative index may be non-zero cannot be expressed by polynomial equations! Therefore we cannot express $k(\!(\epsilon)\!)$ as a closed subscheme of the "doubly infinite" product. On the other hand, writing $k(\!(\epsilon)\!)$ as the union

$$k(\!(\epsilon)\!) = \bigcup_{i \leq 0} \epsilon^i k[\![\epsilon]\!],$$

we find an obvious ind-scheme structure on $k(\!(\epsilon)\!)$.

2.2. The loop group

We now fix a reductive linear algebraic group G over k (for most of this section, it is not important whether G is reductive). The loop group LG of G is the k-space given by the following functor:

$$LG(R) = G(R(\!(\epsilon)\!)), \quad R \text{ a } k\text{-algebra.}$$

The terminology "loop group" refers to the fact that this construction is similar to the construction of the loop group in topology. There one considers the space of continuous maps from the circle S^1 to the given topological group. In the algebraic context, the circle is replaced by an infinitesimal pointed disc, i.e., the spectrum of $k(\!(\epsilon)\!)$. Similarly, we have the *positive loop group* $L^+ G$, defined by

$$L^+ G(R) = G(R[\![\epsilon]\!]), \quad R \text{ a } k\text{-algebra.}$$

The positive loop group is actually a(n) (infinite-dimensional) scheme. Let us first check this for GL_n. The idea is to view $k[\![\epsilon]\!] = \mathbb{A}^1_k(k[\![\epsilon]\!])$ as an infinite product of affine lines over k, as explained above. Via the closed embedding

$$\mathrm{GL}_n \to \mathrm{Mat}_{n \times n} \times \mathrm{Mat}_{n \times n}, \quad A \mapsto (A, A^{-1})$$

we identify $\mathrm{GL}_n(R[\![\epsilon]\!])$ with the set of matrices

$$\{ (A, B) \in \mathrm{Mat}_{n \times n}(R[\![\epsilon]\!]) \times \mathrm{Mat}_{n \times n}(R[\![\epsilon]\!]), \quad AB = 1 \}.$$

Therefore $L^+ G$ is the closed subscheme in $\prod_{i \geq 0}(\mathbb{A}^{n^2} \times \mathbb{A}^{n^2})$ given by the equations obtained from splitting the matrix equality $AB = 1$ into equations for each ϵ-component. Given an arbitrary linear group G, we can embed G as a closed subgroup into some GL_n, and we see that $L^+ G$ is a closed subscheme of $L^+ \mathrm{GL}_n$.

Definition 2.5. The *affine Grassmannian* $\mathcal{G}rass_G$ for G is the quotient k-space LG/L^+G.

The quotient in the category of k-spaces is the sheafification of the presheaf quotient $R \mapsto LG(R)/L^+G(R)$.

Similarly, if we choose a Borel subgroup $B^- \subseteq G$, then we have an Iwahori subgroup $\mathbf{I} \subseteq L^+G$, which by definition is the inverse image of B^- under the projection $L^+G \to G$ (which maps ϵ to 0). Instead of the quotient of LG by the positive loop group L^+G, we can consider the quotient by $\mathbf{I}$:

Definition 2.6. The *affine flag variety* $\mathcal{F}lag_G$ for G is the quotient k-space $LG/\mathbf{I}$.

More generally, by taking quotients by "parahoric subgroups", we can define "partial affine flag varieties". We will show below that the affine Grassmannian and the affine flag variety are ind-schemes over k. In the case of GL_n (and with some more effort, of any classical group) these ind-schemes can be interpreted as parameter spaces of lattices or lattice chains (satisfying certain conditions).

Remark 2.7. The loop and positive loop constructions can be applied to any scheme over $k((\epsilon))$ and $k[\![\epsilon]\!]$, respectively. In particular, one can construct the loop group for a group G over $k((\epsilon))$ which does not come from k by base change. One obtains "twisted" loop groups, and their affine Grassmannians. This is worked out in the paper [63] by Pappas and Rapoport. The basic construction is the same; note however that the notion of parahoric subgroup is more subtle in general than in our case, see [34]. Certain properties from the non-twisted case are shown to carry over to the twisted case in loc. cit., but there are still many open questions.

2.3. Lattices

Let k be a field, and let R be a k-algebra. We denote by $R[\![\epsilon]\!]$ the ring of formal power series over R, and by $R((\epsilon))$ the ring of Laurent series over R, i.e., the localization of $R[\![\epsilon]\!]$ with respect to ϵ. Let r, n be positive integers. The $R[\![\epsilon]\!]$-submodule $R[\![\epsilon]\!]^n \subset R((\epsilon))^n$ is called the *standard lattice*, and is denoted by $\Lambda_R = \Lambda_{0,R}$.

Definition 2.8.

1. A *lattice* $\mathcal{L} \subset R((\epsilon))^n$ is an $R[\![\epsilon]\!]$-submodule such that
   (a) There exists $N \in \mathbb{Z}_{\geq 0}$ with
   $$\epsilon^N \Lambda_R \subseteq \mathcal{L} \subseteq \epsilon^{-N} \Lambda_R, \text{ and}$$
   (b) the quotient $\epsilon^{-N}\Lambda_R/\mathcal{L}$ is locally free of finite rank over R.
2. A lattice $\mathcal{L}$ is called *r-special*, if $\bigwedge^n \mathcal{L} = \epsilon^r \Lambda_R$.

We denote the set of all lattices in $R((\epsilon))^n$ by $\mathcal{L}att_n(R)$, and the set of all 0-special lattices by $\mathcal{L}att_n^0(R)$. Our goal is to equip these sets with an "algebraic-geometric structure". We will see below that $\mathcal{L}att_n$ and $\mathcal{L}att_n^0$ are ind-schemes and can be identified with the affine Grassmannian of GL_n and SL_n respectively.

We also define, for $N \geq 1$, subsets

$$Latt_n^{(N)}(R) \subset Latt_n(R), \qquad Latt_n^{0,(N)}(R) \subset Latt_n^0(R),$$

where the number N in part (a) of the definition of a lattice is fixed. The functors $Latt_n^{(N)}(R), Latt_n^{0,(N)}$ are projective schemes over k. Let us show this for $Latt_n^{0,(N)}$. (Then $Latt_n^{(N)}(R)$ can be obtained as a disjoint union of schemes of the same form.) The morphism of functors

$$Latt_n^{0,(N)}(R) \to \mathrm{Grass}_N(\epsilon^{-N} \Lambda_k / \epsilon^N \Lambda_k)(R), \qquad \mathscr{L} \mapsto \mathscr{L}/\epsilon^N \Lambda_R,$$

defines a closed embedding of $Latt_n^{0,(N)}$ into the Grassmann variety of N-dimensional subspaces of the ($2N$-dimensional) k-vector space $t^{-N} k[\![\epsilon]\!]^n / t^N k[\![\epsilon]\!]^n$. The image is the closed subscheme of all subspaces that are stable under the nilpotent endomorphism induced by ϵ. We have

$$Latt_n(R) = \bigcup_N Latt_n^{(N)}(R), \qquad Latt_n^0(R) = \bigcup_N Latt_n^{0,(N)}(R).$$

We obtain ind-schemes

$$Latt_n = \bigcup Latt_n^{(N)}, \qquad Latt_n^0 = \bigcup Latt_n^{0,(N)}.$$

Example 2.9 (The ind-scheme $Latt_1$ over k). The k-valued points of $Latt_1$ are the finitely generated $k[\![\epsilon]\!]$-submodules of $k(\!(\epsilon)\!)$, i.e., the fractional ideals (ϵ^i), $i \in \mathbb{Z}$. So topologically, $Latt_1$ is simply the disjoint union of countably many points.

Let us determine the ind-scheme structure. An R-valued point $\mathscr{L} \in Latt_1(R)$, where R is an arbitrary ring, is an $R[\![\epsilon]\!]$-submodule $\mathscr{L} \subseteq R(\!(\epsilon)\!)$. The conditions that $\mathscr{L}$ is a lattice are equivalent to saying that $\mathscr{L}$ is generated, over $R[\![\epsilon]\!]$, by an element of $R(\!(\epsilon)\!)^\times$ (cf. Lemma 2.11). This unit of $R(\!(\epsilon)\!)$ is determined by $\mathscr{L}$ up to multiplication by units of $R[\![\epsilon]\!]$.

We have, assuming that $\mathrm{Spec}\, R$ is connected,

$$R(\!(\epsilon)\!)^\times = \left\{ \sum_i a_i \epsilon^i \in R(\!(\epsilon)\!); \, \exists i_0 : a_{i_0} \in R^\times, a_j \text{ nilpotent for all } j < i_0 \right\},$$

so

$$R(\!(\epsilon)\!)^\times / R[\![\epsilon]\!]^\times = \coprod_{i_0} \bigcup_{N \geq 1} \{ (a_1, \ldots, a_N); \, a_i \in R \text{ nilpotent} \}.$$

Here for fixed i_0, the sets in the union over N are embedded into each other in the obvious way (i.e., by extending tuples of smaller length by zeros).

This description shows that as an ind-scheme, $Latt_1$ is highly non-reduced. However, this phenomenon of non-reducedness will be of no importance for us. Note that nevertheless, $Latt_1$ is formally smooth, i.e., it satisfies the infinitesimal lifting criterion for smoothness.

2.4. The affine Grassmannian for GL_n

We prove that the functor $\mathcal{G}rass$ (which we defined above as the quotient of the loop group by the positive loop group) is representable by an ind-scheme in two steps: In this section, we deal with the case of $G = \mathrm{GL}_n$, and in the next section we consider the general case. In the case of the general linear group, we can describe $\mathcal{G}rass$ quite explicitly in terms of lattices. Every element $g \in LG(R)$ gives rise to a lattice $g\Lambda_R$. We obtain:

Proposition 2.10. *The affine Grassmannian for* GL_n *is isomorphic, as a k-space, to $\mathcal{L}att_n$. The affine Grassmannian for* SL_n *is isomorphic to $\mathcal{L}att_n^0$.*

To prove the proposition, one has to show that, fpqc-locally on R, every lattice is free over $R[\![\epsilon]\!]$: choosing a basis for a *free* lattice gives a representation by a matrix in $\mathrm{GL}_n(R(\!(\epsilon)\!))$ which is well defined up to an element of $\mathrm{GL}_n(R[\![\epsilon]\!])$. It is enough to achieve this after a quasi-compact faithfully flat base change $R \to R'$, because the quotient LG/L^+G is the *sheafification* of the presheaf quotient. Therefore the proposition follows from the equivalence of 1. and 4. of the following lemma.

Lemma 2.11. *Let $\mathcal{L} \subset R(\!(\epsilon)\!)^n$ be an $R[\![\epsilon]\!]$-submodule. The following are equivalent:*

1. *The submodule $\mathcal{L}$ is a lattice.*
2. *The submodule $\mathcal{L}$ is a projective $R[\![\epsilon]\!]$-module and $\mathcal{L} \otimes_{R[\![\epsilon]\!]} R(\!(\epsilon)\!) = R(\!(\epsilon)\!)^n$.*
3. *Zariski-locally on R, $\mathcal{L}$ is a free $R[\![\epsilon]\!]$-module of rank n (i.e., there exist $f_1, \ldots, f_r \in R$ such that $(f_1, \ldots, f_r) = (1)$ and for all i, $\mathcal{L} \otimes_{R[\![\epsilon]\!]} R_{f_i}[\![\epsilon]\!]$ is a free $R_{f_i}[\![\epsilon]\!]$-module of rank n) and $\mathcal{L} \otimes_{R[\![\epsilon]\!]} R(\!(\epsilon)\!) = R(\!(\epsilon)\!)^n$.*
4. *Fpqc-locally on R, $\mathcal{L}$ is a free $R[\![\epsilon]\!]$-module of rank n (i.e., there exists a faithfully flat ring homomorphism $R \to R'$ such that $\mathcal{L} \otimes_{R[\![\epsilon]\!]} R'[\![\epsilon]\!]$ is a free $R'[\![\epsilon]\!]$-module) and $\mathcal{L} \otimes_{R[\![\epsilon]\!]} R(\!(\epsilon)\!) = R(\!(\epsilon)\!)^n$.*

Note that in 4., usually one has $R'[\![\epsilon]\!] \neq R' \otimes_R R[\![\epsilon]\!]$, and similarly in 3.

Proof. 1. $\Rightarrow$ 2. To simplify the notation, we assume that $\epsilon^N \Lambda_R \subseteq \mathcal{L} \subseteq \Lambda_R$. First note that for $s \in R$, we have

$$\Lambda_R \otimes_{R[\![\epsilon]\!]} R[\![\epsilon]\!]_s / \mathcal{L} \otimes_{R[\![\epsilon]\!]} R[\![\epsilon]\!]_s = (\Lambda_R/\mathcal{L}) \otimes_R R_s$$

as R-modules, so although $R_s[\![\epsilon]\!]$ usually differs from $R[\![\epsilon]\!]_s$, this difference does not matter for us. To prove 2., one shows that locally on R, there exists a basis $f_1, \ldots, f_n$ of Λ_R over $R[\![\epsilon]\!]$ and $i_1, \ldots, i_n \in \mathbb{Z}_{\geq 0}$, such that the elements $\epsilon^j f_l$, $l = 1, \ldots, n$, $j = i_l, \ldots, N-1$, form an R-basis of $\mathcal{L}/\epsilon^N \Lambda_R$. This can be achieved by successively choosing suitable bases of the subquotients $\ker \epsilon^i / \ker \epsilon^{i-1}$ (locally on R) that are compatible with $\mathcal{L}$.

2. $\Rightarrow$ 3. Since $\mathcal{L}$ is projective of finite rank over $R[\![\epsilon]\!]$, we find elements $g_1, \ldots, g_r \in R[\![\epsilon]\!]$ which generate the unit ideal and such that for all i, $\mathcal{L} \otimes_{R[\![\epsilon]\!]} R[\![\epsilon]\!]_{g_i}$ is free of rank n over $R[\![\epsilon]\!]_{g_i}$. Then the absolute terms $f_i := g_i(0) \in R$ generate the unit ideal of R, and for all i with $f_i \neq 0$, we have that g_i is a unit in

$R_{f_i}[\![\epsilon]\!]$, so

$$\mathscr{L} \otimes_{R[\![\epsilon]\!]} R_{f_i}[\![\epsilon]\!] = \mathscr{L} \otimes_{R[\![\epsilon]\!]} R[\![\epsilon]\!]_{g_i} \otimes_{R[\![\epsilon]\!]_{g_i}} R_{f_i}[\![\epsilon]\!]$$

is free of rank n over $R_{f_i}[\![\epsilon]\!]$.

3. $\Rightarrow$ 4. Trivial.

4. $\Rightarrow$ 1. Let $R \to R'$ be as in 4., and write $\mathscr{L}' = \mathscr{L} \otimes_{R[\![\epsilon]\!]} R'[\![\epsilon]\!]$. By assumption, $\mathscr{L}'$ is free over $R'[\![\epsilon]\!]$, and in particular, for suitable N we have $t^N R'[\![\epsilon]\!]^n \subseteq \mathscr{L}' \subseteq t^{-N} R'[\![\epsilon]\!]^n$. By intersecting with $R[\![\epsilon]\!]^n$, we obtain the analogous property for $\mathscr{L}$. Furthermore,

$$(\epsilon^{-N} R[\![\epsilon]\!]/\mathscr{L}) \otimes_R R' = \epsilon^{-N} R'[\![\epsilon]\!]/\mathscr{L}'$$

is locally free over R', and since R' is faithfully flat over R, $\epsilon^{-N} R[\![\epsilon]\!]/\mathscr{L}$ is locally free over R. $\qquad\square$

Example 2.9 shows that the affine Grassmannian for the multiplicative group $\mathbb{G}_m = \mathrm{GL}_1$ is not reduced. On the other hand, the affine Grassmannian for SL_n is integral (see [4], and [63], Corollary 5.3 and Theorem 6.1, which includes the case of positive characteristic and also deals with other groups).

Similarly as for the affine Grassmannian, we can now show that the affine flag variety $L\,\mathrm{GL}_n\,/\mathbf{I}$ is an ind-scheme. Here we let A be the diagonal torus in GL_n, we let B be the Borel subgroup of upper triangular matrices, so that B^- is the subgroup of lower triangular matrices. To describe $\mathcal{F}\!lag_{\mathrm{GL}_n}$ in terms of lattices, we use the notion of lattice chain:

Definition 2.12. Let R be a k-algebra. A *(full periodic) lattice chain* inside $R(\!(\epsilon)\!)^n$ is a chain

$$\mathscr{L}_0 \supset \mathscr{L}_1 \supset \cdots \supset \mathscr{L}_{n-1} \supset \epsilon\mathscr{L}_0,$$

such that each $\mathscr{L}_i$ is a lattice in $R(\!(\epsilon)\!)^n$, and such that each quotient $\mathscr{L}_{i+1}/\mathscr{L}_i$ is a locally free R-module of rank 1.

Each element of $LG(R)$ gives rise to a lattice chain inside $R(\!(\epsilon)\!)^n$ by applying it to the standard lattice chain

$$\Lambda_{i,R} = \bigoplus_{j=0}^{n-i-1} R[\![\epsilon]\!]e_{j+1} \oplus \bigoplus_{j=n-i}^{n-1} \epsilon R[\![\epsilon]\!]e_{j+1},$$

where $e_1, \ldots, e_n$ denotes the standard basis of $R(\!(\epsilon)\!)^n$. Because the Iwahori subgroup $\mathbf{I}(R) \subset \mathrm{GL}_n(R[\![\epsilon]\!])$ is precisely the stabilizer of the standard lattice chain $\Lambda_{\bullet,R}$, we get:

Proposition 2.13. *Let $\mathcal{F}\!lag$ be the affine flag variety for GL_n, and let R be a k-algebra. Then $\mathcal{F}\!lag(R)$ is the space of lattice chains in $R(\!(\epsilon)\!)^n$. In particular, $\mathcal{F}\!lag$ is an* ind-*scheme. The affine flag variety for SL_n is the closed sub-ind-scheme of all lattice chains $(\mathscr{L}_\bullet)_\bullet$ such that $\mathscr{L}_0$ is 0-special.*

2.5. The affine Grassmannian for an arbitrary linear algebraic group

To establish the existence of the affine Grassmannian as an ind-scheme in the general case, we will embed a given linear algebraic group into a general linear group in a suitable way. We use the following lemma due to Beilinson and Drinfeld.

Lemma 2.14 ([7, Proof of Theorem 4.5.1]). *Let $G_1 \subset G_2$ be linear algebraic groups over k such that the quotient $U := G_2/G_1$ is quasi-affine. Suppose that the quotient LG_2/L^+G_2 is an ind-scheme of ind-finite type. Then the same holds for LG_1/L^+G_1, and the natural morphism $LG_1/L^+G_1 \to LG_2/L^+G_2$ is a locally closed immersion. If U is affine, then this immersion is a closed immersion.*

See also [63, Theorem 1.4] for a version which includes the twisted case. As an application of the lemma, we obtain:

Proposition 2.15. *Let G be a linear algebraic group over k. Then the quotient LG/L^+G is an ind-scheme over k. If G is reductive, then it is ind-projective.*

Proof. First assume that G is a reductive group, the case which will be relevant for us. Choose an embedding $G \to \mathrm{GL}_n$ of G into some general linear group. Since G is reductive, the quotient GL_n/G is affine. In fact, a quotient of a reductive group by a closed subgroup is affine if (and only if) the subgroup is reductive; see [70]. The lemma above, together with Proposition 2.10, then shows that LG/L^+G is an ind-projective ind-scheme over k.

In the general case, one shows that there exists an embedding $G \to \mathrm{GL}_n \times \mathbb{G}_m$ such that the quotient is quasi-affine. See [7, Theorem 4.5.1] or [63, Proposition 1.3]. □

Remark 2.16.

1. Note that in general, given a closed immersion of algebraic groups $G_1 \to G_2$, the induced morphism of the affine Grassmannians is not a closed immersion. For an example consider the inclusion of a Borel subgroup B into a reductive group G. The morphism $LB/L^+B \to LG/L^+G$ is a bijection on k-valued points, but except for trivial cases is far from being an isomorphism.
2. For a different method of constructing the affine Grassmannian in all relevant cases, see Faltings' paper [21].

Similarly, we obtain that the affine flag variety $\mathcal{F}lag_G = LG/\mathbf{I}$ is an ind-scheme. We can either again use an embedding of G into GL_n or $\mathrm{GL}_n \times \mathbb{G}_m$, or show that it is an ind-scheme by considering the natural projection $\mathcal{F}lag_G \to \mathcal{G}rass_G$ which is a fiber bundle whose fibers are all isomorphic to the usual flag variety of G.

Another interesting and important description of the affine Grassmannian for a semisimple group G can be given in terms of G-bundles on a (smooth, projective) curve. See the surveys by Kumar [49] and Sorger [75] for introductions to this point of view. Here, we just sketch how the relationship is established. The basic idea is to glue vector bundles on the curve C by gluing the trivial bundle on the pointed curve $C \setminus \{\mathrm{pt}\}$ and the trivial bundle on the formal neighborhood $\mathrm{Spec}\, k[\![\epsilon]\!]$ of the point to obtain a vector bundle (or more generally, a G-bundle) on C. Note that this gluing is not an instance of faithfully flat descent, because in general the homomorphism $R \to R[\![\epsilon]\!]$ is not flat. It is flat if R is noetherian, or more generally if R is a coherent ring (see [12, I §2, Exercice 12], [24]). For locally free modules, Beauville and Laszlo [5] have proved directly that descent holds in this situation. To prove that in this way one can construct all bundles, one uses the result of Drinfeld and Simpson who have shown that G-bundles on $C \setminus \{\text{pt}\} \times S$ are trivial fpqc-locally on S. Here the assumption that G is semisimple is obviously crucial (since for instance for $G = \mathbb{G}_m$, where G-bundles are just line bundles, this usually fails).

2.6. Decompositions

Below we need the following decompositions. Write $K = G(\mathcal{O})$, and denote by $X_{*,+}$ the subset of $X_*(A)$ consisting of all dominant coweights.

Theorem 2.17 (Cartan decomposition). *The affine Grassmannian decomposes as a disjoint union*

$$\mathcal{G}rass(k) = \bigcup_{\lambda \in X_{*,+}} K\epsilon^\mu K/K.$$

The closure of each "Schubert cell" $K\epsilon^\mu K/K$ is a union of cells, and the closure relations are given by the partial order on dominant coweights introduced in Section 1.1:

$$\overline{K\epsilon^\mu K/K} = \bigcup_{\lambda \leq \mu} K\epsilon^\lambda K/K.$$

In the case of $G = \mathrm{GL}_n$, the translation element λ such that $g \in K\epsilon^\lambda K$ is simply given by the elementary divisors of the lattice $g\Lambda_k$ with respect to the standard lattice Λ_k.

The corresponding result for the affine flag variety is the following. In fact, in this case the geometric structure of each cell is very simple. We denote by

$$I = \mathbf{I}(k) \subset G(\mathcal{O})$$

the Iwahori subgroup of $G(k(\!(\epsilon)\!))$ given by $\mathbf{I}$.

Theorem 2.18 (Iwahori–Bruhat decomposition). *The affine flag variety decomposes as a disjoint union*

$$\mathcal{F}lag(k) = \bigcup_{x \in \widetilde{W}} IxI/I.$$

The closure of each "Schubert cell" IxI/I is a union of Schubert cells, and the closure relations are given by the Bruhat order:

$$\overline{IxI/I} = \bigcup_{y \leq x} IyI/I.$$

For every $x \in \widetilde{W}$, the cell IxI/I is isomorphic to $\mathbb{A}^{\ell(x)}$.

2.7. Connected components

As we have seen in the example of $G = \mathrm{GL}_n$, the affine Grassmannian (and in fact, the loop group) of G may have several connected components.

Definition 2.19. The *algebraic fundamental group* $\pi_1(G)$ of G is the quotient of the cocharacter lattice $X_*(A)$ by the coroot lattice.

Similarly as for the loop group of a topological space, we have

Proposition 2.20 ([7, Proposition 4.5.4], [63, Theorem 0.1]). *The set of connected components of the loop group can be identified with $\pi_1(G)$.*

We denote the map which maps a point to its connected component by $\kappa = \kappa_G \colon G(L) \to \pi_1(G)$. We can describe the map κ explicitly as follows: given $g \in G(L)$, the Cartan decomposition as stated above says that there exists a unique $\lambda \in X_*(A)_+$ such that $g \in G(\mathcal{O})\epsilon^\lambda G(\mathcal{O})$. Then $\kappa(g)$ is the image of λ under the natural projection $X_*(A) \to \pi_1(G)$. The reason is that the positive loop group is connected. For the same reason, the map κ factors through the quotients $\mathcal{G}rass_G$ and $\mathcal{F}lag_G$, and we denote the resulting maps again by κ, or κ_G if we want to indicate which group we refer to.

Example 2.21. If $G = \mathrm{GL}_n$, then we can identify $\pi_1(G)$ with the quotient of $\mathbb{Z}^n$ by the subgroup of elements $(x_1, \ldots, x_n) \in \mathbb{Z}^n$ with $\sum x_i = 0$, and hence $\pi_1(G) \cong \mathbb{Z}$. The map κ maps an element $g \in \mathrm{GL}_n(k((\epsilon)))$ to the valuation of its determinant.

The map κ_G is sometimes called the "Kottwitz map" because of its appearance in Kottwitz' papers [44], [45] on isocrystals. Cf. also the section about the classification of σ-conjugacy classes in Section 4.2 below. Kottwitz works in the p-adic situation and therefore has to define this map in a different way, and also defines it for non-split groups; cf. [63] for a discussion of the relationship of the different definitions. Note that in [26], [27], the same map is denoted η_G.

2.8. The Bruhat–Tits building

We give the definition of the Bruhat–Tits building for $G = \mathrm{PGL}_n$ (or $G = \mathrm{SL}_n$) which is sometimes useful to visualize subsets of the affine Grassmannian or of the affine flag variety. Bruhat and Tits developed this theory for arbitrary reductive groups over local fields. See the books by Garrett [22], and by Abramenko and Brown [1] (where the relevant buildings are called Euclidean and affine buildings, respectively).

We let K denote a complete discretely-valued field, and denote by k its residue class field. For us, the relevant cases are either $K = L = k((\epsilon))$, where k is algebraically closed, and $K = F = \mathbb{F}_q((\epsilon))$. But in contrast to the ind-scheme structure on the set $G(L)/G(\mathcal{O})$, the theory of the building works equally well over fields of mixed characteristic, say the field $\mathbb{Q}_p$ of p-adic numbers, or the completion of its maximal unramified extension $\widehat{\mathbb{Q}}_p^{\mathrm{ur}}$. We denote by $\mathcal{O}_K$ the valuation ring of K. As above, we have the notion of $\mathcal{O}_K$-lattice inside K^n.

Definition 2.22. The *Bruhat–Tits building for* PGL_n over K is the simplicial complex $\mathcal{B}$ where

- The set $\mathcal{B}_0$ of vertices of $\mathcal{B}$ is the set of equivalence classes of $\mathcal{O}_K$-lattices $\mathscr{L} \subseteq K^n$, where the equivalence relation is given by homothety, i.e., $\mathscr{L} \sim \mathscr{L}'$ if and only if there exists $c \in K^\times$ such that $\mathscr{L}' = c\mathscr{L}$.
- A set $\{ L_1, \ldots, L_m \}$ of m vertices is a simplex if there exist representatives $\mathscr{L}_i$ of L_i such that

$$\mathscr{L}_1 \supset \mathscr{L}_2 \supset \cdots \supset \mathscr{L}_m \supset \epsilon \mathscr{L}_1.$$

The n-dimensional simplices are called alcoves. Clearly, every simplex is contained in an alcove. We say that a lattice $\mathscr{L} \subset K^n$ is adapted to a basis $f_1, \ldots, f_n$ of K^n, if $\mathscr{L}$ has an $\mathcal{O}_K$-basis of the form $\epsilon^{i_1} f_1, \ldots, \epsilon^{i_n} f_n$. The apartment corresponding to the basis f_i is the subcomplex of $\mathcal{B}$ whose simplices consist of vertices given by lattices adapted to this basis. The elementary divisor theorem shows that given any two vertices of $\mathcal{B}$, there exists an apartment containing both. In fact, one can show that given any two simplices, there exists an apartment containing both. The apartment corresponding to the standard basis is called the standard apartment.

The standard apartment is closely connected with the affine root system of G. The geometric realization of the standard apartment can be naturally identified with $X_*(A)_{\mathbb{R}}$. The set of vertices lying in the standard apartment can be identified with $X_*(A) = \mathbb{Z}^n/\mathbb{Z} \cdot (1, \ldots, 1)$, the residue class of $(i_1, \ldots, i_n)$ corresponding to the homothety class of $\bigoplus_\nu \epsilon^{i_\nu} \mathcal{O}_K e_\nu$. The grid of affine root hyperplanes gives rise to the simplicial structure; see the figures on pages 41, 42, and 43 for the irreducible root systems of rank 2. In fact, the set of alcoves is equal to the set of connected components of the complement of the union of all affine root hyperplanes

$$\{ x \in X_*(A)_{\mathbb{R}}; \ \langle x, \alpha \rangle = i \}, \quad \alpha \in \Phi, \ i \in \mathbb{Z}.$$

Our choice of a standard lattice chain, or equivalently the choice of an Iwahori subgroup, gives us a distinguished alcove, called the *base alcove*. The extended affine Weyl group acts on the set of alcoves in the standard apartment. The affine Weyl group acts simply transitively, and hence can be identified with the set of alcoves, using the base alcove as a base point. On the other hand, the elements of length 0 are exactly those elements which fix the base alcove. (For instance, in the case of PGL_3, there are 2 non-trivial rotations with center the barycenter of the base alcove, fixing the base alcove.) The length of an element x of the (extended) affine Weyl group is the number of affine root hyperplanes separating the base alcove from the image of the base alcove under x. For a vertex of $\mathcal{B}$ represented by $\mathscr{L} = g\Lambda$, $g \in \mathrm{PGL}_n(K)$, we call the residue class of $\mathrm{val}(\det g)$ in $\mathbb{Z}/n$ the type of the vertex. This number is independent of the choice of representative.

For $n = 2$, the situation is particularly simple:

Example 2.23. Let $G = \mathrm{PGL}_2$. In this case, the simplicial complex $\mathcal{B}$ is a tree, as is easily seen using the notion of distance below. The notion of type gives $\mathcal{B}$

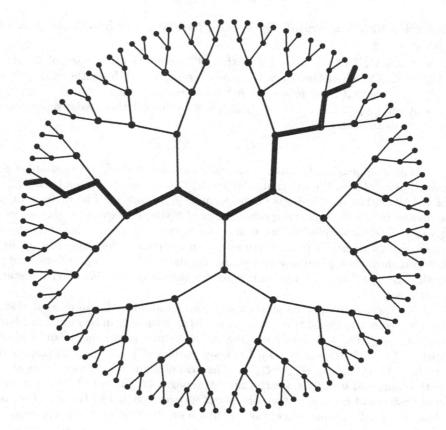

FIGURE 1. Part of the Bruhat–Tits tree for PGL_2 over $\mathbb{F}_2((\epsilon))$ (or over $\mathbb{Q}_2$). One apartment is marked by thick lines.

the structure of a bipartite graph: all neighbors of a vertex of type 0 have type 1, and conversely. The set of neighbors of a point represented by $\mathscr{L}$ can obviously be identified with the projective line $\mathbb{P}^1(k)$. We have an obvious notion of *distance* between two vertices: Given vertices represented by $\mathscr{L} \subseteq \mathscr{L}'$, we can choose bases of the form b_1, b_2, and $\epsilon^{d_1} b_1, \epsilon^{d_2} b_2$ of $\mathscr{L}$ and $\mathscr{L}'$, and the distance is given by $|d_1 - d_2|$ (which is independent of the choice of bases). See Figure 1 for an illustration.

The action of $\mathrm{PGL}_n(K)$, or $\mathrm{SL}_n(K)$, on $\mathcal{B}$ induces identifications

- $\mathcal{F}lag_{\mathrm{SL}_n}(k) =$ set of all alcoves in $\mathcal{B}$,
- $\mathcal{G}rass_{\mathrm{PGL}_n}(k) =$ set of all vertices in $\mathcal{B}$,
- $\mathcal{G}rass_{\mathrm{SL}_n}(k) =$ set of all vertices of type 0 in $\mathcal{B}$.

3. Affine Springer fibers

3.1. Springer fibers

We start with a brief review of the classical theory of Springer fibers which provides important motivation for the notion of affine Springer fiber. A survey of this topic was given by Springer [74].

Let k be an algebraically closed field of characteristic 0, and let G be a connected semisimple algebraic group over k. One can also work in positive characteristic, provided one makes an assumption that the characteristic is sufficiently large with respect to the group.

We fix a maximal torus A of G and a Borel subgroup $B \subset G$ which contains A. We denote the Lie algebras of these groups by $\mathfrak{g}$, $\mathfrak{b}$, $\mathfrak{a}$, respectively. We have $\mathfrak{b} = \mathfrak{a} \oplus \mathfrak{n}$, where $\mathfrak{n}$ is the Lie algebra of the unipotent radical of B.

Denote by $\tilde{\mathfrak{g}}$ the quotient $G \times^B \mathfrak{b}$ of $G \times \mathfrak{b}$ by the B-action given by $b.(g, x) = (gb^{-1}, \mathrm{Ad}(b)x)$.

Theorem 3.1 (Grothendieck, see [74, Theorem 1.4]). *The diagram*

$$
\begin{array}{ccc}
\tilde{\mathfrak{g}} & \xrightarrow{\varphi} & \mathfrak{g} \\
\Big\downarrow{\vartheta} & & \Big\downarrow{\chi} \\
\mathfrak{a} & \xrightarrow{\psi} & \mathfrak{a}/W
\end{array}
$$

where

- *φ maps (g, x) to $\mathrm{Ad}(g)(x)$,*
- *ϑ maps (g, x) to the $\mathfrak{a}$-component of $x \in \mathfrak{b} = \mathfrak{a} \oplus \mathfrak{n}$,*
- *χ is the map induced by the homomorphism $k[\mathfrak{a}/W] = k[\mathfrak{a}]^W \cong k[\mathfrak{g}]^G \subset k[\mathfrak{g}]$,*
- *ψ is the canonical projection from $\mathfrak{a}$ to the quotient of $\mathfrak{a}$ by W,*

is a simultaneous resolution of χ, i.e.,

1. *φ is proper, ϑ is smooth, ψ is finite, and*
2. *for all $a \in \mathfrak{a}$, the morphism $\vartheta^{-1} \to \chi^{-1}(\psi(a))$ is a resolution of singularities (and in particular induces an isomorphism over the smooth locus of $\chi^{-1}(\psi(a))$).*

If $G = \mathrm{GL}_n$, then we can identify $\mathfrak{a}/W$ with affine space $\mathbb{A}^n$, and χ with the map which sends $x \in \mathfrak{gl}_n = \mathrm{Mat}_{n \times n}(k)$ to the coefficients of its characteristic polynomial. Then $\chi^{-1}(0)$ is the nilpotent cone, the subset of all nilpotent matrices.

Definition 3.2. The fibers of φ are called *Springer fibers*. For $x \in \mathfrak{g}$, we write $\varphi^{-1}(x) = Y_x$ (considered as a reduced scheme).

Note that for $x \in \mathfrak{g}$, we can rewrite the Springer fiber Y_x as

$$Y_x = \left\{ g \in G/B; \ \mathrm{Ad}(g^{-1})(x) \in \mathfrak{b} \right\}, \quad (g, z) \mapsto g, \tag{1}$$

the inverse map being given by $g \mapsto (g, \mathrm{Ad}(g^{-1})(x))$.

We note some properties of Springer fibers:

Proposition 3.3 (see [74, Theorem 1.8]). *Let $x \in \mathfrak{g}$, and denote by $H = Z_G(x_s)^0$ the identity component of the centralizer of the semisimple part x_s of x.*

1. *The number of connected components of Y_x is $\#W/W_x$, where W_x is the Weyl group of H.*
2. *Each connected component of Y_x is isomorphic to $Y_{x_n}^H$, where x_n is the nilpotent part of x.*
3. *Y_x is equidimensional of dimension $\frac{1}{2}(\dim Z_G(x) - \operatorname{rk} G)$.*

Part 3., which is the hardest, is due to Spaltenstein. It implies in particular that φ is a "small morphism". An interesting application of Springer fibers is the construction of the irreducible representations of the Weyl group of G.

Springer fibers are usually very singular varieties. All the more surprising is the following purity theorem (see Section 3.5 for a brief discussion of purity):

Theorem 3.4 (Spaltenstein [72]). *Let $G = \mathrm{GL}_n$. Then every Springer fiber Y_x admits a paving by affine spaces. In particular, its (ℓ-adic) cohomology is concentrated in even degrees and is pure.*

If one works over the field of complex numbers, one can replace ℓ-adic cohomology by singular cohomology. Shimomura [71] has generalized the theorem to Springer fibers for GL_n in partial flag varieties, see also the paper [39] by Hotta and Shimomura. On the other hand, the Springer fibers have severe singularities, and in particular Poincaré duality fails for these varieties, even on the level of Betti numbers.

3.2. Affine Springer fibers

Now let k be an algebraically closed field, let $\mathcal{O} = k[\![\epsilon]\!]$, and let $L = k(\!(\epsilon)\!)$ be the field of Laurent series. (The hypothesis that k is algebraically closed is not necessary, and not even desirable for some applications; we make it here to simplify the situation a little bit.) We fix a connected reductive linear algebraic group G over k, and a maximal torus $A \subseteq G$. We denote by $\mathfrak{g}$ and $\mathfrak{a}$ the Lie algebras of G and A, respectively. There is no widely accepted notation for affine Springer fibers. Often they are denoted by $\mathcal{G}rass_\gamma$ (which is of course more appealing when the affine Grassmannian is denoted by a shorter symbol like X (as in [41]) or $\mathfrak{X}^G$ (as in [15])). We will denote affine Springer fibers by $\mathcal{F}(\gamma)$, a notation which is close to the notation of [32], and make the following definition (cf. (1)):

Definition 3.5. The *affine Springer fiber* associated with $\gamma \in \mathfrak{g}(L)$ is

$$\mathcal{F}(\gamma) = \left\{ x \in G(L); \ \mathrm{Ad}(x^{-1})\gamma \in \mathfrak{g}(\mathcal{O}) \right\}/G(\mathcal{O}),$$

a locally closed subset of $\mathcal{G}rass(k)$. We view $\mathcal{F}(\gamma)$ as an ind-scheme over k by giving it the reduced ind-scheme structure.

Remark 3.6.

1. Note that all of the following cases can occur, depending on γ: $\mathcal{F}(\gamma) = \varnothing$, $\mathcal{F}(\gamma)$ is a scheme of finite type over k, $\mathcal{F}(\gamma)$ is a scheme locally of finite type (but not of finite type) over k, $\mathcal{F}(\gamma)$ is not a scheme (but only an ind-scheme).
2. In [61], Ngô gives a functorial definition of affine Springer fibers in terms of G-bundles, and hence obtains a natural ind-scheme structure.
3. There is a variant of the definition where the Lie algebra $\mathfrak{g}(\mathcal{O})$ of the maximal compact subgroup $G(\mathcal{O})$ is replaced by the Lie algebra of an Iwahori subgroup. Then we obtain affine Springer fibers in the affine flag variety of G.
4. Another obvious variant is to replace the Lie algebra with the group itself, and to replace the adjoint action by the conjugation action. At least if the characteristic is 0 or sufficiently large, then one can switch back and forth between these two points of view using a quasi-logarithm, see, e.g., the work of Kazhdan and Varshavsky, [42], 1.8.

Similarly as in Proposition 3.3 2., one can often reduce to the case that γ is a *topologically nilpotent* element, i.e., that γ^n converges to 0 in the ϵ-adic topology, using the topological Jordan decomposition. See Spice's paper [73] for details in the group (rather than the Lie algebra) case.

3.3. General properties

First note that multiplication by g induces an isomorphism $\mathcal{F}(\gamma) \cong \mathcal{F}(\mathrm{Ad}(g)\gamma)$, so we can study all non-empty affine Springer fibers, up to isomorphism, by considering $\gamma \in \mathfrak{g}(O)$.

Recall that a semisimple element in $\mathfrak{g}(L)$ is called regular, if its centralizer is a maximal torus (i.e., if the centralizer is "as small as possible"). In the case of GL_n this just means that all eigenvalues over an algebraic closure are different. Although G is split, the centralizer of a regular semisimple element γ will not be a *split* maximal torus in general, because the field $k((\epsilon))$ is not algebraically closed. If char $k = 0$, then the algebraic closure of $k((\epsilon))$ is the field of Puiseux series,

$$\overline{k((\epsilon))} = \bigcup_{e \in \mathbb{Z}_{\geq 0}} k((\epsilon^{\frac{1}{e}})).$$

If char $k > 0$, then the field on the right-hand side is the perfect closure of the maximal tamely ramified extension of $k((\epsilon))$. See Kedlaya's paper [43] for a description of the algebraic closure and for further references.

For the remainder of Section 3, we make the assumption that the order $\#W$ of the Weyl group of G is invertible in k. This implies that, even if char $k > 0$, every maximal torus of G splits over a tamely ramified extension of L, i.e., over an extension of the form $k((\epsilon^{\frac{1}{e}}))$, where char k does not divide e. See [32] for a more thorough discussion of the situation in positive characteristic.

The following proposition shows that only affine Springer fibers for γ regular semisimple are "reasonable" geometric objects, at least for our purposes:

Proposition 3.7 ([41, §2 Corollary, Lemma 6]). *Let $\gamma \in \mathfrak{g}(\mathcal{O})$. We have*

$$\dim \mathcal{F}(\gamma) < \infty \iff \gamma \in \mathfrak{g}(L) \text{ regular semisimple.}$$

From now on, γ will always be assumed to be regular semisimple. In this case, we get more precise information:

Theorem 3.8 ([41], [9]). *Let $\gamma \in \mathfrak{g}(\mathcal{O})$ be regular semisimple (as an element of $\mathfrak{g}(L)$).*

1. *Let T be the centralizer of γ in $G(k((\epsilon)))$, a maximal torus. Let A_γ be the maximal split subtorus of T. Then $\mathrm{Hom}_{k((\epsilon))}(\mathbb{G}_m, T) = X_*(T) = X_*(A_\gamma)$ acts freely on $\mathcal{F}(\gamma)$, and the quotient $X_*(A_\gamma)\backslash \mathcal{F}(\gamma)$ is a projective k-scheme.*
2. *In particular, if γ is elliptic, i.e., $X_*(A_\gamma) = 0$, then $\mathcal{F}(\gamma)$ is a projective k-scheme.*
3. *Let $\mathfrak{z}(\gamma) \subseteq \mathfrak{g}(L)$ be the centralizer of γ, and let δ_γ be the map*

$$\delta_\gamma = \mathrm{ad}(\gamma) \colon \mathfrak{g}(L)/\mathfrak{z}(\gamma) \to \mathfrak{g}(L)/\mathfrak{z}(\gamma).$$

   Then

$$\dim \mathcal{F}(\gamma) = \frac{1}{2}\big(\mathrm{val}(\det(\delta_\gamma)) - \mathrm{rk}\,\mathfrak{g} + \dim \mathfrak{a}^w\big),$$

where w denotes the type of $\mathfrak{z}(N)$ (see [41], §1, Lemma 2 or [32], 5.2) and $\mathfrak{a}^w$ is the fix point locus of w in the Lie algebra $\mathfrak{a}$ of the fixed maximal torus A.

A different way of obtaining the dimension formula is given by Ngô, [61] 3.8. It was proved by Kazhdan and Lusztig ([41], §4 Proposition 1) that affine Springer fibers in the Iwahori case are equidimensional. Equidimensionality in the Grassmannian case was proved by Ngô, see [61] Proposition 3.10.1.

For a moment, let us consider affine Springer fibers over a finite base field. One reason why this is interesting is that the number of points of a quotient $X_*(A_\gamma)\backslash \mathcal{F}(\gamma)$ over a finite field can be expressed as an orbital integral:

$$\#(X_*(A_\gamma)\backslash \mathcal{F}(\gamma))(\mathbb{F}_q) = \#X_*(A_\gamma)\backslash \{\, g \in G(F);\ \mathrm{Ad}(g^{-1})\gamma \in \mathfrak{g}(\mathfrak{o}) \,\}/G(\mathcal{O})$$

$$= \int_{T(F)\backslash G(F)} \mathbf{1}_{\mathfrak{g}(\mathfrak{o})}(\mathrm{Ad}(g^{-1})\gamma)\, dg/dt =: O_\gamma(\mathbf{1}_{\mathfrak{g}(\mathfrak{o})})$$

for measures dg, dt such that $G(\mathcal{O})$ and $T(\mathcal{O})$ have volume 1. Such orbital integrals are of great interest from the point of view of the Langlands program, and more specifically of the fundamental lemma, a long-standing conjecture of Langlands and Shelstad which was recently proved by Ngô, [61]. Ngô does use affine Springer fibers along the way of his proof, but as Example 3.4.2 by Bernstein and Kazhdan suggests, it is hopeless to "compute" the number of points of an affine Springer fiber directly.

By the Grothendieck–Lefschetz fix point formula, one can express the number of points of a variety over a finite field in terms of the trace of the Frobenius morphism on the ℓ-adic cohomology. We will report on several results on the cohomology of affine Springer fibers in Sections 3.5–3.7.

For classical groups, one can also express these numbers in a completely elementary way, as numbers of lattices satisfying certain conditions.

3.4. Examples

3.4.1. SL$_2$. For more details on the example in this section see [31] 6. We fix an algebraically closed base field k of characteristic $\neq 2$, let $G = \mathrm{SL}_2$, denote by A the diagonal torus, by $\mathcal{G}rass$ the corresponding affine Grassmannian, and by $x_0 \in \mathcal{G}rass$ the base point corresponding to the standard lattice $\Lambda = k[\![\epsilon]\!]^2$.

Denote by α the unique positive root (with respect to the Borel subgroup of upper triangular matrices). We regard α as the morphism $A \to \mathbb{G}_m$, $\mathrm{diag}(a, a^{-1}) \mapsto a^2$. We denote by α' its "differential", i.e., the homomorphism

$$\alpha' : \mathfrak{a}(L) \to L, \qquad \begin{pmatrix} a & \\ & -a \end{pmatrix} \mapsto 2a.$$

For $n \leq -1$, we set $x_n = \begin{pmatrix} 1 & \epsilon^n \\ & 1 \end{pmatrix}$.

Lemma 3.9 (Nadler, [31, Lemma 6.2]).

1. *The affine Grassmannian is the disjoint union*

$$\mathcal{G}rass = \bigcup_{n \leq 0} A(L)x_n,$$

and for each $n \leq 0$, we have $\dim A(L)x_n = |n|$.

2. *For $\gamma \in \mathfrak{a}(\mathcal{O})$, setting $v = \mathrm{val}(\alpha'(\gamma))$, we have*

$$\mathcal{F}(\gamma) = \bigcup_{n=-v}^{0} A(L)x_n.$$

Proof. The proof is elementary; see [31]. In terms of the Bruhat–Tits tree for SL_2, we can understand the lemma as follows: The action of $A(L)$ on the tree fixes the standard apartment, and hence preserves the distance to the standard apartment. One checks that x_n has distance $|n|$ to the standard apartment, and to prove the first part, one has to prove that $A(L)$ acts transitively on the set of points of a fixed distance to the standard apartment.

It is clear that $A(L)$ acts on $\mathcal{F}(\gamma)$, so $\mathcal{F}(\gamma)$ is a union of A-orbits, and the lemma says, from the point of view of the building, that $\mathcal{F}(\gamma)$ is the set of all points of type 0 of distance $\leq \mathrm{val}(\alpha'(\gamma))$ from the standard apartment. $\square$

We see that for $\mathrm{val}(\alpha'(\gamma)) = 0$, $\mathcal{F}(\gamma) \cong X_*(A)$ is a discrete set, while for $\mathrm{val}(\alpha'(\gamma)) = 1$ it is a chain of countably many projective lines where the ith chain and the $(i+1)$th chain intersect transversally in a single point, and no other intersections occur. We can take the quotient $\mathbb{Z}\backslash\mathcal{F}(\gamma)$ and obtain a nodal rational curve. This illustrates Theorem 3.8, 1.

Of course, one checks immediately that the above is consistent with the general dimension formula stated above: for $\gamma \in \mathfrak{a}(\mathcal{O})$ regular, with the notation of Theorem 3.8, $\mathfrak{z}(\gamma) = \mathfrak{a}$, and $w = \mathrm{id}$. The factor $\frac{1}{2}$ arises because δ_γ is defined on the whole (positive and negative) root space.

One can now go on to describe the $A(k)$-fix points and orbits in $\mathcal{F}(\gamma)$, and hence compute its T-equivariant homology; see [31] 7 and Section 3.7 below. Since there is only one positive root, for SL_2 one is always in the "equivaluation" case, see below.

3.4.2. The example of Bernstein and Kazhdan.
In the appendix to [41], J. Bernstein and D. Kazhdan give an example of an affine Springer fiber $\mathcal{F}(\gamma)$ in the affine flag variety of $G = \mathrm{Sp}_6$ which is not a rational variety. More precisely, it has an irreducible component which admits a dominant morphism to an elliptic curve (whose isomorphism class depends on the element γ). This shows that one cannot expect to have a closed formula for the number of points of an affine Springer fiber over a finite field. Furthermore, this affine Springer fiber cannot have a paving by affine spaces.

3.5. Purity

In this section, k denotes an algebraic closure of $\mathbb{F}_p$. Whenever we consider ℓ-adic cohomology, we assume that ℓ is a prime different from p.

Let X_0 be a separated $\mathbb{F}_q$-scheme of finite type, and let $X = X_0 \otimes_{\mathbb{F}_q} k$. The geometric Frobenius $\mathrm{Fr} \in \mathrm{Gal}(k/\mathbb{F}_q)$, i.e., the inverse of the usual ("arithmetic") Frobenius morphism $x \mapsto x^q$, acts on X via its action on the second factor of the product $X_0 \otimes_{\mathbb{F}_q} k$, and hence on the ℓ-adic cohomology groups $H^i(X) := H^i(X, \overline{\mathbb{Q}}_\ell)$. The cohomology is called pure, if for every integer i, the space $H^i(X, \overline{\mathbb{Q}}_\ell)$ is pure of weight n in the sense of Deligne: For every embedding $\iota \colon \overline{\mathbb{Q}}_\ell \to \mathbb{C}$ and every eigenvalue α of Fr on $H^i(X)$, $|\iota(\alpha)| = q^{i/2}$. Note that this is really a property of X; it is independent of the choice of X_0 and q. Every X of finite type over k is defined over some finite field.

There is a large class of varieties with pure cohomology:

Theorem 3.10 (Deligne). *Let X be a smooth and proper k-scheme. Then the cohomology of X is pure.*

On the other hand, the cohomology of a singular variety will usually not be pure. Springer fibers and affine Springer fibers are (expected to be) exceptions to this rule. Another standard method of checking purity is to show that the variety in question admits a paving by affine spaces. Since affine Springer fibers can have cohomology in odd degrees, they cannot have such a paving in general, however; cf. also Example 3.4.2. See Section 3.7 for positive results.

3.6. (Co-)Homology of Ind-schemes

In the sequel, it will sometimes be more useful to use homology rather than cohomology. One defines

$$H_i(X) = H_i(X, \overline{\mathbb{Q}}_\ell) = \mathrm{Hom}_{\overline{\mathbb{Q}}_\ell}(H^i(X), \overline{\mathbb{Q}}_\ell) = H_c^{-i}(X, K_X),$$

where K_X denotes the dualizing complex of X, and the final equality is given by Poincaré duality. Cf. [15] 3.3.

If X is an ind-scheme, say $X = \bigcup_n X_n$, with X_n of finite type and separated, then we set

$$H^i(X) = \varprojlim H^i(X_n), \qquad \text{and} \qquad H_i(X) = \varinjlim H_i(X_n).$$

These groups are independent of the choice of representation of X as a union of finite-dimensional schemes.

3.7. Equivariant cohomology

One of the important tools in studying cohomological properties of affine Springer fibers is equivariant cohomology. The centralizer T of γ acts on the affine Springer fiber $\mathcal{F}(\gamma)$, and equivariant cohomology takes into account the additional structure given by this action. Under a purity assumption, the equivariant cohomology is completely encoded by the 0- and 1-dimensional orbits of T; see the Lemma of Chang and Skjelbred (Proposition 3.15). Furthermore, in favorable situations, for instance if the cohomology is pure, the usual cohomology can easily be recovered from the equivariant one. We sketch the definition of equivariant cohomology in the ℓ-adic setting. Though elegant, it is not easy to digest because it uses ℓ-adic cohomology of algebraic stacks. As long as one works over the field of complex numbers, one can also use the classical topological version of equivariant cohomology, see [30] and Tymoczko's introductory paper [76]. The reference we follow in the ℓ-adic setting is the paper [15] by Chaudouard and Laumon.

Let k be an algebraic closure of the field $\mathbb{F}_q$ with q elements, let $p = \mathrm{char}\, k$. Let X be a separated k-scheme of finite type, and let T be an algebraic torus acting on X.

Definition 3.11. The *T-equivariant ℓ-adic cohomology groups* of X are

$$H_T^n(X) := H_T^n(X, \overline{\mathbb{Q}}_\ell) := H^n([X/T], \overline{\mathbb{Q}}_\ell),$$

where $[X/T]$ denotes the stack quotient of X by the action of T.

For $X = \mathrm{Spec}\, k$, the Chern–Weil isomorphism describes the equivariant cohomology (with respect to the trivial action by T). This is particularly important, because for any X, writing $H_T^*(X) := \bigoplus_{n\geq 0} H_T^n(X)$, cup-product induces on $H_T^*(X)$ the structure of a graded algebra, and of a $H_T^*(\mathrm{Spec}\, k)$-module. To state the Chern–Weil isomorphism, we define

$$\mathscr{D}^* := \mathrm{Sym}^*(X^*(T) \otimes \overline{\mathbb{Q}}_\ell(-1)).$$

Here $\overline{\mathbb{Q}}_\ell(-1)$ is the Tate twist, i.e., the vector space $\overline{\mathbb{Q}}_\ell$ where the geometric Frobenius acts by multiplication by q.

Theorem 3.12 (Chern–Weil isomorphism). *There is a natural isomorphism*

$$\mathscr{D}^* \longrightarrow H_T^*(\operatorname{Spec} k)$$

doubling the degree, i.e.,

$$H_T^i(\operatorname{Spec} k) = \begin{cases} \operatorname{Sym}^j(X^*(T) \otimes \overline{\mathbb{Q}}_\ell(-1)) & \text{if } i = 2j \text{ is even} \\ 0 & \text{if } i \text{ is odd} \end{cases}.$$

Proof. We give a sketch of the proof; see Behrend's paper [8], 2.3, for details. Using the Künneth formula, one reduces to the case $T = \mathbb{G}_m$. Now $[\mathbb{A}_k^{N+1}/\mathbb{G}_m]$ has the same cohomology as $[\operatorname{Spec} k/\mathbb{G}_m]$, and the open immersion

$$\mathbb{P}^N = (\mathbb{A}^{N+1} \setminus \{0\})/\mathbb{G}_m \to [\mathbb{A}^{N+1}/\mathbb{G}_m]$$

gives us, by purity, that

$$H^i([\mathbb{A}^{N+1}/\mathbb{G}_m]) \cong H^i(\mathbb{P}^N) \cong \overline{\mathbb{Q}}_\ell(-i)$$

for all $i \leq 2N$. $\qquad\qquad\square$

Now we return to the case of an arbitrary separated k-scheme X of finite type on which T acts. The Leray spectral sequence for the natural morphism $[X/T] \to [\operatorname{Spec} k/T]$ has the form

$$E_2^{p,q} = H_T^p(\operatorname{Spec} k) \otimes H^q(X) \Longrightarrow H_T^{p+q}(X). \tag{2}$$

Definition 3.13. The scheme X (together with the given T-action) is called *equivariantly formal*, if the spectral sequence (2) degenerates at the E_2 term.

If X is equivariantly formal, then

$$H_T^*(X) \cong H^*(X) \otimes H_T^*(\operatorname{Spec} k),$$
$$H^*(X) \cong H_T^*(X) \otimes_{H_T^*(\operatorname{Spec} k)} \overline{\mathbb{Q}}_\ell,$$

so the usual and the equivariant cohomology determine each other in a simple way. Because the differentials in the spectral sequence respect the Frobenius action, we obtain

Proposition 3.14. *Let X be as above, and assume that $H^*(X) = H^*(X, \overline{\mathbb{Q}}_\ell)$ is pure. Then X is equivariantly formal.*

Similarly as for usual cohomology, we define

$$H_i^T(X) = \operatorname{Hom}_{\overline{\mathbb{Q}}_\ell}(H_T^n(X), \overline{\mathbb{Q}}_\ell),$$

and if $X = \bigcup_n X_n$ is an ind-scheme with X_n separated, of finite type, we define

$$H_T^i(X) = \varprojlim H_T^i(X_n), \qquad H_i^T(X) = \varinjlim H_i^T(X_n).$$

Equivariant cohomology incorporates the additional structure given by the torus action of T on X.

One way to make use of this is the following:

Proposition 3.15 (Lemma of Chang–Skjelbred, [15, Lemme 3.1]). *Let V be a finite-dimensional vector space on which T acts algebraically, and let $X \subseteq \mathbb{P}(V)$ be a T-stable closed subscheme of the projective space of lines in V. Suppose that the cohomology of X is pure. Denote by X_0 the set of T-fix points in X, and by X_1 the union of all orbits of T of dimension ≤ 1.*

There is an exact sequence

$$H_\bullet^T(X_1, X_0) \to H_\bullet^T(X_0) \to H_\bullet^T(X) \to 0,$$

where $H_\bullet^T(X_1, X_0)$ denotes the relative equivariant homology (see [15], 3.5).

Therefore for pure varieties with torus action we can compute the equivariant homology, and hence the homology, once we understand the 1-dimensional and 0-dimensional T-orbits. This makes the following purity conjecture a central topic in the theory of affine Springer fibers.

Conjecture 3.16 (Goresky, Kottwitz, MacPherson [31, Conjecture 5.3]). *For every $n \geq 0$, the homology group $H_n(\mathcal{F}(\gamma))$ is pure of weight $-n$.*

Assuming the conjecture, the usual cohomology is related to the equivariant cohomology as explained above, and since the equivariant cohomology can be described in terms of the fixed points and one-dimensional orbits of the torus, a detailed study of the torus action yields an explicit description of the cohomology of affine Springer fibers. In the case where γ is unramified (i.e., its centralizer is split over L, and hence can be assumed to be equal to A), Goresky, Kottwitz and MacPherson have proved:

Theorem 3.17 ([31, Theorem 9.2]). *Let $\gamma \in \mathfrak{a}(\mathcal{O})$, and assume that $\mathcal{F}(\gamma)$ is pure. Then Proposition 3.15 induces an isomorphism*

$$H_\bullet^A(\mathcal{F}(\gamma)) \cong (k[X_*(A)] \otimes \mathscr{D}^*) / \sum_{\alpha \in \Phi^+} L_{\alpha,\gamma},$$

where

$$L_{\alpha,\gamma} = \sum_{d=1}^{\mathrm{val}(\alpha'(\gamma))} (1 - \alpha^\vee) k[X_*(A)] \otimes \mathscr{D}^* \{\partial_\alpha^d\}.$$

Here ∂_α is the differential operator of degree 1 on $\mathscr{D}^*$ corresponding to α. Note that the set of A-fixed points in $\mathcal{F}(\gamma)$ is equal to the set of all A-fixed points in $\mathcal{G}rass$, and hence can be identified with $X_*(A)$. Therefore its homology is just $k[X_*(A)] \otimes \mathscr{D}^*$. See loc. cit. for details.

In the "equivalued" case, Goresky, Kottwitz and MacPherson have proved the purity conjecture. Let $\gamma \in \mathfrak{g}(L)$ be a regular semisimple element with centralizer T. The element γ is called *integral*, if $\mathrm{val}(\lambda'(\gamma)) \geq 0$ for every $\lambda \in X^*(T)$, and it is called *equivalued*, if for every root α of T (over an algebraic closure of L), the valuation $\mathrm{val}(\alpha'(\gamma))$ is equal to some constant s independent of α, and $\mathrm{val}(\lambda'(\gamma)) \geq s$ for every $\lambda \in X^*(T)$.

Theorem 3.18 ([32, Theorem 1.1]). *Assume that p does not divide the order of W. Let γ be an integral equivalued regular element of $(\operatorname{Lie} T)(L)$, where $T \subset G_L$ is a maximal torus. Then the affine Springer fiber $\mathcal{F}(\gamma)$ has pure cohomology.*

The theorem is proved by showing that the affine Springer fibers in question admit a paving by varieties which are fiber bundles in affine spaces over certain smooth, projective varieties, and invoking Deligne's Theorem 3.10. V. Lucarelli [56] has provided examples of affine Springer fibers for PGL_3 and elements of unequal valuation which admit pavings by affine spaces, thus proving the purity conjecture in these cases.

3.8. The fundamental lemma

The fundamental lemma, sometimes called, more appropriately, the matching conjecture of Langlands and Shelstad, is a family of combinatorial identities which relate orbital integrals (of different sorts) for different groups, which was conjectured by Langlands and Shelstad [50]. The precise statement is quite complicated; see loc. cit. and Hales' paper [35]. The relationship between the groups occurring is given by "endoscopy". Endoscopic groups for G are described in terms of the root datum of G, and there is no simple relationship in terms of the groups themselves. As an example, if $G = U(n)$ is a quasi-split unitary group, then the products $U(n_1) \times U(n_2)$, $n_1 + n_2 = n$, are endoscopic groups for n. Note however that in general an endoscopic group is not a subgroup of the given group.

Before the work of Ngô, see below, the fundamental lemma had been proved in several special cases; see [35] for references. Furthermore, results of Langlands–Shelstad, Hales and Waldspurger allowed to reduce the originally p-adic statement to a Lie algebra version in the function field case over fields of high positive characteristic. The idea of translating the fundamental lemma into an algebro-geometric statement and using the highly developed machinery of algebraic geometry to prove it has been around for some time. For instance, see the paper [54] by Laumon and Rapoport. It was not clear until quite recently, though, how to translate the complicated combinatorics to "simple" geometry, rather than to intractable geometry.

We have seen above that the number of points of an affine Springer fiber can be expressed as an orbital integral. In fact, the orbital integrals occurring in the statement of the fundamental lemma can be expressed in terms of affine Springer fibers, and one can say that the fundamental lemma predicts some (totally unexpected) kind of relationship between affine Springer fibers for different groups.

One can try to prove the fundamental lemma by studying affine Springer fibers. Goresky, Kottwitz and MacPherson [31] have proved the fundamental lemma in the "equivaluation case", using their theorems about affine Springer fibers that we stated above (Theorem 3.17, Theorem 3.18). Laumon had the idea of using deformations of affine Springer fibers to make the problem more accessible, see [51], but still only obtained the result (for unitary groups) assuming the purity conjecture of Goresky, Kottwitz and MacPherson.

A break-through occurred with the work of Laumon and Ngô [53] who realized that the "Hitchin fibration" is a suitable global situation into which (slight modifications of) the relevant affine Springer fibers can be embedded. This provides a geometric interpretation of the theory of endoscopy. Heuristically, one gets a natural way to deform, and eventually "get rid of", the singularities. The work of Laumon and Ngô dealt with the simpler case of unitary groups, where the endoscopic groups in question are actually subgroups of the original group, so that it is easier than in the general case to relate the intervening geometric objects to each other. Recently, Ngô in his celebrated paper [61] was able to overcome the big remaining difficulties and to prove the fundamental lemma in the general case.

In addition to the original papers cited above we mention the surveys written by Dat [16], Laumon [52] and Ngô [60]. Neither of these includes the most recent and complete results by Ngô [61], but see the preprint [17] by Dat and Ngo Dac.

4. Affine Deligne–Lusztig varieties

4.1. Deligne–Lusztig varieties

We start with a short reminder about usual Deligne–Lusztig varieties in order to put the theory described below into context. Let k be an algebraic closure of the finite field $\mathbb{F}_q$, let G be a connected reductive group over $\mathbb{F}_q$, let B be a Borel subgroup defined over $\mathbb{F}_q$, and let $T \subseteq B$ be a maximal torus defined over $\mathbb{F}_q$. We denote by σ the Frobenius morphism on k, $G(k)$, etc. Let W be the absolute Weyl group of the pair (G, T), i.e., the Weyl group for $T_k \subset G_k$. If G is split, then W is equal to the Weyl group "over $\mathbb{F}_q$", but here it is unnecessary to make this assumption. Recall the Bruhat decomposition $G(k) = \bigcup_{w \in W} B(k)wB(k)$.

Definition 4.1 (Deligne–Lusztig, [18]). The *Deligne–Lusztig variety* associated with $w \in W$ is the locally closed subvariety $X_w \subset G_k/B_k$ with

$$X_w(k) = \left\{ g \in G(k); \; g^{-1}\sigma(g) \in B(k)wB(k) \right\}/B(k).$$

Given $g, h \in (G/B)(k) = G(k)/B(k)$, one says that the relative position $\mathrm{inv}(g, h)$ of g, h is the unique element $w \in W$, such that $g^{-1}h \in B(k)wB(k)$. The latter condition should be understood as a condition on representatives of g, h in $G(k)$, but is independent of the choice of representatives. With this notion, we can say that $X_w(k)$ is the set of all elements $g \in (G/B)(k)$ such that g and $\sigma(g)$ have relative position w.

Example 4.2. If $w = \mathrm{id}$ is the identity element, then $X_{\mathrm{id}} = (G/B)(\mathbb{F}_q)$ is the set of $\mathbb{F}_q$-rational points in G_k/B_k, i.e., the set of fix points of σ.

Example 4.3. If $G = \mathrm{GL}_n$, then we can identify $G(k)/B(k)$ with the set of full flags of subvector spaces in k^n. We identify the Weyl group with the subgroup of permutation matrices, and hence with the symmetric group S_n on n letters.

The relative position of flags $\mathscr{F}_\bullet$, $\mathscr{G}_\bullet$ can be described as follows: It is the unique permutation $\gamma \in S_n$ such that for all i, j,

$$\dim(\mathscr{F}_i \cap \mathscr{G}_j) = \#\{\, 1 \le l \le j;\ \gamma(l) \le i \,\}.$$

We record some foundational properties of Deligne–Lusztig varieties:

Proposition 4.4. *Let $w \in W$.*

1. *The Deligne–Lusztig variety X_w is smooth and of dimension $\ell(w)$, the length of w.*
2. *The closure $\overline{X}_w$ of X_w in G_k/B_k is normal, and*

$$\overline{X}_w = \bigcup_{v \le w} X_v,$$

   where $\le$ denotes the Bruhat order in W.
3. *The Deligne–Lusztig variety X_w is connected if and only if w is not contained in a σ-stable standard parabolic subgroup of W.*

In fact, generalizing 3., one can easily give a formula for the number of connected components of a Deligne–Lusztig variety.

Proof. See [18], [10], [25]. □

Deligne–Lusztig varieties play an important role in the representation theory of finite groups of Lie type, i.e., of groups of the form $G(\mathbb{F}_q)$, where G is as above. The reason is that $G(\mathbb{F}_q)$ acts on X_w, and hence on its cohomology. Deligne and Lusztig [18] have shown that all irreducible representations of $G(\mathbb{F}_q)$ can be realized inside the (ℓ-adic) cohomology of Deligne–Lusztig varieties, with suitable $G(\mathbb{F}_q)$-equivariant local systems as coefficients.

Deligne–Lusztig varieties also occur in many other situations. As one example we mention the results of C.-F. Yu and the author [29] which show that all "Kottwitz–Rapoport" strata that are entirely contained in the supersingular locus of a Siegel modular variety with Iwahori level structure are disjoint unions of copies of a Deligne–Lusztig variety.

Remark 4.5. There is a natural relationship between the Springer representation and the representations associated with Deligne–Lusztig varieties. This was first proved by Kazhdan; for details see the appendix of the paper [42] of Kazhdan and Varshavsky.

4.2. σ-conjugacy classes

Now and for the following sections we fix a finite field $\mathbb{F}_q$, and let k be an algebraic closure of $\mathbb{F}_q$. The Frobenius $\sigma \colon x \mapsto x^q$ acts on k, and also (on the coefficients) on $L = k((\epsilon))$: $\sigma(\sum a_i \epsilon^i) = \sum a_i^q \epsilon^i$. We write $F = \mathbb{F}_q((\epsilon))$, the fixed field of σ in L. As usual, we fix an algebraic group G over $\mathbb{F}_q$. We assume, since that is the case we will consider below, that G is a *split* connected reductive group (see Kottwitz' paper for the classification of σ-conjugacy classes in the general case).

Before we come to the definition of affine Deligne–Lusztig varieties, we discuss Kottwitz' classification of σ-conjugacy classes in $G(L)$. The (right) action of $G(L)$ on itself by σ-conjugation is given by $h \cdot g = g^{-1}h\sigma(g)$. Correspondingly, the σ-conjugacy class of b in $G(L)$ is the subset $\{g^{-1}b\sigma(b); \; g \in G(L)\}$. We denote by $B(G)$ the set of σ-conjugacy classes in $G(L)$.

If we consider $G(k)$ instead of $G(L)$, then the situation is considerably simpler: Lang's theorem (see, e.g., [11], Theorem 16.3) says that $G(k)$ is a single σ-conjugacy class. In fact, this statement is of crucial importance for all three points of Proposition 4.4.

The set $B(G)$ of σ-conjugacy classes was described by Kottwitz [44], [45]. A simple invariant of a σ-conjugacy class is the connected component of the loop group $G(L)$ it lies in. In other words, the map $\kappa\colon G(L) \to \pi_1(G)$ factors through a map $\kappa\colon B(G) \to \pi_1(G)$ (sometimes called the Kottwitz map).

A more interesting invariant of a σ-conjugacy class is its Newton vector, an element in $X_*(A)_\mathbb{Q}/W$. We will not give its definition here (see [44], [45]; see Example 4.6 for the case of GL_n). For practical purposes, the following description is often good enough, however:

The restriction $N_G T(L) \to B(G)$ of the natural map from $G(L)$ to $B(G)$ factors through the extended affine Weyl group $\widetilde{W} = N_G T(L)/T(\mathcal{O})$. This follows from a variant of Lang's theorem. The resulting map $\widetilde{W} \to B(G)$ is surjective (this is implicit in Kottwitz' classification, see, e.g., [27] Corollary 7.2.2). Now if $w \in \widetilde{W}$, its Newton vector ν can be computed as follows. Let n be the order of the finite Weyl group part of w, i.e., the order of the image of w under the projection $\widetilde{W} = X_*(A) \rtimes W \to W$. Then $w^n = \epsilon^\lambda$ for some translation element $\lambda \in X_*(A)$, and $\nu = \frac{1}{n}\lambda \in X_*(A)_\mathbb{Q}/W$. The resulting map $B(G) \to X_*(A)_\mathbb{Q}/W$ is called the *Newton map*.

Kottwitz shows that combining the Newton map with the map κ, one obtains an injection

$$B(G) \to X_*(A)_\mathbb{Q}/W \times \pi_1(G).$$

In the special case that the derived group G^{der} is simply connected, the connected component can be recovered from the Newton vector, so that the Newton map is injective. For instance this is true for GL_n and SL_n, but not for PGL_n. We sometimes identify the quotient $X_*(A)_\mathbb{Q}/W$ with the dominant chamber $X_*(A)_{\mathbb{Q},+}$ of rational coweights λ such that $\langle \alpha, \lambda \rangle \geq 0$ for all roots α, and consider Newton vectors as elements of the latter.

Example 4.6. Let us consider the case $G = \mathrm{GL}_n$. Every σ-conjugacy class contains a representative b of the following form: b is a block diagonal matrix, and each block has the form

$$\begin{pmatrix} 0 & \epsilon^{k_i+1}I_{k_i'} \\ \epsilon^{k_i}I_{n_i-k_i'} & 0 \end{pmatrix} \in \mathrm{GL}_{n_i}(L).$$

Here $n = \sum n_i$, $k_i, k_i' \in \mathbb{Z}$, $0 \leq k_i < n$. The Newton vector of b is the composite of the Newton vectors of the single block, and the Newton vector of each block is

$(k_i + \frac{k_i'}{n_i}, \ldots, k_i + \frac{k_i'}{n_i})$ (where the tuple has n_i entries). This representative is called the standard representative in [27] 7.2.

This shows that the set of elements in $X_*(A)_{\mathbb{Q},+}$ (which we can identify with the set of n-tuples of rational numbers in descending order) is the subset of sequences

$$a_1 = \cdots = a_{i_1} > a_{i_1+1} = \cdots = a_{i_1+i_2} > a_{i_1+i_2+1} \cdots > a_{i_1+\cdots+i_r+1} = \cdots = a_n$$

that satisfy the integrality condition $i_\nu a_{i_1+\cdots+i_{\nu-1}+1} \in \mathbb{Z}$ for each $1 \leq \nu \leq r+1$ (with $i_{r+1} = n - i_1 - \cdots - i_r$).

Given $(a_1, \ldots, a_n) \in X_*(A)_{\mathbb{Q},+}$, we can view the a_i as the slopes of the *Newton polygon* attached to b. (One usually orders the slopes in ascending order, so that the Newton polygon is the lower convex hull of the points $(0,0)$ and $(i, \sum_{j=n-i+1}^{n} a_j)$, $i = 1, \ldots, n$.) The integrality condition says that the break points of this polygon should have integer coefficients.

The classification of σ-conjugacy classes in GL_n is the same as the classification of isocrystals (due to Dieudonné/Manin). More precisely, an isocrystal is a pair (V, Φ) consisting of a finite-dimensional L-vector space V and a σ-linear bijection Φ (i.e., Φ is additive, $\Phi(av) = \sigma(a)v$ for all $a \in L$, $v \in V$, and Φ is bijective). Choosing a basis of V, we can write $\Phi = b\sigma$, $b \in \mathrm{GL}_n(L)$, $n = \dim V$. A change of basis corresponds to σ-conjugating b. See for instance Demazure's book [19], Chapter IV.

Given $b \in G(L)$, its σ-centralizer is the algebraic group J_b over F with

$$J_b(F) = \{ g \in G(L);\ g^{-1}b\sigma(g) = b \}.$$

In fact, J_b is an inner twist of the Levi subgroup $\mathrm{Cent}_G(\nu)$, the centralizer of the Newton vector of ν. See [44] 6.5.

Definition 4.7 ([47]). Let $b \in G(L)$. The *defect of b* is the difference

$$\mathrm{def}(b) = \mathrm{rk}_F(G) - \mathrm{rk}_F(J_b)$$

of the (F-)rank of G and the F-rank of the σ-centralizer J_b.

Example 4.8. If $b = 1$, then $J_b = G$ and $\mathrm{def}(b) = 0$. On the other hand, suppose that $G = \mathrm{GL}_n$, and let b be the generator of the group of length 0 elements in $\widetilde{W}$ with Newton vector $(\frac{1}{n}, \ldots, \frac{1}{n})$. Then $J_b = D_{\frac{1}{n}}^\times$, the group of units of the central division algebra over F with invariant $\frac{1}{n}$. In this case, $\mathrm{def}(b) = n - 1$.

Kottwitz [47] has proved that the defect of an element $b \in G(L)$ can also be expressed in a way which is close to Chai's conjectural formula [14] for the dimension of Newton strata in Shimura varieties, and to Rapoport's conjectural formula ([64] Conjecture 5.10) for the dimension of affine Deligne–Lusztig varieties in the affine Grassmannian (Theorem 4.17 below).

To simplify the discussion, we assume here that the derived group G^{der} is simply connected. We have an exact sequence

$$1 \to G^{\mathrm{der}} \to G \to D \to 1,$$

where $D := G/G^{\mathrm{der}}$ is a split torus, and obtain an exact sequence of character groups

$$0 \to X^*(D) \to X^*(A) \to X^*(A \cap G^{\mathrm{der}}) \to 0.$$

We lift the fundamental weights in $X^*(A \cap G^{\mathrm{der}})$ to elements ω_i, $i = 1, \ldots, l$, in $X^*(A)$, and in addition choose a basis $\omega_{l+1}, \ldots, \omega_n$ of $X^*(D)$. Then the characters $\omega_1, \ldots, \omega_n$ form a $\mathbb{Z}$-basis of $X^*(A)$.

Example 4.9. If $G = \mathrm{GL}_n$, then $G^{\mathrm{der}} = \mathrm{SL}_n$ is simply connected, and a possible choice of the ω_i is

$$\omega_i = (1^{(i)}, 0^{(n-i)}) \in \mathbb{Z}^n = X^*(A), \quad i = 1, \ldots, n,$$

the notation meaning that 1 is repeated i times, and 0 is repeated $n - i$ times.

Proposition 4.10. *Assume that the derived group G^{der} is simply connected, and let $b \in G(L)$ with Newton vector ν_b. Then*

$$\mathrm{def}(b) = 2 \sum_{i=1}^{n} \mathrm{fr}(\langle \omega_i, \nu_b \rangle),$$

where for any rational number α, $\mathrm{fr}(\alpha) \in [0,1)$ denotes its fractional part.

Finally, we make the following important definition:

Definition 4.11. A σ-conjugacy class in $G(L)$ is called *basic*, if the following equivalent conditions are satisfied:

1. The Newton vector ν is central, i.e., lies in the image of $X_*(Z)_{\mathbb{Q}}$, where $Z \subseteq G$ is the center of G.
2. The σ-conjugacy class can be represented by an element $\tau \in \widetilde{W}$ with $\ell(\tau) = 0$.

We call an element $b \in G(L)$ basic, if its σ-conjugacy class is basic.

More precisely, one can show that the restriction of the map $\widetilde{W} \to B(G)$ to the set Ω_G of elements of length 0 is a bijection from Ω_G to the set of basic σ-conjugacy classes ([45] 7.5, [27] Lemma 7.2.1). Looking at σ-conjugacy classes from the point of view of Newton strata in the special fiber of a Shimura variety, the basic locus is the unique closed Newton stratum. In the case of the Siegel modular variety, for instance, this is just the supersingular locus.

4.3. Affine Deligne–Lusztig varieties: the hyperspecial case

Similarly as for usual Deligne–Lusztig varieties, we want to consider all elements g which map under a "Lang map" to a fixed double coset. When we consider the affine Grassmannian, we look at $G(\mathcal{O})$-cosets, which by the Cartan decomposition (Theorem 2.17) are parameterized by the set $X_*(A)_+$ of dominant coweights. Furthermore, in the affine context one has to consider generalizations of the Lang map, i.e., we consider maps of the form $g \mapsto g^{-1} b \sigma(g)$ for an element $b \in G(L)$.

Definition 4.12. The *affine Deligne–Lusztig variety* $X_\mu(b)$ in the affine Grassmannian associated with $b \in G(L)$ and $\mu \in X_*(A)_+$ is given by

$$X_\mu(b)(k) = \left\{ g \in G(L); \ g^{-1} b \sigma(g) \in G(\mathcal{O}) \epsilon^\mu G(\mathcal{O}) \right\} / G(\mathcal{O}).$$

We can view this definition as a σ-linear variant of affine Springer fibers, and as it turns out, there are several analogies between the two theories. For instance, see the discussion of the dimension formula (Theorem 4.17) below. Of course, there are also many differences, and an important difference is that the definition above includes a parameter μ: While in the case of affine Springer fibers we always considered the set of g such that $g^{-1}bg \in G(\mathcal{O})$ (or rather the Lie algebra version of this), here we consider this relationship for an arbitrary $G(\mathcal{O})$-double coset. Furthermore, while in the case of affine Springer fibers the element γ, which corresponds to the b above, was regular semisimple in the most interesting cases, in the case of affine Deligne–Lusztig varieties the most interesting case is where b is a basic element, i.e., up to σ-conjugacy, b is a representative of a length 0 element of $\widetilde{W}$.

The subset $X_\mu(b)(k)$ is locally closed in $\mathcal{G}rass(k)$, so it inherits the structure of a (reduced) sub-ind-scheme. In fact, $X_\mu(b)$ is a scheme locally of finite type over k; see Corollary 5.5 in the paper [36] by Hartl and Viehmann. The key point here is that points x in the building such that the distance from x to $b\sigma(x)$ is bounded (the bound being given by μ) have bounded distance to the building of J_b over a finite unramified extension of F; this was proved by Rapoport and Zink [67]. Compare the corresponding fact for affine Springer fibers, where every point has bounded distance to $X_*(A_\gamma)$ (Proposition 3.8 1.). Usually $X_\mu(b)$ has infinitely many irreducible components:

Proposition 4.13. *Assume that G is simple and of adjoint type. Let $b \in G(L)$, $\mu \in X_*(A)$ with $X_\mu(b) \neq \varnothing$. The following are equivalent:*

1. $X_\mu(b)$ *is of finite type over k.*
2. *The element b is superbasic, i.e., no σ-conjugate of b is contained in a proper Levi subgroup of G.*
3. $G = \mathrm{PGL}_n$ *and the Newton vector of b has the form $(\frac{r}{n}, \dots, \frac{r}{n})$, $r \in \mathbb{Z}$ coprime to n.*

Proof. The implication 2.$\Rightarrow$ 1., which is the most difficult one, follows from Viehmann's detailed study of the superbasic case [77]. The equivalence of 2. and 3. is explained in [26] 5.9. There it is also shown that b is superbasic if and only if J_b is anisotropic. If this is not the case, then the J_b-action on $X_\mu(b)$ shows that $X_\mu(b)$ contains points of arbitrarily high distance to the origin (in the sense of the building), and hence cannot be of finite type. $\qquad\square$

Remark 4.14.

1. Multiplication by $g^{-1} \in G(L)$ induces an isomorphism between $X_\mu(b)$ and $X_\mu(g^{-1}b\sigma(g))$. Since we are only interested in affine Deligne–Lusztig varieties up to isomorphism, we are free to replace b by another representative of its σ-conjugacy class. The σ-centralizer $J_b(F)$ of b acts on $X_\mu(b)$.
2. We may have $X_\mu(b) = \varnothing$. It is one of the basic questions, when this happens. The reason that usual Deligne–Lusztig varieties are always non-empty is that the Lang map $g \mapsto g^{-1}\sigma(g)$ is a surjection $G(k) \to G(k)$: All elements of $G(k)$

are σ-conjugate to the identity element. Therefore in the classical case there is no need to introduce the parameter b which we see above. On the other hand, the Lang map $G(L) \to G(L)$ is not surjective. In fact, we have seen above that $G(L)$ consists of many σ-conjugacy classes. The usual proof for the surjectivity of the Lang map fails in the setting of ind-schemes: although the differential of the Lang map $g \mapsto g^{-1}\sigma(g)$ is an isomorphism, one cannot conclude that the map itself is "étale".

3. One can obviously generalize the definition to cover other parahoric subgroups of $G(L)$. One obtains affine Deligne–Lusztig varieties inside (partial) affine flag varieties. We will consider the case of the Iwahori subgroup in detail in the following section.

4. There is a p-adic variant, where L is replaced by the completion of the maximal unramified extension of $\mathbb{Q}_p$. In this case, the same definition gives an *affine Deligne–Lusztig set*. Although one still uses the term *affine Deligne–Lusztig variety* in the p-adic situation, this is not really justified. One does not have a k-ind-scheme structure on the quotient $G(L)/I$, and hence there is no variety structure on $X_\mu(b)$. On the other hand, this case is particularly interesting because of its connection to the theory of moduli spaces of p-divisible groups and Shimura varieties (see 4.9 below). In fact, in some cases one obtains a variety structure on the affine Deligne–Lusztig set, induced from the scheme structure of the corresponding moduli space.

5. We mention that in [57], Lusztig has considered a different analogue of usual Deligne–Lusztig varieties in an "affine" context (his varieties are infinite-dimensional, they have a pro-structure rather than an ind-structure).

6. We mention in passing that it is also interesting to replace σ by other morphisms. See the papers by Baranovsky and Ginzburg [3] and by Caruso [13] for two other choices of σ. The first one is related to the study of conjugacy classes in Kac–Moody groups, the latter one is motivated by the theory of Breuil, Kisin and others about the classification of finite flat group schemes.

Example 4.15. If $G = \mathrm{SL}_2$, then $X_*(A)_+ = \{\, (a, -a) \in \mathbb{Z}^2;\ a \geq 0 \,\}$, and an element of the affine Grassmannian is in the $G(\mathcal{O})$-orbit corresponding to $(a, -a)$ if and only if it, seen as a vertex in the Bruhat–Tits building over L, has distance a to the rational building (i.e., the building over F). In particular, we find a description of affine Deligne–Lusztig varieties $X_\mu(1)$ which is completely analogous to the description of affine Springer fibers for SL_2 in Section 3.4.1.

There is the following criterion for non-emptiness of $X_\mu(b)$:

Theorem 4.16. *Let $b \in G(L)$ with Newton vector $\nu \in X_*(A)_{\mathbb{Q},+}$, and let $\mu \in X_*(A)$ be dominant. Then*

$$X_\mu(b) \neq \varnothing \quad \Longleftrightarrow \quad \kappa_G(b) = \mu \text{ and } \nu \leq \mu,$$

where we denote the image of μ in $\pi_1(G)$ again by μ, and where $\nu \leq \mu$ means by definition that $\mu - \nu$ is a non-negative linear combination of simple coroots.

The implication $\Rightarrow$ is called Mazur's inequality; for GL_n the statement above boils down to a version of an inequality considered by Mazur in the study of p-adic estimates of the number of points over a finite field of certain algebraic varieties. It was proved by Rapoport and Richartz [65] for general G. The converse, accordingly called the "converse to Mazur's inequality" was proved only recently. It was conjectured to hold, and proved for GL_n and GSp_{2g}, by Kottwitz and Rapoport [48]. C. Lucarelli [55] proved the theorem for classical groups, and finally Gashi [23] proved it for the exceptional groups (and even in the quasi-split case). See also Kottwitz' paper [46].

Theorem 4.17. *Let* $b \in G(L)$ *with Newton vector* $\nu_b \in X_*(A)_{\mathbb{Q},+}$, *and let* $\mu \in X_*(A)$ *be dominant. Assume that* $X_\mu(b) \neq \varnothing$. *Then*

$$\dim X_\mu(b) = \langle \rho, \mu - \nu_b \rangle - \frac{1}{2}\operatorname{def}(b).$$

Recall that we denote by ρ half the sum of the positive roots. The defect in the dimension formula should be seen as a correction term, and should be compared with the term $\frac{1}{2}(\operatorname{rk}\mathfrak{g} - \dim \mathfrak{a}^w)$ in the dimension formula for affine Springer fibers (Proposition 3.8), compare [47], in particular (1.9.1).

The formula for the dimension of $X_\mu(b)$ was conjectured by Rapoport in [64] Conjecture 5.10. and reformulated in the current form by Kottwitz [47], using the notion of defect. In [26], the proof of the dimension formula is reduced to the superbasic case, i.e., to the case where no σ-conjugate of b is contained in a proper Levi subgroup of G. Subsequently the formula was proved in the superbasic case by Viehmann [77].

To reduce the proof of the dimension formula to the superbasic case, one has to compare the affine Deligne–Lusztig varieties $X_\mu^M(b)$ and $X_\mu^G(b)$ for a Levi subgroup $A \subseteq M \subseteq G$ and $b \in M(L)$. If $P = MN$ is a parabolic subgroup, then there is a bijection $P(L)/P(\mathfrak{o}) \cong G(L)/K$, and this defines a map

$$\alpha\colon G(L)/K \cong P(L)/P(\mathfrak{o}) \longrightarrow M(L)/M(\mathfrak{o})$$

from the affine Grassmannian for G to the affine Grassmannian for M. This map is not a morphism of ind-schemes, but for any connected component Y of the affine Grassmannian for M, the restriction of α to $\alpha^{-1}(Y)$ is a morphism of ind-schemes. This map can be used to relate the affine Deligne–Lusztig varieties for M and for G; see [26] 5.6.

It is expected that in the hyperspecial case, all affine Deligne–Lusztig varieties are equidimensional. This has been proved in the two extreme cases:

If $b = \epsilon^\nu$ is a translation element, then Proposition 2.17.1 in [26] shows that $X_\mu(b)$ is equidimensional. The proof relies on a result proved by Mirković and Vilonen as part of their proof of the geometric Satake isomorphism. More precisely, their results about the intersection cohomology of intersections of $U(L)$- and K-orbits imply that these intersections are equidimensional. For a proof of the relevant fact in positive characteristic, which is what one needs in our situation, see the paper [62] by Ngô and Polo.

On the other hand, if b is basic, then it was proved by Hartl and Viehmann in [36] that $X_\mu(b)$ is equidimensional; see Section 4.10.

To prove equidimensionality in general, one might want to apply the strategy of the proof of the dimension formula, that is to reduce to the (super-)basic case. However, to carry out this reduction, a better understanding of the restriction of the map α to $X_\mu^G(b)$ would be needed.

A general result about the relationship between affine Deligne–Lusztig varieties for G and Levi subgroups of G is the Hodge–Newton decomposition. Let M_b be the centralizer of the Newton vector ν of b. There is a unique standard parabolic subgroup $P_b = M_b N_b$ with Levi subgroup M_b (and unipotent radical N_b). Denote by A_{P_b} the identity component of the center of M, and let

$$\mathfrak{a}_{P_b}^+ = \left\{\, x \in X_*(A_{P_b}) \otimes \mathbb{R};\ \langle \alpha, x \rangle > 0 \text{ for every root } \alpha \text{ of } A_{P_b} \text{ in } N_b \,\right\}.$$

Replacing b by a σ-conjugate, we may and will assume that $b \in M_b(L)$, b is basic with respect to M_b, and $\kappa_{M_b}(b) \in X_*(A_{P_b})_\mathbb{R}$ actually lies in $\mathfrak{a}_{P_b}^+$. Given any standard parabolic subgroup $P \subseteq G$, we use analogous notation as for P_b: $P = MN$, A_P, etc. Then we have

Theorem 4.18 ([46], [78, Theorem 1]). *Let $\mu \in X_*(A)$ be dominant, and let b, M_b as above. Let $P = MN \subseteq G$ be a standard parabolic subgroup with $P_b \subseteq P$. If $\kappa_M(b) = \mu$, then the inclusion $X_\mu^M(b) \to X_\mu^G(b)$ is an isomorphism.*

See also the paper [59] by Mantovan and Viehmann for a generalization to unramified groups.

It is hard to determine the set of connected components of an affine Deligne–Lusztig variety $X_\mu(b)$. The problem is that in general the group $J_b(F)$ does not act transitively on the set of connected components, see Viehmann's paper [78], Section 3, for an example. As Viehmann shows in loc. cit., the situation is better if instead one considers the following variant: Given μ, define

$$X_{\leq \mu}(b) = \bigcup_{\lambda \leq \mu} X_\lambda(b).$$

This is a closed subscheme of the affine Grassmannian (equipped with the reduced scheme structure). A priori, it could be bigger than the closure of $X_\mu(b)$ in $\mathcal{G}rass_G$, though. For b basic, Hartl and Viehmann [36] show that $X_{\leq \mu}(b)$ is equal to the closure of $X_\mu(b)$. For these "closed affine Deligne–Lusztig varieties" one has

Theorem 4.19 (Viehmann [78] Theorem 2). *Suppose that the data G, μ, b are indecomposable with respect to a Hodge–Newton decomposition, i.e., there is no standard $P \supseteq P_b$ with $\kappa_M(b) = \mu$. Assume that G is simple.*

1. *If ϵ^μ is central in G, and b is σ-conjugate to ϵ^μ, then*

$$X_\mu(b) = X_{\leq \mu}(b) \cong J_b(F)/(J_b(F) \cap G(\mathcal{O})) \cong G(F)/G(\mathcal{O}_F)$$

is discrete.

2. *Assume that we are not in the situation of* 1. *Then* $\kappa_M(b) \neq \mu$ *for all proper standard parabolic subgroups* $P = MN \subsetneq G$ *with* $b \in M(L)$, *and* κ_G *induces a bijection*

$$\pi_0(X_{\leq \mu}(b)) \cong \pi_1(G).$$

The group $J_b(F)$ *acts transitively on* $\pi_0(X_{\leq \mu}(b))$.

4.4. Affine Deligne–Lusztig varieties: the Iwahori case

Now we come to the Iwahori case. Because of the Iwahori–Bruhat decomposition $G(L) = \bigcup_{x \in \widetilde{W}} IxI$ (Theorem 2.18) we associate affine Deligne–Lusztig varieties inside the affine flag variety with elements $b \in G(L)$ and $x \in \widetilde{W}$:

Definition 4.20. The *affine Deligne–Lusztig variety* $X_x(b)$ in the affine flag variety associated with $b \in G(L)$ and $x \in \widetilde{W}$ is given by

$$X_x(b)(k) = \left\{ g \in G(L); \; g^{-1}b\sigma(g) \in IxI \right\}/I.$$

The same remarks as in the case of the affine Grassmannian apply. In fact, in the Iwahori case it is much harder (and not yet completely settled) to give a criterion for which $X_x(b)$ are non-empty, and a closed formula for their dimensions.

Note that $X_x(b) = \varnothing$ whenever x and b do not lie in the same connected component of $G(L)$. Since for each $x \in \widetilde{W}$ we have a unique basic σ-conjugacy class in the same connected component as x, there is a unique basic σ-conjugacy class for which $X_x(b)$ can possibly be non-empty. Therefore, as long as we talk only about basic σ-conjugacy classes, x practically determines b, and below we sometimes assume implicitly that x and b are in the same connected component of $G(L)$.

Example 4.21. Let us discuss the case of $G = \mathrm{SL}_2$, $b = 1$. For SL_2, the situation is particularly simple. For instance, every element in the affine Weyl group of SL_2 has a unique reduced expression, and there are only two elements of any given length > 0. What are the Schubert cells IxI which can contain an element of the form $g^{-1}\sigma(g)$? Fix $gI \in G(L)/I$ with $g \neq \sigma(g)$. We consider gI as an alcove in the Bruhat–Tits building of SL_2, and denote by d the distance from gI to the rational building (normalizing the distance so that an alcove has distance 0 to the rational building if and only if it is contained in there). Clearly, $\sigma(g)I$ also has distance d to the rational building, and because the rational building is equal to the subcomplex of σ-fix points in the building over L, one sees that the distance from gI to $\sigma(g)I$ is $2d - 1$. This implies that the element $x \in \widetilde{W}$ with $g^{-1}\sigma(g) \in IxI$ has length $2d - 1$, and shows that $X_x(1) = \varnothing$ if $\ell(x)$ is different from 0 and even. It is not hard to show that on the other hand all $X_x(1)$ for x of odd length are non-empty, and that in this case $\dim X_x(1) = \frac{1}{2}(\ell(x) + 1)$. For a more detailed consideration along these lines see Reuman's PhD thesis [68].

To simplify some of the statements below, we assume from now on, for the rest of Section 4.4, that the Dynkin diagram of G is connected. The (non-)emptiness of affine Deligne–Lusztig varieties for simply connected groups of rank 2 is illustrated

FIGURE 2. Left: The shrunken Weyl chambers (gray) in the root system of type C_2.
Right: The set of P-alcoves for the group of type G_2 and
$P = {}^{s_1 s_2 s_1}(B \cap B s_2 B)$.

in Section 4.5 below. Looking at these pictures, or rather at the picture for a fixed b (cf. [26], [27]), one notices that the behavior is more complicated and more difficult to describe close to the walls of the finite root system. The following definition was made by Reuman in order to describe the "good" region.

Definition 4.22. Let $x \in \widetilde{W}$. We say that x lies in the *shrunken Weyl chambers*, if for every finite root α, $U_\alpha(L) \cap xIx^{-1} \neq U_\alpha(L) \cap I$.

In other words, x lies in the shrunken Weyl chambers, if for every finite root α there exists an affine root hyperplane parallel to $\{\alpha = 0\}$ which separates x and the base alcove. We sometimes call the complement of the shrunken chambers the *critical strips*. See the left-hand side picture in Figure 2.

We define maps from the extended affine Weyl group $\widetilde{W}$ to the finite Weyl group W as follows:

$\eta_1 \colon \widetilde{W} = X_*(T) \rtimes W \to W$, the projection

$\eta_2 \colon \widetilde{W} \to W$, where $\eta_2(x)$ is the unique element $v \in W$ such that $v^{-1}x \in {}^S\widetilde{W}$

$\eta(x) = \eta_2(x)^{-1}\eta_1(x)\eta_2(x).$

Let $x \in \widetilde{W}$, and let b be a representative of the unique basic σ-conjugacy class corresponding to the connected component of x. We define the *virtual dimension* (of $X_x(b)$):

$$d(x) = \frac{1}{2}\big(\ell(x) + \ell(\eta(x)) - \mathrm{def}(b)\big).$$

The following conjecture which extends a conjecture made by D. Reuman [69] gives a very simple "closed formula" for non-emptiness and dimension of affine Deligne–Lusztig varieties for b basic and x in the shrunken Weyl chambers. (Note that the conjecture as it stands does not extend to all x. See Conjecture 4.28 below for a more precise, but more technical conjecture about non-emptiness.)

Conjecture 4.23 ([27, Conjecture 9.4.1 (a)]). *Let $b \in G(L)$ be basic. Assume that $x \in \widetilde{W}$ lies in the shrunken Weyl chambers. Then $X_x(b) \neq \varnothing$ if and only if $\kappa_G(x) = \kappa_G(b)$ and $\eta(x) \in W \setminus \bigcup_{T \subsetneq S} W_T$. In this case,*

$$\dim X_x(b) = d(x).$$

Note that this statement easily implies the dimension formula for affine Deligne–Lusztig varieties in the affine Grassmannian. Several partial results towards this conjecture have been obtained. In [27], one direction is proved; in fact, there is the following slightly stronger result which also covers most of the critical strips:

Theorem 4.24 ([27, Proposition 9.4.4]). *Let b be basic. Let $x \in \widetilde{W}$, say $x = \epsilon^\lambda v$, $v \in W$. Assume that $\lambda \neq \nu_b$ and that $\eta(x) \in \bigcup_{T \subsetneq S} W_T$. Then $X_x(b) = \varnothing$.*

As in the hyperspecial case, the key point of the theorem is to relate certain affine Deligne–Lusztig varieties for G to Deligne–Lusztig varieties for Levi subgroups M. Since the group of connected components of the loop group of M is much larger than the one for G, the trivial condition that b and x must belong to the same connected component becomes much stronger, and yields an obstruction for affine Deligne–Lusztig varieties to be non-empty. To single out the elements $x \in \widetilde{W}$, where this can be done, we need the notion of P-alcove introduced in loc. cit.; see the right-hand side picture of Figure 2 for an example.

In the sequel, we consider parabolic subgroups $P \subseteq G$. They need not be standard, i.e., we do not require that $B \subseteq P$, but we only consider *semi-standard* parabolic subgroups, i.e., we ask that $A \subseteq P$. We denote by $P = MN$ the Levi decomposition of such a subgroup; here N denotes the unipotent radical of P, and M is the unique Levi subgroup of P which contains A. Then the (extended affine, or finite) Weyl group of M is contained in the (extended affine, or finite) Weyl group of G.

Definition 4.25. Let $P = MN \subseteq G$ be a semi-standard parabolic subgroup. An element $x \in \widetilde{W}$ is called a *P-alcove*, if it satisfies the following conditions:

1. $x \in \widetilde{W}_M$, the extended affine Weyl group of M,
2. $x(I \cap N(L))x^{-1} \subseteq I \cap N(L)$.

For P-alcoves, one has a "Hodge–Newton decomposition" (see above and [46], [59] for analogues in the hyperspecial case):

Theorem 4.26 ([27, Theorem 2.1.4]). *Suppose that x is a P-alcove for $P = MN \supseteq A$. If $X_x(b) \neq \varnothing$, then the σ-conjugacy class of b meets $M(L)$. Now assume that $b \in M(L)$. Then the closed immersion $X_x^M(b) \to X_x(b)$ induces a bijection*

$$J_b^M(F) \backslash X_x^M(b) \xrightarrow{\cong} J_b(F) \backslash X_x(b).$$

Here $X_x(b)$ denotes the affine Deligne–Lusztig variety for M, and J_b^M denotes the σ-centralizer of b in M.

This is deduced easily from the following, slightly more technical statement, which shows that x being a P-alcove is a strong requirement from the point of view of σ-conjugacy classes occurring inside IxI.

Theorem 4.27 ([27, Theorem 2.1.2]). *Let x be a P-alcove for the semi-standard parabolic subgroup $P = MN \subset G$. Then every element of IxI is σ-conjugate under I to an element of $I_M x I_M$, where $I_M = I \cap M$.*

These results, together with experimental evidence, lead to the following conjecture:

Conjecture 4.28 ([27, Conjecture 9.3.2]). *Let b be basic with Newton vector ν_b, and let $x \in \widetilde{W}$. Then $X_x(b)$ is empty if and only if there exists a semi-standard parabolic subgroup $P = MN \subsetneq G$, such that x is a P-alcove, and $\kappa_M(x) \neq \kappa_M(\nu_b)$.*

On the other hand, still assuming that b is basic, in [28] X. He and the author prove non-emptiness of $X_x(b)$ using the "reduction method of Deligne and Lusztig" (see [18], proof of Theorem 1.6, or [28]) and combinatorial considerations about the affine Weyl group for all elements x that are sufficiently far from the walls and which are expected to give rise to a non-empty $X_x(b)$. More precisely, let us denote by $\rho^\vee$ the sum of all fundamental coweights, and by θ the largest root.

Definition 4.29. An element $\mu \in X_*(A)$ is said to lie in the *very shrunken* Weyl chambers, if

$$|\langle \mu, \alpha \rangle| \geq \langle \rho^\vee, \theta \rangle + 2$$

for every root α.

We then have the following theorem.

Theorem 4.30 ([28]). *Let b be basic, let $x \in \widetilde{W}$ be in the same connected component of $G(L)$ as b, and write $x = t^\mu w$.*

1. *If μ is regular, or $\eta_2(x) = w_0$, the longest element of W, then $\dim X_x(b) \leq d(x)$.*
2. *Assume that $\eta(x) \in W \setminus \bigcup_{T \subsetneq S} W_T$. If μ is in the very shrunken Weyl chambers or $\eta_2(x) = w_0$, then $X_x(b) \neq \varnothing$.*

3. *Let G be a classical group, and let $x \in W_a$ be an element of the affine Weyl group such that $\eta(x) \in W \setminus \bigcup_{T \subsetneq S} W_T$. If μ is in the very shrunken Weyl chambers or $\eta_2(x) = w_0$, then $\dim X_x(1) = d(x)$.*

A crucial ingredient for part 3 is a theorem of He [37] about conjugacy classes in affine Weyl groups. E. Beazley has independently obtained similar results, using the reduction method of Deligne and Lusztig and results by Geck and Pfeiffer about conjugacy classes in finite Weyl groups.

If b is not basic, then because of the experimental evidence we expect the following, see [27] Conjecture 9.4.1 (b). To simplify the statement, let us assume that b is in $\widetilde{W}$, and is of minimal length among all the elements representing this σ-conjugacy class. Let $x \in \widetilde{W}$ be in the same connected component of $G(L)$ as b.

- If $\ell(x)$ is small (with respect to $\ell(b)$), then $X_x(b) = \varnothing$.
- If $\ell(x)$ is large (with respect to $\ell(b)$), $X_x(b) \neq \varnothing$ if and only if $X_x(b_{\text{basic}}) \neq \varnothing$, where b_{basic} represents the unique basic σ-conjugacy class in the same connected component as x. In this case, the dimension of the two affine Deligne–Lusztig varieties differs by a constant (depending on b, but not on x).

It is not easy to give precise bounds for what "small" and "large" should mean. (For the first, one gets approximate information by considering the projection to the affine Grassmannian.) This question can be viewed as the problem of finding a suitable analogue of Mazur's inequality in the Iwahori case. It is hard to describe the pattern of non-emptiness for x of length close to $\ell(b)$. For abundant examples, see the figures for the rank 2 case given in the next section.

One can also study the question of non-emptiness from the slightly different point of view where one fixes x, and asks for the set of b's which give a non-empty affine Deligne–Lusztig variety. See Beazley's paper [6] for an analysis of the case of SL_3 from this standpoint.

In the Iwahori case, affine Deligne–Lusztig varieties are not equidimensional in general. An example with $G = SL_4$ is given in [28].

4.5. The rank 2 case

There are several ways of computing, in specific cases, whether an affine Deligne–Lusztig variety is non-empty, and what its dimension is. Let us illustrate one approach to eliminate the Frobenius morphism from the problem and hence to reduce it to a purely combinatorial statement, in the case $b = 1$. By definition, $X_x(1) \neq \varnothing$ if and only if there exists $g \in G(L)$ such that $g^{-1}\sigma(g) \in IxI$. Now we decompose $G(L) = \bigcup_{w \in \widetilde{W}} IwI$, and we see that the existence of g as before is equivalent to the existence of $w \in \widetilde{W}$ and $i \in I$ with $w^{-1}i^{-1}\sigma(i)w \cap IxI \neq \varnothing$. Now the group I is a single σ-conjugacy class: $I = \{i^{-1}\sigma(i); \ i \in I\}$. Therefore the existence of g is equivalent to the existence of $i' \in I$ with $w^{-1}i'w \cap IxI \neq \varnothing$, or in other words:

$$X_x(1) \neq \varnothing \iff x \in Iw^{-1}IwI \text{ for some } w \in \widetilde{W}.$$

For given x and w, the condition on the right-hand side can easily be translated into a combinatorial statement about the Bruhat–Tits building or the affine Weyl group, respectively. See [26] 6,7, and [27] 10–13, for a detailed discussion. In the case of rank 2 it is in principle feasible to do such computations "by hand"; see Reuman's papers [68], [69]. Using a computer program, one can assemble a large number of examples.

Here we give examples for simply connected groups of type A_2 (i.e., SL$_3$), C_2 (i.e., Sp$_4$) and G_2, for $b = 1$. We identify the affine Weyl group W_a with the set of alcoves in the standard apartment, i.e., with the set of small triangles in the figures below. The base alcove is marked by a thick border.

We denote the σ-conjugacy classes by letters, according to the tables given below. The letters associated with σ-conjugacy classes which meet an Iwahori double coset are printed inside the corresponding alcove. To save space, sequences of letters without gaps are abbreviated as follows: ABCDE is abbreviated to A-E, etc. When denoting elements in $\widetilde{W}$ as products in the generators, we write s_{021} instead of $s_0 s_2 s_1$, etc.

4.6. Type A_2

In this case, the set of σ-conjugacy classes is well known, and we just need to say which σ-conjugacy classes occur in the figure, and how they are named.

$b =$		$\bar{\nu}_b =$	$b =$		$\bar{\nu}_b =$
A	$\epsilon^{(0,0,0)} = 1$	$(0,0,0)$	Q	$\epsilon^{(4,1,-5)}$	$(4,1,-5)$
B	$\epsilon^{(1,0,-1)}$	$(1,0,-1)$	R	$\epsilon^{(3,2,-5)}$	$(3,2,-5)$
C	$\epsilon^{(2,0,-2)}$	$(2,0,-2)$	S	$\epsilon^{(6,0,-6)}$	$(6,0,-6)$
D	$\epsilon^{(2,-1,-1)}$	$(2,-1,-1)$	T	$\epsilon^{(6,-1,-5)}$	$(6,-1,-5)$
E	$\epsilon^{(1,1,-2)}$	$(1,1,-2)$	U	$\epsilon^{(6,-2,-4)}$	$(6,-2,-4)$
F	$\epsilon^{(3,0,-3)}$	$(3,0,-3)$	V	$\epsilon^{(6,-3,-3)}$	$(6,-3,-3)$
G	$\epsilon^{(3,-1,-2)}$	$(3,-1,-2)$	W	$\epsilon^{(5,1,-6)}$	$(5,1,-6)$
H	$\epsilon^{(2,1,-3)}$	$(2,1,-3)$	X	$\epsilon^{(4,2,-6)}$	$(4,2,-6)$
I	$\epsilon^{(4,0,-4)}$	$(4,0,-4)$	Y	$\epsilon^{(3,3,-6)}$	$(3,3,-6)$
J	$\epsilon^{(4,-1,-3)}$	$(4,-1,-3)$	a	s_{021}	$(1,-\frac{1}{2},-\frac{1}{2})$
K	$\epsilon^{(4,-2,-2)}$	$(4,-2,-2)$	b	s_{012}	$(\frac{1}{2},\frac{1}{2},-1)$
L	$\epsilon^{(3,1,-4)}$	$(3,1,-4)$	c	$s_{021021021}$	$(3,-\frac{3}{2},-\frac{3}{2})$
M	$\epsilon^{(2,2,-4)}$	$(2,2,-4)$	d	$s_{012012012}$	$(\frac{3}{2},\frac{3}{2},-3)$
N	$\epsilon^{(5,0,-5)}$	$(5,0,-5)$	e	$s_{021021021021021}$	$(5,-\frac{5}{2},-\frac{5}{2})$
O	$\epsilon^{(5,-1,-4)}$	$(5,-1,-4)$	f	$s_{012012012012012}$	$(\frac{5}{2},\frac{5}{2},-5)$
P	$\epsilon^{(5,-2,-3)}$	$(5,-2,-3)$			

4.7. Type C_2

Again, it is well known what the σ-conjugacy classes are, so we just list those which we consider, and under which names they appear.

	$b =$	$\bar{\nu}_b =$		$b =$	$\bar{\nu}_b =$
A	$\epsilon^{(0,0)} = 1$	$(0,0)$	K	$\epsilon^{(4,0)}$	$(4,0)$
B	$\epsilon^{(1,0)}$	$(1,0)$	L	$\epsilon^{(4,1)}$	$(4,1)$
C	$\epsilon^{(1,1)}$	$(1,1)$	M	$\epsilon^{(4,2)}$	$(4,2)$
D	$\epsilon^{(2,0)}$	$(2,0)$	a	$\epsilon^{(6,5)}$	$(6,5)$
E	$\epsilon^{(2,1)}$	$(2,1)$	b	$\epsilon^{(6,6)}$	$(6,6)$
F	$\epsilon^{(2,2)}$	$(2,2)$	c	s_{012}	$(\frac{1}{2},\frac{1}{2})$
G	$\epsilon^{(3,0)}$	$(3,0)$	d	$s_{012012012}$	$(\frac{3}{2},\frac{3}{2})$
H	$\epsilon^{(3,1)}$	$(3,1)$	e	$s_{012012012012}$	$(\frac{5}{2},\frac{5}{2})$
I	$\epsilon^{(3,2)}$	$(3,2)$	f	$s_{012012012012012012}$	$(\frac{7}{2},\frac{7}{2})$
J	$\epsilon^{(3,3)}$	$(3,3)$	g	$s_{0120120120120120120120120}$	$(\frac{9}{2},\frac{9}{2})$

4.8. Type G_2

The set $B(G)$ is the union of the set of dominant translation elements, and the following two families, each coming from one of the two standard parabolic subgroups:

$$\left\{ n \cdot \left(-\frac{1}{6}, -\frac{1}{6}, \frac{1}{3}\right) ; \; n \in \mathbb{Z}_{>0} \right\}, \quad \left\{ n \cdot \left(0, -\frac{1}{2}, \frac{1}{2}\right) ; \; n \in \mathbb{Z}_{>0} \right\}.$$

FIGURE 3. Dimensions of affine Deligne–Lusztig varieties, type A_2, $b = 1$.

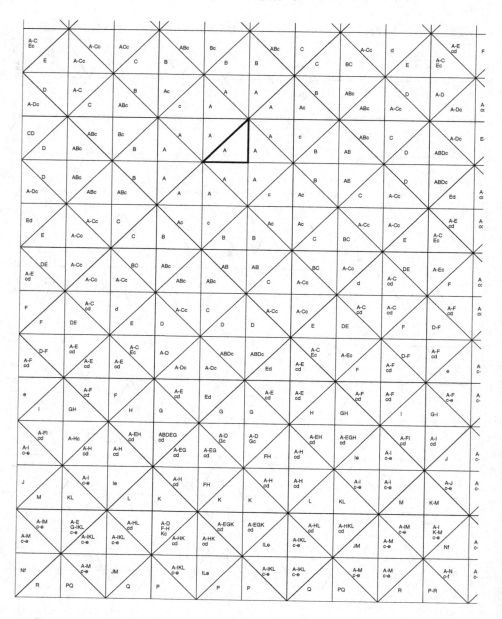

FIGURE 4. Dimensions of affine Deligne–Lusztig varieties, type C_2, $b = 1$.

FIGURE 5. Dimensions of affine Deligne–Lusztig varieties, type G_2, $b = 1$.

To find the alcove corresponding to a certain reduced expression, recall that with our normalization the shortest edge of the alcoves is of type 0, the medium edge is of type 1, and the longest edge is of type 2.

Here is the list of σ-conjugacy classes considered for the figure.

	$b =$	$\bar{\nu}_b =$		$b =$	$\bar{\nu}_b =$
A	$\epsilon^{(0,0,0)} = 1$	$(0,0,0)$	K	$\epsilon^{(-1,-2,3)}$	$(-1,-2,3)$
B	$\epsilon^{(-\frac{1}{3},-\frac{1}{3},\frac{2}{3})}$	$(-\frac{1}{3},-\frac{1}{3},\frac{2}{3})$	L	$\epsilon^{(-\frac{1}{3},-\frac{7}{3},\frac{8}{3})}$	$(-\frac{1}{3},-\frac{7}{3},\frac{8}{3})$
C	$\epsilon^{(-\frac{2}{3},-\frac{2}{3},\frac{4}{3})}$	$(-\frac{2}{3},-\frac{2}{3},\frac{4}{3})$	a	s_{021}	$(-\frac{1}{6},-\frac{1}{6},\frac{1}{3})$
D	$\epsilon^{(0,-1,1)}$	$(0,-1,1)$	b	s_{02121}	$(0,-\frac{1}{2},\frac{1}{2})$
E	$\epsilon^{(-1,-1,2)}$	$(-1,-1,2)$	c	$s_{021210212}$	$(-\frac{1}{2},-\frac{1}{2},1)$
F	$\epsilon^{(-\frac{1}{3},-\frac{4}{3},\frac{5}{3})}$	$(-\frac{1}{3},-\frac{4}{3},\frac{5}{3})$	d	$s_{021210212102121}$	$(0,-\frac{3}{2},\frac{3}{2})$
G	$\epsilon^{(-\frac{4}{3},-\frac{4}{3},\frac{8}{3})}$	$(-\frac{4}{3},-\frac{4}{3},\frac{8}{3})$	e	$s_{021212021210212}$	$(-\frac{5}{6},-\frac{5}{6},\frac{5}{3})$
H	$\epsilon^{(-\frac{2}{3},-\frac{5}{3},\frac{7}{3})}$	$(-\frac{2}{3},-\frac{5}{3},\frac{7}{3})$	f	$s_{021212021210212021212}$	$(-\frac{7}{6},-\frac{7}{6},\frac{7}{3})$
I	$\epsilon^{(0,-2,2)}$	$(0,-2,2)$	g	$s_{0212102121021210212102121}$	$(0,-\frac{5}{2},\frac{5}{2})$
J	$\epsilon^{(-\frac{5}{3},-\frac{5}{3},\frac{10}{3})}$	$(-\frac{5}{3},-\frac{5}{3},\frac{10}{3})$	h	$s_{021212021210212021212021212}$	$(-\frac{3}{2},-\frac{3}{2},3)$

4.9. Relationship to Shimura varieties

Affine Deligne–Lusztig varieties are related to the reduction of certain Shimura varieties, or more directly to moduli spaces of p-divisible groups. To establish this relationship, we consider the p-adic variant of affine Deligne–Lusztig varieties, cf. Remark 4.14 4. The relation relies on Dieudonné theory (see for instance [19]), which classifies p-divisible groups over a perfect field in terms of their Dieudonné modules. A Dieudonné module is a free module of finite rank over the ring of Witt vectors W together with a σ-linear operator F (Frobenius) and a σ^{-1}-linear operator V (Verschiebung) such that $FV = VF = p$ (so that V is uniquely determined by F).

Now fix a p-divisible group $\mathbb{X}$ over k, and denote by M its Dieudonné module, and by $N = M \otimes_W W[\frac{1}{p}]$ its rational Dieudonné-module, or *isocrystal*. We fix a basis of M over W and write F as $b\sigma$, $b \in \mathrm{GL}_n(W[\frac{1}{p}])$, where $n = \mathrm{rk}_W M$. Lattices inside N which are stable under F and V correspond to quasi-isogenies $\mathbb{X} \to X$ of p-divisible groups over k. A lattice $\mathscr{L} = gM$, $g \in \mathrm{GL}_n(W[\frac{1}{p}])$ is stable under F and V if and only if

$$p\mathscr{L} \subseteq F\mathscr{L} \subseteq \mathscr{L},$$

i.e., $g^{-1}b\sigma(g) \in G(\mathcal{O})\epsilon^\mu G(\mathcal{O})$ for some μ of the form $(1,\ldots,1,0,\ldots,0)$. In other words, μ is a minuscule dominant coweight. Since μ must have the same image as b under the Kottwitz map κ_G, it is determined uniquely by $\mathbb{X}$. Therefore we can identify the set of k-valued points of the moduli space of quasi-isogenies attached to $\mathbb{X}$ with the affine Deligne–Lusztig set attached to GL_n, b and μ over $L = \widehat{\mathbb{Q}_p^{\mathrm{un}}}$. See the book [66] by Rapoport and Zink for more information about these moduli spaces, which are often called Rapoport–Zink spaces nowadays. One can also consider variants for other groups, associated with EL- or PEL-data; see loc. cit. See Viehmann's papers [79], [80] for results about the structure of such moduli spaces of p-divisible groups from this point of view. For instance, Viehmann determines the sets of connected and irreducible components, and the dimensions.

Similarly, one obtains a relationship to Shimura varieties, or more precisely to the Newton strata in the special fiber of the corresponding moduli space of abelian varieties. Restricting to a Newton stratum corresponds to fixing an isogeny type of p-divisible groups, i.e., to choosing b. Roughly speaking, the Newton stratum splits up, up to a finite morphism, as a product of a truncated Rapoport–Zink space and a "central leaf". See Mantovan's paper [58] and [26], 5.10 for details and further references.

One can also consider the Iwahori case from this point of view. Again, choosing b corresponds to fixing a Newton stratum (or to considering a Rapoport–Zink space instead of a moduli space of abelian varieties). The choice of $x \in \widetilde{W}$ corresponds to the choice of a Kottwitz–Rapoport stratum. The affine Deligne–Lusztig set $X_x(b)$ is related to the intersection of these two strata. For instance, $X_x(b) \neq \varnothing$ if and only if Newton stratum for b and KR stratum for x intersect (see Haines' survey [33] Proposition 12.6).

One can show using an algorithmic description of the non-emptiness question (see [27]) that the p-adic variant of $X_x(b)$ is non-empty if and only if the function field variant $X_x(b)$ is non-empty (to formulate this properly, we assume that $b \in \widetilde{W}$). In particular, all of the results above in this direction yield information about the intersections of Newton strata and Kottwitz–Rapoport strata and hence about the geometric structure of these moduli spaces of abelian varieties. This is used by Viehmann [81] to obtain results about Shimura varieties from considerations about the function field case. On the other hand, there is no good a priori notion of dimension for the p-adic affine Deligne–Lusztig sets. It seems, however, that once one has a reasonable dimension theory for spaces of this kind, then the dimensions of $X_x(b)$ should agree in the p-adic and function field case. In the supersingular case, the dimension of the affine Deligne–Lusztig variety and the corresponding intersection of a Newton and a Kottwitz–Rapoport stratum are expected to be equal; in general there should be a non-trivial central leaf which governs the difference between the affine Deligne–Lusztig variety and the intersection.

For many more details along these lines see the survey papers by Haines [33] and Rapoport [64].

4.10. Local Shtuka

In [36], Hartl and Viehmann relate affine Deligne–Lusztig varieties to deformations of local shtuka. One could call this the function field version of Section 4.9. In the function field case, the theory works in full generality (whereas in the context of p-divisible groups one is limited to minuscule cocharacters, and also has limitations on which groups one can consider). Using their theory, they prove that for basic b, all affine Deligne–Lusztig varieties $X_\mu(b) \subset \mathcal{G}rass$ are equidimensional, and that the closure of $X_\mu(b)$ is equal to $\bigcup_{\lambda \leq \mu} X_\lambda(b)$.

4.11. Cohomology of affine Deligne–Lusztig varieties

Consider an affine Deligne–Lusztig variety $X_w(b)$. The σ-centralizer J_b acts on $X_w(b)$, and hence on its cohomology with compact support, and on its (Borel–Moore) homology. Since usual Deligne–Lusztig varieties are nowadays an indispensable tool in the representation theory of finite groups of Lie type, one expects that the representations of J_b occurring in the homology of affine Deligne–Lusztig varieties are also of great interest. However, because the geometric properties in the affine case are so much harder to understand, at the moment not much is known about representation theoretic properties. Let us give an overview about the results obtained so far.

Zbarsky [82] considered the following case: $G = \mathrm{SL}_3$, $b = \epsilon^\nu$ where ν is dominant regular. In this case, $J_b = A(F) \cong \mathbb{Z}^2 \times A(\mathcal{O}_F)$. Zbarsky shows that the subgroup $A(\mathcal{O}_F)$ acts trivially on the Borel–Moore homology of $X_w(b)$, and that the action of $\mathbb{Z}^2$ corresponds to permutation of the homology spaces of disjoint closed subsets of $X_w(b)$. A strategy to show that the integral part of the torus acts trivially is to extend the action to an action of $A(\mathcal{O}_L)$. An action of the latter must be trivial because of a "homotopy argument"; therefore the action of the subgroup $A(\mathcal{O}_F)$ is a fortiori trivial. However, it is not possible to extend the action in this way in general. Zbarsky defines a stratification of $X_w(b)$ such that on each stratum the action extends to an action of the larger torus, which is enough to reach the desired conclusion.

The Iwahori case for $G = \mathrm{GL}_2$ has been worked out in detail by Ivanov [40]. In this case, one can determine the geometric structure of the affine Deligne–Lusztig varieties completely. They are disjoint unions of copies of a product of some affine space with the complement of finitely many points on a projective line. As a consequence, one reads off directly the (co-)homology groups (with constant coefficients), and by analyzing the action of J_b on the set of connected components one can determine the representations of J_b which one gets; see loc. cit. Ivanov identifies these representations in terms of compact inductions, and also analyzes them from the point of view of the Langlands classification. There are no non-trivial morphisms to supercuspidal representations.

As in the finite-dimensional case, in addition to the homology with constant coefficients, one should also consider coefficients in certain local systems, or in other words, one should consider the homology of certain coverings of these affine Deligne–Lusztig varieties. Finally we mention the results of He [38] who, at least for $G = \mathrm{PGL}_n$ and $G = \mathrm{PSp}_{2n}$ identifies a subset of $\widetilde{W}$ such that all representations in the homology of affine Deligne–Lusztig varieties occur already in the homology of affine Deligne–Lusztig varieties $X_x(b)$ with x in this subset.

Acknowledgment.

It is a pleasure to thank the organizers of the workshop, Jochen Heinloth and Alexander Schmitt for the invitation; the workshop was a very interesting and enjoyable event. I would also like to thank the audience of the course for their active participation, questions and remarks. Over the years, I learned a lot about these topics from many people, in particular Thomas Haines, Robert Kottwitz, Michael Rapoport, and Eva Viehmann, whom I want to thank as well.

Furthermore I am grateful to the Deutsche Forschungsgemeinschaft for its support through a Heisenberg research grant and the Sonderforschungsbereich SFB/TR45.

References

[1] P. Abramenko, K. Brown, *Buildings*, Springer Graduate Texts in Mathematics **248**, 2008.

[2] M. Artin, *Grothendieck topologies*, Seminar notes, Harvard University 1962.

[3] V. Baranovsky, V. Ginzburg, *Conjugacy classes in loop groups and G-bundles on elliptic curves*, Int. Math. Res. Not. **1996**, no. 15, 733–751.

[4] A. Beauville, Y. Laszlo, *Conformal blocks and generalized theta functions*, Comm. Math. Phys. **164** (1994), no. 2, 385–419.

[5] A. Beauville, Y. Laszlo, *Un lemme de descente*, C. R. A. S. **320** (1995), no. 3, 335–340.

[6] E.T. Beazley, *Codimensions of Newton strata for* $SL_3(F)$ *in the Iwahori case*, Math. Z. **263** (2009), 499–540.

[7] A. Beilinson, V. Drinfeld, *Quantization of Hitchin's integrable system and Hecke eigensheaves*, Preprint, available at
http://www.math.uchicago.edu/~mitya/langlands.html

[8] K. Behrend, *The Lefschetz trace formula for algebraic stacks*, Invent. math. **112** (1993), 127–149.

[9] R. Bezrukavnikov, *The dimension of the fixed point set on affine flag manifolds*, Math. Res. Lett. **3** (1996), 185–189.

[10] C. Bonnafé, R. Rouquier, *On the irreducibility of Deligne–Lusztig varieties*, C.R.A.S. **343** (2006), 37–39.

[11] A. Borel, *Linear Algebraic Groups*, Springer Graduate Text in Mathematics **126**, 1991.

[12] N. Bourbaki, *Algèbre commutative*, Chapitres 1 à 4, Masson 1985.

[13] X. Caruso, *Dimension de certaines variétés de Deligne–Lusztig affines généralisées*, Preprint 2008,
http://perso.univ-rennes1.fr/xavier.caruso/articles/dimension.pdf

[14] C.-L. Chai, *Newton polygons as lattice points*, Amer. J. Math. **122** (2000), no. 5, 967–990.

[15] P.H. Chaudouard, G. Laumon, *Sur l'homologie des fibres de Springer affines tronquées*, Duke Math. J. **145**, no. 3 (2008), 443–535.

[16] J.F. Dat, *Lemme fondamental et endoscopie, une approche géométrique (d'après Gérard Laumon et Ngô Báo Châu)*, Séminaire Bourbaki 2004/2005. Astérisque **307** (2006), Exposé 940, 71–112.

[17] J.F. Dat, Ngo Dac T., *Lemme fondamental pour les algèbres de Lie (d'après Ngô Bao-Châu)*, Preprint, available at
http://people.math.jussieu.fr/~dat/recherche/publis/proj_livre.pdf

[18] P. Deligne, G. Lusztig, *Representations of reductive groups over finite fields*, Ann. of Math. **103** (1976), 103–161.

[19] M. Demazure, *Lectures on p-divisible groups*, Springer Lecture Notes in Mathematics **302** (1972).

[20] V. Drinfeld, *Infinite-dimensional vector bundles in algebraic geometry (an introduction)*, in: The unity of mathematics, Progr. Math. **244**, Birkhäuser 2006, 263–304.

[21] G. Faltings, *Algebraic loop groups and moduli spaces of bundles*, J. Europ. Math. Soc. **5** (2003), 41–68.

[22] P. Garrett, *Buildings and classical groups*, Chapman and Hall 1997.

[23] Q. Gashi, *On a conjecture of Kottwitz and Rapoport*, Preprint arXiv:0805.4575v2 (2008).

[24] S. Glaz, *Commutative coherent rings*, Lecture Notes in Mathematics **1371**, Springer-Verlag, 1989.

[25] U. Görtz, *On the connectedness of Deligne–Lusztig varieties*, Represent. Theory **13** (2009), 1–7.

[26] U. Görtz, T. Haines, R. Kottwitz, D. Reuman, *Dimensions of some affine Deligne–Lusztig varieties*, Ann. sci. École Norm. Sup. 4e série, t. **39** (2006), 467–511.

[27] U. Görtz, T. Haines, R. Kottwitz, D. Reuman, *Affine Deligne–Lusztig varieties in affine flag varieties*, Preprint arXiv:math/0805.0045v2 (2008).

[28] U. Görtz, X. He, *Dimension of affine Deligne–Lusztig varieties in affine flag varieties*, in preparation.

[29] U. Görtz, C.-F. Yu, *Supersingular Kottwitz–Rapoport strata and Deligne–Lusztig varieties*, to appear in J. Inst. Math. Jussieu, arXiv:math/0802.3260v2.

[30] M. Goresky, R. Kottwitz, R. MacPherson, *Equivariant cohomology, Koszul duality, and the localization theorem*, Invent. math. **131** (1998), 25–83.

[31] M. Goresky, R. Kottwitz, R. MacPherson, *Homology of affine Springer fibers in the unramified case*, Duke Math. J. **121** (2004), 509–561.

[32] M. Goresky, R. Kottwitz, R. MacPherson, *Purity of equivalued affine Springer fibers*, Represent. Theory **10** (2006), 130–146.

[33] T. Haines, *Introduction to Shimura varieties with bad reduction of parahoric type*, in: Harmonic analysis, the trace formula, and Shimura varieties, Clay Math. Proc. **4**, Amer. Math. Soc. 2005, 583–642.

[34] T. Haines, M. Rapoport, *On parahoric subgroups*, Appendix to [63], Adv. in Math. **219** (2008), 188–198.

[35] T. Hales, *A statement of the fundamental lemma*, in: Harmonic analysis, the trace formula, and Shimura varieties, Clay Math. Proc. **4**, Amer. Math. Soc. 2005, 643–658.

[36] U. Hartl, E. Viehmann, *The Newton stratification on deformations of local G-shtuka*, Preprint arXiv:0810.0821v2.

[37] X. He, *Minimal length elements in conjugacy classes of extended affine Weyl groups*, in preparation.

[38] X. He, *A class of equivariant sheaves on some loop groups*, in preparation.

[39] R. Hotta, N. Shimomura, *The fixed-point subvarieties of unipotent transformations on generalized flag varieties and the Green functions*, Math. Ann. **241**, no. 3 (1979), 193–208.

[40] A. Ivanov, *The cohomology of the affine Deligne–Lusztig varieties in the affine flag manifold of* GL$_2$, Diploma thesis Bonn 2009.

[41] D. Kazhdan, G. Lusztig, *Fixed point varieties on affine flag manifolds*, Isr. J. Math. **62**, no. 2, (1988), 129–168.

[42] D. Kazhdan, Y. Varshavsky, *On endoscopic decomposition of certain depth zero representations*, in: Studies in Lie theory, Progr. Math. **243**, Birkhäuser 2006, 223–301.

[43] K. Kedlaya, *The algebraic closure of the power series field in positive characteristic*, Proc. Amer. Math. Soc. **129**, no. 12 (2001), 3461–3470.

[44] R. Kottwitz, *Isocrystals with additional structure*, Compositio Math. **56** (1985), 201–220.

[45] R. Kottwitz, *Isocrystals with additional structure. II*, Compositio Math. **109** (1997), 255–339.

[46] R. Kottwitz, *On the Hodge–Newton decomposition for split groups*, Int. Math. Res. Not. **26** (2003), 1433–1447.

[47] R. Kottwitz, *Dimensions of Newton strata in the adjoint quotient of reductive groups*, Pure Appl. Math. Q. **2** (2006), no. 3, 817–836.

[48] R. Kottwitz, M. Rapoport, *On the existence of F-crystals*, Comment. Math. Helv. **78** (2003), no. 1, 153–184.

[49] S. Kumar, *Infinite Grassmannians and moduli spaces of G-bundles*, in: Vector bundles on curves – new directions (Cetraro, 1995), Lecture Notes in Mathematics **1649**, Springer 1997, 1–49.

[50] R. Langlands, D. Shelstad, *On the definition of transfer factors*, Math. Ann. **278** (1987), 219–271.

[51] G. Laumon, *Sur le lemme fondamental pour les groupes unitaires*, Preprint math.AG/0212245 (2002).

[52] G. Laumon, *Aspects géométriques du Lemme Fondamental de Langlands–Shelstad*, in: Proc. ICM Madrid 2006, Vol. II, Eur. Math. Soc., 2006, 401–419.

[53] G. Laumon, B.C. Ngô, *Le lemme fondamental pour les groupes unitaires*, Ann. of Math. (2) **168** (2008), no. 2, 477–573.

[54] G. Laumon, M. Rapoport, *A geometric approach to the fundamental lemma for unitary groups*, Preprint math.AG/9711021 (1997).

[55] C. Lucarelli, *A converse to Mazur's inequality for split classical groups*, J. Inst. Math. Jussieu **3** (2004), no. 2, 165–183.

[56] V. Lucarelli, *Affine pavings for affine Springer fibers for split elements in* PGL(3), math.RT/0309132 (2003).

[57] G. Lusztig, *Some remarks on the supercuspidal representations of p-adic semisimple groups*, in: Automorphic forms, representations and L-functions, Proc. Symp. Pure Math. **33** Part 1 (Corvallis 1977), Amer. Math. Soc. 1979, 171–175.

[58] E. Mantovan, *On the cohomology of certain PEL type Shimura varieties*, Duke Math. J. **129** (3) (2005), 573–610.

[59] E. Mantovan, E. Viehmann, *On the Hodge–Newton filtration for p-divisible O-modules*, to appear in Math. Z., arXiv:0701.4194 (2007).

[60] B.C. Ngô, *Fibration de Hitchin et structure endoscopique de la formule des traces*, Proc. ICM Madrid, Vol. II, Eur. Math. Soc. (2006), 1213–1225.

[61] B.C. Ngô, *Le lemme fondamental pour les algèbres de Lie*, Preprint arXiv:0801.0446v3 (2008).

[62] B.C. Ngô, P. Polo, *Résolutions de Demazure affines et formule de Casselman–Shalika géométrique*, J. Algebraic Geom. **10** (2001), no. 3, 515–547.

[63] G. Pappas, M. Rapoport, *Twisted loop groups and their affine flag varieties*, Adv. Math. **219** (2008), no. 1, 118–198.

[64] M. Rapoport, *A guide to the reduction modulo p of Shimura varieties*, Astérisque (2005), no. 298, 271–318.

[65] M. Rapoport, M. Richartz, *On the classification and specialization of F-isocrystals with additional structure*, Compositio Math. **103** (1996), 153–181.

[66] M. Rapoport, Th. Zink, *Period spaces for p-divisible groups*, Ann. Math. Studies **141**, Princeton Univ. Press 1996.

[67] M. Rapoport, Th. Zink, *A finiteness theorem in the Bruhat–Tits building: an application of Landvogt's embedding theorem*, Indag. Mathem. N. S. **10** 3 (1999), 449–458.

[68] D. Reuman, *Determining whether certain affine Deligne–Lusztig sets are non-empty*, Thesis Chicago 2002, math.NT/0211434.

[69] D. Reuman, *Formulas for the dimensions of some affine Deligne–Lusztig varieties*, Michigan Math. J. **52** (2004), no. 2, 435–451.

[70] R. Richardson, *Affine coset spaces of reductive algebraic groups*, Bull. London Math. Soc. **9** (1977), no. 1, 38–41.

[71] N. Shimomura, *A theorem on the fixed point set of a unipotent transformation on the affine flag manifold*, J. Math. Soc. Japan **32**, no. 1 (1980), 55–64.

[72] N. Spaltenstein, *Classes unipotentes et sous-groupes de Borel*, Lecture Notes in Mathematics **946**, Springer 1982.

[73] L. Spice, *Topological Jordan decompositions*, J. Alg. **319** (2008), 3141–3163.

[74] T. Springer, *On representations of Weyl groups*, in: Proc. Hyderabad Conference on Algebraic Groups (Hyderabad, 1989), Manoj Prakashan (1991), 517–536.

[75] Ch. Sorger, *Lectures on moduli of principal G-bundles over algebraic curves*, in: Proc. School on Algebraic Geometry (Trieste, 1999), ICTP Lect. Notes **1** (2000), 1–57.

[76] J. Tymoczko, *An introduction to equivariant cohomology and homology, following Goresky, Kottwitz, and MacPherson*, in: Snowbird Lectures on Algebraic Geometry, ed. R. Vakil, Contemp. Math. **388**, Amer. Math. Soc. 2005, 169–188.

[77] E. Viehmann. *The dimension of affine Deligne–Lusztig varieties*, Ann. sci. École Norm. Sup. 4e série, t. **39** (2006), 513–526.

[78] E. Viehmann, *Connected components of closed affine Deligne–Lusztig varieties*, Math. Ann. **340**, no. 2 (2008), 315–333.

[79] E. Viehmann, *Moduli spaces of p-divisible groups*, J. Alg. Geom. **17** (2008), 341–374.

[80] E. Viehmann, *Moduli spaces of polarized p-divisible groups*, Documenta Math. **13** (2008), 825–852.

[81] E. Viehmann, *Truncations of level 1 of elements in the loop group of a reductive group*, Preprint arXiv:0907.2331v1 (2009).

[82] B. Zbarsky, *On some stratifications of affine Deligne–Lusztig varieties for* SL_3, Preprint arXiv:0906.0186v1 (2009).

Ulrich Görtz
Institut für Experimentelle Mathematik
Universität Duisburg-Essen
Ellernstr. 29
D-45326 Essen, Germany
e-mail: ulrich.goertz@uni-due.de

Affine Flag Manifolds and Principal Bundles
Trends in Mathematics, 51–90
© 2010 Springer Basel AG

Quantization of Hitchin's Integrable System and the Geometric Langlands Conjecture

Tomás L. Gómez

Abstract. This is an introduction to the work of Beilinson and Drinfeld [6] on the Langlands program.

Mathematics Subject Classification (2000). Primary 11R39; Secondary 14D20.

Keywords. Langlands program, Chiral algebras, G-opers, Hitchin integrable system, $\mathcal{D}$-modules.

Let G be a complex reductive group, and let $T \subset G$ be a maximal torus. Let Λ and $\Lambda^\vee$ be the lattices of characters and 1-parameter subgroups of T. We have natural inclusions $\Lambda \subset \mathfrak{t}^*$ and $\Lambda^\vee \subset \mathfrak{t}$, $\mathfrak{t}$ the Lie algebra of T. Let $\Delta \subset \Lambda$ and $\Delta^\vee \subset \Lambda^\vee$ be the sets of roots and coroots. We write the root and coroot data as follows

$$\left(\Delta \subset \Lambda \subset \mathfrak{t}^*, \ \Delta^\vee \subset \Lambda^\vee \subset \mathfrak{t} \right).$$

Let G' be another reductive complex group with a maximal torus T' and root and coroot data

$$\left(\Delta' \subset \Lambda' \subset \mathfrak{t}'^*, \ \Delta'^\vee \subset \Lambda'^\vee \subset \mathfrak{t}' \right).$$

Assume that there is an isomorphism $\varphi \colon \mathfrak{t} \to \mathfrak{t}'^*$ interchanging the root and coroot data. More precisely, φ should identify Δ with $\Delta'^\vee$ and Λ with $\Lambda'^\vee$, and the isomorphism $\mathfrak{t}' \to \mathfrak{t}^*$ induced by φ should identify Δ' with $\Delta^\vee$ and Λ' with $\Lambda^\vee$. Then we say that G' is the Langlands dual group of G, and we denote it by ^LG. We have $^L(^LG) = G$.

For example, $^L\mathrm{GL}_n \cong \mathrm{GL}_n$, $^L\mathrm{SL}_n \cong \mathrm{PSL}_n$, $^L\mathrm{SO}_{2n} \cong \mathrm{SO}_{2n}$, $^L\mathrm{SO}_{2n+1} \cong \mathrm{Sp}_{2n}$ and if T is a complex torus, then ^LT is the torus dual to T. In general, the center $Z(G)$ of G is naturally isomorphic to $\mathrm{Hom}(\pi_1(^LG), \mathbb{C}^*)$, the Pontrjagin dual of the fundamental group of ^LG.

Let X be a smooth projective curve over $\mathbb{C}$ with genus $g > 1$, and let Bun_G be the moduli stack of principal G-bundles. An ^LG-local system on X is a differentiable principal ^LG-bundle together with a flat connection. Equivalently, it is a

Supported by the Spanish Ministerio de Educación y Ciencia [MTM2007-63582].

pair (P, ∇) where P is a holomorphic ${}^L G$-bundle and ∇ a holomorphic connection (the flat connection decomposes into a $(0,1)$ part, which is the holomorphic
structure on P, and a $(1,0)$ part, which is the holomorphic connection).

Conjecture 0.1 (Geometric Langlands). *For each irreducible ${}^L G$-local system E on
X, there is an irreducible holonomic $\mathcal{D}$-module $\mathcal{F}_E$ on Bun_G, such that $\mathcal{F}_E$ is a
"Hecke eigensheaf" with "eigenvalue" E (cf. Section 4).*

This conjecture is inspired by the Langlands correspondence between Galois
${}^L G$-representations and automorphic functions on $G(\mathbb{A})$, where $\mathbb{A}$ is the ring of
adèles.

In these notes we only consider the case of curves over $\mathbb{C}$, but we should
mention here that there is a version of the geometric Langlands conjecture for
curves over $\mathbb{F}_q$, but using perverse sheaves instead of $\mathcal{D}$-modules. For $G = \mathrm{GL}_n$ this
conjecture is due to Drinfeld and Laumon [26] generalizing Drinfeld's construction
for GL_2 [12] and Deligne's proof for GL_1, and it is was proved by Lafforgue [25].
The version over a field of characteristic zero (or also a finite field of l elements,
with l sufficiently large) was proved by Frenkel, Gaitsgory and Vilonen in [15, 18].

For a complex reductive group, the geometric Langlands conjecture is due to
Beilinson and Drinfeld, and they proved it (over $\mathbb{C}$) when G is semisimple and E
is an ${}^L G$-oper [6] (cf. Section 5).

For a geometric motivation, consider the group $G = \mathrm{GL}_1$ and the Picard
scheme $\mathrm{Pic}(X)$ (note that we are using the Picard scheme, not the Picard stack).
As we have said, there is a proof in this case due by Deligne, which works both
over $\mathbb{F}_q$ and over $\mathbb{C}$ (see [14, 4.1]). In the complex case there is a direct construction
(see [14, 4.3]) which we explain now.

Let's look at $\mathbb{C}^*$-local systems on X. They are in bijection with 1-dimensional
representations of $\pi_1(X)$, or $\pi_1^{\mathrm{ab}}(X)$ (since the representations are 1-dimensional,
they factor through the abelianization). Consider the embedding $i \colon X \subset \mathrm{Pic}^1(X)$
given by the Abel–Jacobi map. This map induces an isomorphism between the
abelianization $\pi_1^{\mathrm{ab}}(X)$ of the fundamental group of X and the fundamental group
$\pi_1(\mathrm{Pic}^1(X))$ of $\mathrm{Pic}^1(X)$ (this is already Abelian), and hence a bijection between
1-dimensional local systems. This bijection gives the following theorem (which is
a reformulation of the geometric Langlands conjecture)

Theorem 0.2. *For each 1-dimensional local system E on X, there exists a 1-
dimensional local system $\mathcal{F}_E$ on $\mathrm{Pic}(X)$ such that*

1. *$m^*(\mathcal{F}_E) \cong \mathcal{F}_E \boxtimes \mathcal{F}_E$, where $m \colon \mathrm{Pic}(X) \times \mathrm{Pic}(X) \longrightarrow \mathrm{Pic}(X)$ is tensor
   multiplication of line bundles.*
2. *$i^* \mathcal{F}_E \cong E$, where $i \colon X \longrightarrow \mathrm{Pic}^1(X)$ is the Abel–Jacobi map.*

Therefore the geometric Langlands conjecture can be seen as a (rather non-
trivial!) generalization of this theorem. For simplicity, in these notes we will assume
that G is semisimple, simply connected. It follows that ${}^L G$ is of adjoint type.

A very good introduction to the geometric Langlands program, with an explanation of how it is related to the Langlands program in number theory, see

Frenkel's lectures [14]. Kapustin and Witten [24] have found a relationship between the geometric Langlands program and the S-duality appearing in four-dimensional gauge theories. There is a different version of Langlands duality (cf. [21, 11]) which can be considered as a "classical limit" of the Langlands duality discussed in this article. It states that the fibers of the Hitchin map for the moduli of principal Higgs G-bundles should be dual to the fibers of the Hitchin map for $^L G$.

Defining $\mathcal{D}$-modules on an arbitrary smooth stack is technically difficult, but we will deal with a special class of stacks (DG-free stacks), and this leads to some technical simplifications (Section 1). The stack of G-modules is constructed using an idea that goes back to A. Weil (Section 3). In Section 4 we define the Hecke eigenproperty, and with this, the statement of the Geometric Langlands conjecture is complete.

Beilinson and Drinfeld construct the eigensheaf $\mathcal{F}_\sigma$ for a certain class of local systems, called opers (Section 5). There is a general way of producing $\mathcal{D}$-modules using an integrable quantum system (cf. Section 6). The quantum integrable system they use is defined in Section 9, using the formalism of the localization functor for Harish-Chandra modules (Section 8). An important point is that the quantum system used by Beilinson and Drinfeld is a quantization of the classical integrable system of Hitchin. This system is described in Section 7, and it is recast in the language of chiral algebras in Section 10 (see Section 2 for the definition of a chiral algebra). This second description is needed to show that Beilinson–Drinfeld's system is indeed a quantization of Hitchin's system (cf. 11). The fact that Beilinson–Drinfeld's system is a quantization of Hitchin's system is important, because it tells us that the $\mathcal{D}$-module $\mathcal{F}_\sigma$ we have constructed is nonzero. Finally, in Section 12, we state a few words about how to prove that $\mathcal{F}_\sigma$ is a Hecke eigensheaf.

Notation. Given an affine scheme Z, we will denote by $\mathcal{O}(Z)$ its coordinate ring. Let P be a principal H-bundle on M, and let H act on F. Then we denote by

$$P \times_H F = (P \times F)/H$$

the associated fiber bundle on M with fiber F. Unless otherwise noted, X will be a fixed smooth projective curve over $\mathbb{C}$ with genus $g > 1$. All the derived categories that appear in this article are assumed to be bounded derived categories. For more details about sheaves on stacks, see [33, 28, 29, 30]

1. $\mathcal{D}$-modules on stacks

Given a smooth scheme, the cotangent bundle is a vector bundle of rank equal to the dimension of the scheme. The correct generalization of this notion, when working with smooth Artin stacks, is not a vector bundle, but rather the *cotangent complex* [22, 23, 27]. Fortunately, the Artin stacks we will consider are "good" (or DG-free, see definition below), and for this class of stacks one can give a simplified definition. For this section, see [6, Section 1 and 7].

Let $\mathcal{Y}$ be a smooth algebraic stack, let Z be a scheme, and let $\pi\colon Z \longrightarrow \mathcal{Y}$ be a smooth morphism. Let $T_{Z/\mathcal{Y}}$ be the relative tangent sheaf on Z. This sheaf is defined as $T_{Z/\mathcal{Y}} := \Delta^*(T_{(Z\times_{\mathcal{Y}} Z)/Z})$, where $\Delta\colon Z \longrightarrow Z \times_{\mathcal{Y}} Z$ is the diagonal.

For all smooth morphisms $Z \longrightarrow \mathcal{Y}$ there is a complex $T_{Z/\mathcal{Y}} \to T_Z$. This morphism is not necessarily a bundle morphism. Given smooth morphisms

$$Z' \xrightarrow{g} Z \longrightarrow \mathcal{Y},$$

the canonical map

$$g^*(T_{Z/\mathcal{Y}} \to T_Z) \xrightarrow{\sim} (T_{Z'/\mathcal{Y}} \to T_{Z'})$$

is a quasi-isomorphism, and hence the following definition makes sense.

Definition 1.1. The sheaf $T_{\mathcal{Y}}$ on $\mathcal{Y}$ is defined as follows: For any smooth morphism $\pi\colon Z \to \mathcal{Y}$ we set

$$\pi^*(T_{\mathcal{Y}}) := T_Z/T_{Z/\mathcal{Y}}.$$

Define the stack $T^*\mathcal{Y}$ as

$$T^*\mathcal{Y} := \operatorname{Spec}_{\mathcal{Y}}(\operatorname{Sym}(T_{\mathcal{Y}})).$$

In general we have $\dim T^*\mathcal{Y} \geq 2\dim \mathcal{Y}$.

Definition 1.2 (cf. [6, Section 1.1.1]). A smooth algebraic stack is called good (or DG free) if one of the following equivalent conditions holds

1. $\dim T^*\mathcal{Y} = 2\dim \mathcal{Y}$
2. $\operatorname{codim}\{\, y \in \mathcal{Y} \mid \dim \operatorname{Aut}(y) = n\,\} \geq n,\ \forall\, n > 0$
3. The complex $\operatorname{Sym}(T_{Z/\mathcal{Y}} \longrightarrow T_Z)$, defined as

$$\cdots \longrightarrow \operatorname{Sym}(T_Z) \otimes \bigwedge^2 T_{Z/\mathcal{Y}} \longrightarrow \operatorname{Sym}(T_Z) \otimes T_{Z/\mathcal{Y}} \longrightarrow \operatorname{Sym}(T_Z),$$

is exact except in degree 0, and the 0th cohomology of this complex is $\operatorname{Sym}(T_Z)/\operatorname{Sym}(T_Z)T_{Z/\mathcal{Y}}$.
4. The morphism $T^*Z \longrightarrow T^*Z/\mathcal{Y}$ is flat.

Note that item 2 implies that there is a dense open substack of $\mathcal{Y}$ where it is Deligne–Mumford. Note that the stack $T^*\mathcal{Y}$ is well defined for an arbitrary smooth stack, but it is only a reasonable definition for the cotangent stack if $\mathcal{Y}$ is good. By this I mean that, for an arbitrary smooth stack, the complex in item 3 has cohomology in several degrees, but our definition only encodes the 0th cohomology. For an arbitrary smooth stack, instead of a cotangent bundle we have a cotangent complex, instead of the symmetric algebra we have a DG-algebra, and instead of a scheme (locally on the stack) we have a DG-scheme [3].

Definition 1.3. A left $\mathcal{D}$-module on $\mathcal{Y}$ is a $\mathcal{D}_Z$-module M_Z for each smooth morphism, $Z \to \mathcal{Y}$, with the obvious compatibility conditions. More precisely, given a pair $f_i\colon Z_i \to \mathcal{Y}$ and $f_j\colon Z_j \to \mathcal{Y}$ of smooth morphisms, and denoting $Z_{ij} = Z_i \times_{\mathcal{Y}} Z_j$, an isomorphism between $p_i^* M_{Z_i}$ and $p_j^* M_{Z_j}$ is given, satisfying a cocycle condition on triples.

The sheaf $\mathcal{D}_{\mathcal{Y}}$ is defined as follows: For any smooth morphism $\pi \colon Z \to \mathcal{Y}$ we set

$$\pi^*(\mathcal{D}_{\mathcal{Y}}) := \mathcal{D}_Z/(\mathcal{D}_Z \cdot T_{Z/\mathcal{Y}}).$$

Again, this definition for the sheaf of differential operators on $\mathcal{Y}$ is reasonable if $\mathcal{Y}$ is DG-free. For a general smooth scheme we should have considered the relative de Rham complex

$$\cdots \longrightarrow \mathcal{D}_Z \otimes \overset{2}{\bigwedge} T_{Z/\mathcal{Y}} \longrightarrow \mathcal{D}_Z \otimes T_{Z/\mathcal{Y}} \longrightarrow \mathcal{D}_Z. \tag{1.1}$$

The point is that if $\mathcal{Y}$ is DG-free, then this complex is exact except in degree 0, and hence it is enough to take only the 0th cohomology instead of the whole complex. For the definition of $\mathcal{D}$-module on an arbitrary smooth stack, see [6, 7.3, 7.5]. For the following lemma, see [6, 1.1.4].

Lemma 1.4. *If $\mathcal{Y}$ is DG-free, then there is an isomorphism*

$$p_*\mathcal{O}_{T^*\mathcal{Y}} \overset{\cong}{\longrightarrow} \operatorname{gr}\mathcal{D}_{\mathcal{Y}},$$

where $p \colon T^*\mathcal{Y} \longrightarrow \mathcal{Y}$ is the natural projection.

For an arbitrary smooth stack, this morphism is only a surjection. The inverse of this isomorphism gives the symbol map σ. The following composition is denoted $\sigma_{\mathcal{Y}}$:

$$\operatorname{gr}\Gamma(\mathcal{Y}, \mathcal{D}_{\mathcal{Y}}) \longrightarrow \Gamma(\mathcal{Y}, \operatorname{gr}\mathcal{D}_{\mathcal{Y}}) \overset{\Gamma(\sigma^{-1})}{\longrightarrow} \Gamma(\mathcal{Y}, p_*\mathcal{O}_{T^*\mathcal{Y}}).$$

Analogously, given a line bundle L on $\mathcal{Y}$, we can define the category of L-twisted $\mathcal{D}$-modules on $\mathcal{Y}$, and the sheaf $\mathcal{D}_{\mathcal{Y}}^L$.

Proposition 1.5. *If G is semisimple, then the moduli stack of principal G-bundles on a smooth curve X of genus $g > 0$ is DG-free.*

This is proved in [6, 2.10.5].

2. Chiral algebras

In this section we will recall some definitions and constructions which will be used in Section 9. Chiral algebras first appeared in Mathematical Physics, in the study of conformal field theory [8]. From the mathematical point of view, chiral algebras can be considered as the geometric approach to the vertex algebras introduced in [9]. For an introduction to vertex algebras, and its relationship with chiral algebras, see [16]. For a reference on chiral algebras, see [7], [17] or [1]. For the theory of $\mathcal{D}$-modules, see [10].

Let X be a complex curve. Let $\mathcal{D}_X$ be the sheaf of differential operators on X. Unless otherwise stated, by $\mathcal{D}_X$-module we mean a left $\mathcal{D}_X$-module, i.e., a quasi-coherent $\mathcal{O}_X$-module endowed with a left action of $\mathcal{D}_X$.

Explicitly, M is a $\mathcal{D}_X$-module if there is an action of T_X on M such that

1. $\xi(fm) = \xi(f)m + f\xi(m)$
2. $(f\xi)(m) = f\xi(m)$
3. $[\xi_1, \xi_2](m) = \xi_1(\xi_2(m)) - \xi_2(\xi_1(m))$

for all $\xi \in T_X$, $f \in \mathcal{O}_X$ and $m \in M$.

Let ω_X be the dualizing sheaf. Recall that $M \mapsto M \otimes_{\mathcal{O}_X} \omega_X$ gives an equivalence of categories from the category of left $\mathcal{D}_X$-modules to right $\mathcal{D}_X$-modules. Indeed, ω_X is a right $\mathcal{D}_X$-module (a tensor field $\xi \in T_X \subset \mathcal{D}_X$ acts on $\nu \in \omega_X$ as $\nu \cdot \xi = -\mathrm{Lie}_\xi(\nu)$), and ξ acts on $m \otimes \nu \in M \otimes_{\mathcal{O}_X} \omega_X$ as $m \otimes (\nu \cdot \xi) - \xi(m) \otimes \nu$. The inverse is given by $M \mapsto M \otimes_{\mathcal{O}_X} \omega_X^{-1}$.

The category $(\mathcal{D}_X\text{-mod})$ is a tensor category: Given two $\mathcal{D}_X$-modules M and N, the product $M \otimes_{\mathcal{O}_X} N$ gets a $\mathcal{D}_X$-module structure by the Leibniz formula

$$\xi(m \otimes n) = \xi(m) \otimes n + m \otimes \xi(n).$$

Let $f: Y \to Z$ be a morphism between smooth schemes. Let N be a right $\mathcal{D}_Z$-module. Let f^{-1} be the inverse image in the category of sheaves. The sheaf

$$\mathcal{D}_{Y \to Z} = \mathcal{O}_Y \otimes_{\pi^{-1}\mathcal{O}_Z} \pi^{-1}\mathcal{D}_Z$$

is a left $\mathcal{D}_Y$-module and right $\pi^{-1}\mathcal{D}_Z$-module. The sheaf $f^{-1}N$ is a left $\pi^{-1}\mathcal{D}_Z$-module, and therefore

$$f^! N = \mathcal{D}_{Y \to Z} \otimes_{\pi^{-1}\mathcal{D}_Z} f^{-1}N$$

is a right $\mathcal{D}_Y$-module. Now let M be a right $\mathcal{D}_Y$-module. The sheaf

$$\mathcal{D}_{Z \leftarrow Y} = \mathcal{D}_{Y \to Z} \otimes_{\mathcal{O}_Y} \omega_Y \otimes_{\pi^{-1}\mathcal{O}_Z} \pi^{-1}\omega_X^{-1}$$

is a right $\mathcal{D}_Y$-module and left $\pi^{-1}\mathcal{D}_Z$-module, hence

$$f_! M = M \otimes_{\mathcal{D}_Y} \mathcal{D}_{Z \leftarrow Y}$$

is a left $\pi^{-1}\mathcal{D}_Z$ module, and hence also a left $\mathcal{D}_Z$-module. If f is an open embedding, then $f^!$ is just restriction, and is therefore denoted f^*, and, on the other hand, $\mathcal{D}_{Z \leftarrow Y}$ is just $\mathcal{D}_Y$, so, for an open embedding, $f_!$ is denoted f_*.

A $\mathcal{D}_X$-algebra is a commutative algebra with unit in the tensor category of $\mathcal{D}_X$-modules. For example, $\mathcal{O}_X$ is a $\mathcal{D}_X$-algebra, and if F is a commutative $\mathbb{C}$-algebra, then $F \otimes_{\mathbb{C}} \mathcal{O}_X$ is a $\mathcal{D}_X$-algebra.

Consider the forgetful functor from $(\mathcal{D}_X\text{-alg})$ to $(\mathcal{O}_X\text{-alg})$. It has a left adjoint functor, called the jet construction $J(\cdot)$

$$\mathrm{Hom}_{\mathcal{D}_X\text{-alg}}(J(C), B) = \mathrm{Hom}_{\mathcal{O}_X\text{-alg}}(C, \mathrm{Forget}(B)) \qquad (2.1)$$

Explicitly, $J(C)$ is the $\mathcal{D}$-algebra defined as the quotient of $\mathrm{Sym}_{\mathcal{O}_X}(\mathcal{D}_X \otimes_{\mathcal{O}_X} C)$ by the ideal generated by $(1 \otimes c_1) \cdot (1 \otimes c_2) - (1 \otimes c_1 \cdot c_2)$ and $(1 \otimes 1) - 1$.

From now on we will assume that X is a projective curve. Consider the functor from $(\mathbb{C}\text{-alg})$ to $(\mathcal{D}_X\text{-alg})$ sending F to $F \otimes \mathcal{O}_X$. It has a left adjoint, called the coinvariants construction $H_\nabla(X, \cdot)$

$$\mathrm{Hom}_{\mathbb{C}\text{-alg}}(H_\nabla(X, A), F) = \mathrm{Hom}_{\mathcal{D}_X\text{-alg}}(A, F \otimes_{\mathcal{O}_X} \mathcal{O}_X).$$

It is given by the formula

$$H_\nabla(X, A) = A_x/\mathrm{DR}^0(X - x, A) \otimes A_x$$

where x is a point on X, $A_x = A \otimes_{\mathcal{O}_X} \mathcal{O}_X/\mathfrak{m}_x$ is the fiber at x, and $\mathrm{DR}^0(X - x, A)$ is the set of sections on $X - x$ of coinvariants of the action of T_X on A^r, where $A^r = A \otimes \Omega_X$ is the right $\mathcal{D}$-module corresponding to the left $\mathcal{D}$-module A. The algebra $H_\nabla(X, A)$ is independent of the point x chosen. This formula shows that for any $x \in X$, there is a surjection

$$A_x \twoheadrightarrow H_\nabla(X, A).$$

Let $\Delta\colon X \longrightarrow X \times X$ be the diagonal and $j\colon X \times X - \Delta(X) \longrightarrow X \times X$ the inclusion of the complement of the diagonal.

An algebra structure on an $\mathcal{O}_X$-module M can be described by a homomorphism

$$M \boxtimes M \to \Delta_* M.$$

Chiral algebras are "meromorphic" generalizations of this notion for $\mathcal{D}_X$-modules, allowing poles along the diagonal. Therefore, instead of $M \boxtimes M$ we consider $j_* j^* M \boxtimes M$. A local section of this sheaf is of the form $f(x, y) a \boxtimes b$ where $f(x, y)$ is a rational function on an open subset of $X \times X$, where we allow poles along the diagonal, and a and b are local sections of M. Also, instead of $\Delta_* M$ ("extension of M by zero"), we will use $\Delta_! M$. Denote by σ_{12} the automorphism of $X \times X$ which permutes the factors. It follows from the definition of $\Delta_! M$ that there is a canonical lift to $\Delta_! M$, which we denote

$$\widetilde{\sigma}_{12}\colon \Delta_! M \longrightarrow \sigma_{12}^* \Delta_! M.$$

A chiral algebra is a right $\mathcal{D}_X$-module A together with a right $\mathcal{D}_{X \times X}$-module homomorphism

$$\{\ \ \}\colon j_* j^*(A \boxtimes A) \longrightarrow \Delta_!(A)$$

such that (antisymmetry)

$$\{f(x, y) a \boxtimes b\} = -\widetilde{\sigma}_{12}\{f(y, x) b \boxtimes a\},$$

where $f(y, x)$ is the transposition σ_{12} composed with $f(x, y)$, and if $f(x, y, z) a \boxtimes b \boxtimes c$ is a section on the complement of all the diagonals in $X \times X \times X$, then (Jacobi identity)

$$\{\{f(x, y, z) a \boxtimes b\} \boxtimes c\} + \widetilde{\sigma}_{123}\{\{f(x, y, z) b \boxtimes c\} \boxtimes a\} + \widetilde{\sigma}_{123}^2\{\{f(y, z, x) c \boxtimes a\} \boxtimes b\} = 0$$

as a section of $\Delta_{x=y=z*}(A)$, where $\widetilde{\sigma}_{123}$ is the lift of the cyclic permutation. A unit of a chiral algebra is a morphism $u\colon \Omega_X \longrightarrow A$ such that the following diagram is commutative

$$\begin{CD}
j_* j^*(\Omega_X \boxtimes A) @>u \boxtimes \mathrm{id}>> j_* j^*(A \boxtimes A) \\
@VfVV @VV\{\ \}V \\
\Delta_!(A) @= \Delta_!(A)
\end{CD} \qquad (2.2)$$

where the morphism f comes from the short exact sequence

$$0 \longrightarrow \Omega_X \boxtimes A \longrightarrow j_* j^* (\Omega_X \boxtimes A) \longrightarrow \Delta_!(A) \longrightarrow 0.$$

A chiral algebra is called commutative if the restriction of the bracket $\{\ \}$ to $A \boxtimes A \subset j_* j^* (A \boxtimes A)$ is zero

$$\{\ \}|_{A \boxtimes A} = 0.$$

Alternatively, the bracket factors as

$$
\begin{array}{ccc}
j_* j^* (A \boxtimes A) & \xrightarrow{\ \ \{\}\ \ } & \Delta_! A \\
& \searrow{\scriptstyle p} & \nearrow \\
& \Delta_! \Delta^! (A \boxtimes A) & \cong \ \Delta_!(A \otimes A) .
\end{array}
\tag{2.3}
$$

Hence, in a commutative chiral algebra, the bracket map comes from a morphism $A \otimes A \longrightarrow A$.

Proposition 2.1. *If A is a commutative chiral algebra, then $A \otimes \Omega_X^{-1}$ is a $\mathcal{D}_X$-algebra. Conversely, if A is a $\mathcal{D}_X$-algebra, then $A^r := A \otimes \Omega_X$ is a commutative chiral algebra, with the bracket defined using the algebra structure $A^r \otimes A^r \longrightarrow A^r$ and (2.3).*

A Lie* algebra is a right $\mathcal{D}_X$-module A together with a right $\mathcal{D}_{X \times X}$-module morphism

$$[\ ]: A \boxtimes A \longrightarrow \Delta_! A$$

which is antisymmetric and satisfies the Jacobi identity, in a similar way as in the definition of chiral algebra. Given a chiral algebra, we define a Lie* algebra by composition

$$[\ ]: A \boxtimes A \hookrightarrow j_* j^* (A \boxtimes A) \xrightarrow{\{\}} \Delta_! A.$$

This gives a "forgetful" functor from (Ch_X) to (Lie_X^*). It has a left adjoint functor, called the chiral envelope

$$\mathrm{Hom}_{\mathrm{Ch}_X} (\mathcal{U}(L), A) \ = \ \mathrm{Hom}_{\mathrm{Lie}_X^*} (L, \mathrm{Forget}(A)).$$

Let $\mathfrak{g}$ be a Lie algebra, and let q be an invariant symmetric bilinear form on $\mathfrak{g}$. Consider the right $\mathcal{D}_X$-module

$$\mathfrak{g} \otimes \mathcal{D}_X \oplus \Omega_X \tag{2.4}$$

with bracket

$$[g_1 \otimes 1 \boxtimes g_2 \otimes 1] = [g_1, g_2] \otimes 1 \oplus q(g_1, g_2)\mathbf{1}'$$

where $\mathbf{1}'$ is the canonical antisymmetric section of $\Delta_!(\Omega_X)$. This gives (2.4) the structure of a Lie* algebra, called the Kac–Moody Lie* algebra. The fiber of $\mathcal{U}(\mathfrak{g} \otimes \mathcal{D}_X \oplus \Omega_X)$ over x is

$$\mathcal{U}(\mathfrak{g} \otimes \mathcal{D}_X \oplus \Omega_X)_x \ = \ \mathrm{Ind}_{\mathfrak{g} \otimes \hat{\mathcal{O}} \oplus \mathbb{C}\mathbf{1}}^{\hat{\mathfrak{g}}_q} \ \mathbb{C}. \tag{2.5}$$

See Section 9 for the definition. For a proof of this formula, and further details about the chiral envelope, see [7, Section 3.7].

3. Geometry of the affine Grassmannian

For a reference for this section, see [34, Section 5] and the references therein. Given a ring R, $R[[t]]$ will be the ring of formal power series with coefficients in R, and $R((t))$ will be the ring of formal Laurent series. Let $\widehat{\mathcal{O}} = \mathbb{C}[[t]]$, and let $\widehat{\mathcal{K}} = \mathbb{C}((t))$ be its quotient field. Let Z be an affine scheme. We denote by $Z[[t]]$ (or $Z(\widehat{\mathcal{O}})$) the functor defined as

$$\mathrm{Hom}\big(S, Z[[t]]\big) := \mathrm{Hom}_{\mathrm{alg}}\big(\mathcal{O}(Z), \mathcal{O}(S)[[t]]\big)\,,$$

where S is an affine scheme. It can be shown that $Z[[t]]$ is representable by a scheme. Note that the $\mathbb{C}$-valued points of $Z[[t]]$ are the $\widehat{\mathcal{O}}$-valued points of Z. We will denote by $Z((t))$ (or $Z(\widehat{\mathcal{K}})$) the functor

$$\mathrm{Hom}\big(S, Z((t))\big) := \mathrm{Hom}_{\mathrm{alg}}\big(\mathcal{O}(Z), \mathcal{O}(S)((t))\big).$$

Note that the $\mathbb{C}$-valued points of $Z((t))$ are the $\widehat{\mathcal{K}}$-valued points of Z. It can be shown that $Z((t))$ is an ind-scheme. Recall that a functor is called an ind-scheme if is representable by a direct limit of closed embeddings. More precisely a functor $F\colon (\mathrm{Sch}) \to (\mathrm{Sets})$ is called an ind-scheme if there are schemes Y_i, $i \in \mathbb{N}$, closed embeddings $Y_i \to Y_{i+1}$ and

$$F = \varinjlim Y_i,$$

where this functor is defined as

$$(\varinjlim Y_i)(S) = \varinjlim \mathrm{Hom}(S, Y_i).$$

If Y_i and Y_j' are two inductive systems that are cofinal (i.e., for all i there is an j such that $Y_i \subset Y_j'$ and for all j there is an i such that $Y_i \supset Y_j'$), then the functors $\varinjlim Y_i$ and $\varinjlim Y_i'$ are canonically isomorphic. We say that an ind-scheme is of ind-finite type if the schemes Y_i can be chosen of finite type. We say that it is ind-complete if the schemes can be chosen to be complete, and analogously, for any property P that is stable under restriction to a closed subscheme, we say that an ind-scheme is ind-P if the schemes Y_i have the property P.

A vector bundle over the disk $\mathbb{D} = \mathrm{Spec}\,\widehat{\mathcal{O}}$ is a finitely generated free $\widehat{\mathcal{O}}$-module. A family of vector bundles over $\mathbb{D}$ parameterized by an affine scheme S is a finitely generated $\mathcal{O}(S)[[t]]$-module, such that locally in the Zariski topology of S, it is isomorphic to the trivial bundle.

Replacing $\mathbb{D}$ with $\mathbb{D}^{\times} = \mathrm{Spec}\,\widehat{\mathcal{K}}$ and $\widehat{\mathcal{O}}$ with $\widehat{\mathcal{K}}$, we obtain the analogous notions for the punctured disk.

Definition 3.1. A family of principal G-bundles on $\mathbb{D}$ parameterized by an affine scheme S is a tensor functor from the category of representations of G to the category of S-families of vector bundles on $\mathbb{D}$

$$\mathrm{Rep}(G) \longrightarrow (S\text{-families of vector bundles on } \mathbb{D}).$$

We also have the analogous notion for the punctured formal disk.

Let G be an affine algebraic group. Then $G[[t]]$ is a group scheme, and $G((t))$ is a group ind-scheme. We define the affine Grassmannian Gr_G to be the quotient (as fpqc sheaves) $G((t))/G[[t]]$.

Lemma 3.2. *The affine Grassmannian is naturally isomorphic to the functor*

$$S \longmapsto (P_G, \beta \colon P_G|_{\mathbb{D}^\times \times S} \longrightarrow P_G^0|_{\mathbb{D}^\times \times S})$$

where P_G is a family of G-bundles on the formal disk $\mathbb{D}$ parameterized by S, P_G^0 is the trivial G-bundle, and β is an isomorphism.

Theorem 3.3. *The affine Grassmannian Gr_G is an ind-scheme of ind-finite type. If G is reductive, then Gr_G is ind-complete.*

A pro-algebraic group H is an affine group scheme that is represented by a projective limit of affine algebraic groups of finite type, i.e., there are algebraic groups H_i of finite type, group morphisms $H_{i+1} \to H_i$ and

$$H = \varprojlim H_i.$$

For example, $G[[t]]$ is a pro-algebraic group. The group of automorphisms of $\mathbb{C}[[t]] = \widehat{\mathcal{O}}$, denoted $\mathrm{Aut}(\widehat{\mathcal{O}})$, is also a pro-algebraic group

$$\mathrm{Aut}(\widehat{\mathcal{O}}) = \varprojlim \Big(\mathrm{Aut}\left(\mathbb{C}[t]/(t^i)\right) \Big).$$

Definition 3.4. An action of a pro-algebraic H group on an ind-scheme Y is nice if we can write $Y = \lim Y_i$ such that

- Y_i is H-invariant for all i.
- The H-action on Y_i factors through a finite-dimensional quotient of H.

Lemma 3.5. *The natural action of $G(\widehat{\mathcal{O}}) = G[[t]]$ on $\mathrm{Gr}_G = G((t))/G[[t]]$ is nice. The natural action of $\mathrm{Aut}(\widehat{\mathcal{O}})$ on Gr_G is also nice.*

There are countably many orbits of $G(\widehat{\mathcal{O}})$ on Gr_G, and these orbits are enumerated by $\Lambda^+ = \Lambda/W$, where Λ are the coweights of G, Λ^+ is the semi-group of dominant coweights, and W is the Weyl group. The dimension of the orbit corresponding to the coweight λ is

$$\dim(\mathrm{Gr}_G^\lambda) = 2\rho(\lambda)$$

where ρ is half the sum of the positive roots (see [31, Section 2]).

Now we will explain the relationship between the affine Grassmannian and the moduli stack of principal G-bundles on a curve X. We choose a point $x \in X$ and fix an isomorphism between $\widehat{\mathcal{O}} = \mathbb{C}[[t]]$ and the completion $\widehat{\mathcal{O}}_x$ of the local ring of X at x. This induces an isomorphism between $\widehat{\mathcal{K}} = \mathbb{C}((t))$ and the quotient field $\widehat{\mathcal{K}}_x$ of $\widehat{\mathcal{O}}_x$. In other words, we have chosen a local parameter t at the point x.

Define the functor $L_X G$ as

$$\mathrm{Hom}(S, L_X G) := \mathrm{Hom}_{\mathrm{alg}}\big(\mathcal{O}(G), \mathcal{O}(S) \otimes \mathcal{O}(X - x)\big),$$

so that $\mathbb{C}$-points of $L_X G$ correspond to morphisms from $X - x$ to G.

The intuitive idea of uniformization is the following: A principal bundle is trivial when restricted to a disk $\mathbb{D}$ or the complement of a point $X^* = X - x$ (for the latter, we need G to be semisimple). Therefore, to describe a principal bundle we have to give the transition function, which is a map from the intersection, i.e., the pointed disk $\mathbb{D}^\times$, to G, in other words, a $\mathbb{C}$-valued point of $G((t))$. Then we have to "forget" the trivializations, so the set of isomorphism classes of principal bundles will correspond to the double quotient $L_X G \setminus G((t))/G[[t]]$. If we take this quotient in the sense of stacks, we obtain the moduli stack of principal G-bundles on X.

Of course, we have to work with families of bundles, and to make this rigorous we need two technical theorems. The first one tells us that the restriction to $X - x$ is trivial [13].

Theorem 3.6 (Drinfeld–Simpson). *Suppose G is semisimple. Let S be an affine scheme and let P be a principal G-bundle on $X \times S$. Then the restriction of P to $(X - x) \times S$ is trivial, locally for the étale topology on S.*

In positive characteristic we would need the fppf topology. The second theorem tells us that we can glue trivial G-bundles on $\mathbb{D}$ and $X - x$ to obtain a principal G-bundle on X. In [2] it is proved for vector bundles, but it is easy to generalize to principal G-bundles (see also [32]).

Theorem 3.7 (Beauville–Laszlo). *Let γ be an S-valued point of $G((t))$. Then there exists a principal G-bundle P on $X \times S$ and trivializations σ and τ on $\mathbb{D} \times S$ and $(X - x) \times X$ whose difference on the intersection $\mathbb{D}^\times \times S$ is γ. Moreover, the triple (P, σ, τ) is unique up to unique isomorphism.*

Using these theorems, the uniformization theorem for principal G-bundles follows (recall that we are assuming that G is semisimple).

Theorem 3.8.

1. *The* ind-*scheme $G(\widehat{\mathcal{K}})$ represents the functor of principal bundles with a trivialization on the formal disk and on $X^* = X - x$ (recall Definition 3.1)*
$$\mathrm{Hom}\big(S, G(\widehat{\mathcal{K}})\big) = \big\{ P, \alpha \colon P|_{\mathbb{D}_S} \to P^0|_{\mathbb{D}_S}, \beta \colon P|_{X_S^*} \to P^0|_{X_S^*} \big\},$$
   where P is a principal G-bundle on $X \times S$, $\mathbb{D}_S = \mathbb{D} \times S$, $X_S^* = (X - x) \times S$ and α, β are isomorphisms.

2. *The affine Grassmannian $G(\widehat{\mathcal{K}})/G(\widehat{\mathcal{O}})$ represents the functor of principal G-bundles with a trivialization on $X - x$*
$$\mathrm{Hom}(S, \mathrm{Gr}_G) = \big\{ P, \beta \colon P|_{X_S^*} \longrightarrow P^0|_{X_S^*} \big\}.$$

3. *The quotient $L_X G \setminus G(\widehat{\mathcal{K}})$ represents the functor $\mathrm{Bun}_{G,x}$ of principal G-bundles with a trivialization on the formal disk $\mathbb{D}$*
$$\mathrm{Hom}(S, \mathrm{Bun}_{G,x}) = \big\{ P, \alpha \colon P|_{\mathbb{D}_S} \to P^0|_{\mathbb{D}_S} \big\}. \tag{3.1}$$

4. *The space $\mathrm{Bun}_{G,x}$ is a principal $G(\widehat{\mathcal{O}})$-bundle on Bun_G, and we have*
$$\mathrm{Bun}_G \cong [\mathrm{Bun}_{G,x}/G(\widehat{\mathcal{O}})]. \tag{3.2}$$

4. Hecke eigenproperty

4.1. Convolution product

In this section we consider the category Sph_G of $G(\widehat{\mathcal{O}})$-equivariant perverse sheaves on Gr_G. We have chosen to use perverse sheaves, following [31], but we could have worked with $\mathcal{D}$-modules as in [6]. Both approaches are equivalent, thanks to the Riemann–Hilbert correspondence.

We define a convolution product $\mathcal{S}_1 * \mathcal{S}_2$ in this category, that will give a structure of symmetric tensor category. This symmetric tensor category will be equivalent to the category of representations of the group $^L G$, the Langlands dual group of G. Since a group is defined by the symmetric tensor category of its representations, this gives a geometric definition of the Langlands dual group.

We now recall the definition of perverse sheaf and equivariant perverse sheaf (see [4] and [31, Section 2]). Let H be an algebraic group acting on a scheme Y of finite type. Let $m \colon H \times Y \to Y$ be the action and let p be the projection from $H \times Y$ to Y. Fix a Whitney stratification $\mathcal{T}$ of Y such that the action of H preserves the strata (in our application, the strata will be the orbits of H). Let $D_{\mathcal{T}}(Y)$ be the bounded derived category of $\mathcal{T}$-constructible $\mathbb{C}$-sheaves. That is, the full subcategory of the derived category of $\mathbb{C}$-sheaves whose objects $\mathcal{S}$ have $H^k(Y, \mathcal{S}) = 0$ unless $k = 0$ and the restriction of the cohomology $H^k(\mathcal{S})|_T$ for any $T \in \mathcal{T}$ is a local system of finite-dimensional $\mathbb{C}$-vector spaces.

An object $\mathcal{S} \in D_{\mathcal{T}}(Y)$ is called perverse if, for all $i \colon T \hookrightarrow Y$, $T \in \mathcal{T}$,

1. $H^k(i^* \mathcal{S}) = 0$ for $k > -\dim_{\mathbb{C}} T$
2. $H^k(i^! \mathcal{S}) = 0$ for $k < \dim_{\mathbb{C}} T$.

The full subcategory (of $D_{\mathcal{T}}(Y)$) of perverse sheaves $\mathrm{Perv}_{\mathcal{T}}(Y)$ is an abelian category. An H-equivariant perverse sheaf on Y is a pair $(\mathcal{S}, \varphi)$, where $\mathcal{S}$ is a perverse sheaf $\mathcal{S}$ on Y, and φ is an isomorphism

$$\varphi \colon m^* \mathcal{S} \longrightarrow p^* \mathcal{S}$$

such that

1. (Identity) φ is the identity map when restricted to $e \times Y$ (where e is the identity element of H).
2. (Associativity) The two isomorphisms induced by the two natural maps from $H \times H \times Y$ to Y coincide.

We denote the category of equivariant perverse sheaves as $\mathrm{Perv}_H(Y)$. If H is connected, then the isomorphism φ, if it exists, is unique, and then $\mathrm{Perv}_H(Y) \subset \mathrm{Perv}_{\mathcal{T}}(Y)$. If H is a pro-algebraic group and Y is an ind-scheme and if the action is nice (cf. Definition 3.4), then we define

$$\mathrm{Perv}_H(Y) = \varinjlim \mathrm{Perv}_H(Y_i).$$

This allows us to define:

Definition 4.1. The category Sph_G of spherical sheaves is set to be $\mathrm{Perv}_{G(\hat{\mathcal{O}})}(\mathrm{Gr}_G)$, where the strata are taken to be the $G(\hat{\mathcal{O}})$-orbits in Gr_G.

A homomorphism $\lambda\colon \mathbb{C}^* \to T$ to a maximal torus of G determines a coset $\lambda \cdot G(\hat{\mathcal{O}}) \subset G(\hat{\mathcal{K}})$ and hence a point in Gr_G. Let Gr_G^λ be the $G(\hat{\mathcal{O}})$-orbit of this point. We have $\mathrm{Gr}_G^\lambda = \mathrm{Gr}_G^\mu$ if and only if λ and μ are conjugate by the Weyl group. The category Sph_G is semisimple ([31, Lemma 7.1]), and since the $G(\hat{\mathcal{O}})$-orbits Gr_G^λ are simply connected, it follows that every $G(\hat{\mathcal{O}})$-equivariant perverse sheaf on Gr_G is a direct sum of intersection cohomology sheaves $\mathrm{IC}_{\overline{\mathrm{Gr}_G^\lambda}}$, where $\mathrm{IC}_{\overline{\mathrm{Gr}_G^\lambda}}$ is the Goresky–MacPherson extension of the trivial local system on Gr_G^λ. Therefore (see [31, Proposition 2.2]), we obtain the following

Lemma 4.2. *Every object in* Sph_G *is automatically equivariant with respect to* $\mathrm{Aut}(\hat{\mathcal{O}})$.

By definition, $G(\hat{\mathcal{K}})$ is a principal $G(\hat{\mathcal{O}})$-bundle on Gr_G. Define the convolution diagram Conv_G to be the associated fiber bundle on Gr_G with fiber Gr_G

$$\mathrm{Conv}_G = G(\hat{\mathcal{K}}) \times_{G(\hat{\mathcal{O}})} \mathrm{Gr}_G .$$

Let p_1 and p be defined as follows

$$\begin{array}{ccc}
\mathrm{Conv}_G & \longrightarrow & \mathrm{Gr}_G \\
p_1\colon \quad (g_1, \overline{g_2}) & \longmapsto & \overline{g_1} \\
p\colon \quad (g_1, \overline{g_2}) & \longmapsto & \overline{g_1 g_2}.
\end{array}$$

Note that p_1 is the structure morphism of Conv_G as a Gr_G-fiber bundle over Gr_G. This fiber bundle is not trivial, but still we have an isomorphism

$$(p, p_1)\colon \mathrm{Conv}_G \xrightarrow{\cong} \mathrm{Gr}_G \times \mathrm{Gr}_G .$$

Now we give a more geometric description of Conv_G. It represents the functor

$$S \longmapsto \{P, P', \beta\colon P|_{X_S^*} \longrightarrow P'|_{X_S^*}, \beta'\colon P'|_{X_S^*} \longrightarrow P^0|_{X_S^*}\},$$

where P and P' are principal G-bundles on $X \times S$, P^0 is the trivial G-bundle, $X_S^* = (X - x) \times S$, and $\tilde{\beta}$ and β' are isomorphisms. The morphisms p_1 and p are defined as follows:

$$\begin{aligned}
p_1(P, P', \beta, \beta') &= (P', \beta') \\
p(P, P', \beta, \beta') &= (P, \beta' \circ \beta).
\end{aligned}$$

The convolution diagram is used to define a product

$$\mathrm{Perv}(\mathrm{Gr}_G) \times \mathrm{Perv}_{G(\hat{\mathcal{O}})}(\mathrm{Gr}_G) \longrightarrow \mathrm{Perv}(\mathrm{Gr}_G)$$

$$(\mathcal{S}_1, \mathcal{S}_2) \longmapsto \mathcal{S}_1 * \mathcal{S}_2.$$

To define this product, we will use the following useful construction. Let $\pi \colon P \to Y$ be a principal H-bundle on a scheme Y, and let H act on a scheme F. Let $\mathcal{S}_1$ be a perverse sheaf on Y, and let $\mathcal{S}_2$ be an H-equivariant perverse sheaf on F. Consider the sheaf $\pi^*\mathcal{S}_1 \boxtimes \mathcal{S}_2$ on $P \times F$. It is H-equivariant because $\mathcal{S}_2$ is H-equivariant, and hence it descends to a sheaf on $(P \times F)/H = P \times_H F$, i.e., the F-fiber bundle associated with the principal H-bundle

$$\operatorname{Perv}(Y) \times \operatorname{Perv}_H(F) \longrightarrow \operatorname{Perv}(P \times_H F)$$
$$(\mathcal{S}_1, \mathcal{S}_2) \longmapsto \mathcal{S}_1 \widetilde{\boxtimes} \mathcal{S}_2. \tag{4.1}$$

If P, Y and F are ind-schemes, H is a pro-algebraic group, and the action is nice, then this construction still makes sense.

Applying this to our situation, we obtain a perverse sheaf $\mathcal{S}_1 \widetilde{\boxtimes} \mathcal{S}_2$ on Conv_G. Let $p_!$ be the push-forward in the derived category (i.e., $p_! = Rp_*$, where p_* is the push-forward in the category of $\mathbb{C}$-sheaves). Apply the functor $p_!$ to define

$$\mathcal{S}_1 * \mathcal{S}_2 = p_!(\mathcal{S}_1 \widetilde{\boxtimes} \mathcal{S}_2).$$

If $\mathcal{S}_2$ is $G(\widehat{\mathcal{O}})$-equivariant, then $\mathcal{S}_1 * \mathcal{S}_2$ is also $G(\widehat{\mathcal{O}})$-equivariant. The sheaf $\mathcal{S}_1 * \mathcal{S}_2$ is perverse [31, Proposition 4.2], i.e.,

$$\mathcal{S}_1 * \mathcal{S}_2 \in \operatorname{Sph}_G .$$

Theorem 4.3. *There are functorial isomorphisms $\mathcal{S}_1 * \mathcal{S}_2 \cong \mathcal{S}_2 * \mathcal{S}_1$ and $(\mathcal{S}_1 * \mathcal{S}_2) * \mathcal{S}_3 \cong \mathcal{S}_1 * (\mathcal{S}_2 * \mathcal{S}_3)$, giving the category Sph_G the structure of a unital rigid commutative associative tensor category (in particular, the "hexagon axiom" holds, cf. [19, Section 3.7]).*

The associativity is [31, Proposition 4.5], and the commutativity is [31, Section 5]. The unit for the convolution product is $\operatorname{IC}_{\overline{\operatorname{Gr}_G^0}}$. Rigidity means that duals exist, and it is proved in [19, Proposition 1.3.1(ii)].

Now we will define Gr_X and $\operatorname{Gr}_{X \times X}$ and use them to give an alternative definition of the product $\mathcal{S}_1 * \mathcal{S}_2$. We will use this to define the commutativity isomorphism.

Let $\pi \colon \mathfrak{X} \longrightarrow X$ be the canonical principal $\operatorname{Aut}(\mathbb{C}[\![t]\!])$-bundle on X. A point of $\mathfrak{X}$ is a pair

$$(x, \varphi \colon \widehat{\mathcal{O}}_x \xrightarrow{\cong} \mathbb{C}[\![t]\!]), \tag{4.2}$$

where $x \in X$, and φ is an isomorphism between $\mathbb{C}[\![t]\!]$ and the completion $\widehat{\mathcal{O}}_x$ of the local ring at x. In general, if $\operatorname{Aut}(\widehat{\mathcal{O}})$ acts on F, we can form the associated fiber bundle

$$\mathfrak{X}(F) := \mathfrak{X} \times_{\operatorname{Aut}(\mathbb{C}[\![t]\!])} F,$$

and using (4.1), a functor

$$\operatorname{Perv}_{\operatorname{Aut}(\widehat{\mathcal{O}})}(F) \longrightarrow \operatorname{Perv}(\mathfrak{X}(F))$$
$$\mathcal{S} \longmapsto \mathcal{S}_X := \mathbb{C}_X \widetilde{\boxtimes} \mathcal{S} \tag{4.3}$$

where $\mathbb{C}_X$ is the constant sheaf on X.

We define Gr_X as the associated Gr_G-fiber bundle

$$\mathrm{Gr}_X = \mathfrak{X} \times_{\mathrm{Aut}(\mathbb{C}[t])} \mathrm{Gr}_G .$$

It is an ind-scheme. The fiber of Gr_X over a point $x \in X$ is canonically isomorphic to $G(\mathcal{K}_x)/G(\widehat{\mathcal{O}}_x)$, where $\mathcal{K}_x$ is its quotient field of $\widehat{\mathcal{O}}_x$. The ind-scheme Gr_X represents the functor

$$S \longmapsto \left(f \colon S \to X, P, \beta \colon P|_{X \times S - \Gamma_f} \xrightarrow{\cong} P^0|_{X \times S - \Gamma_f} \right),$$

where P is a principal G-bundle on $X \times S$, and β is a trivialization on $X \times S - \Gamma_f$, where Γ_f is the graph of f.

Define $\mathrm{Gr}_{X \times X}$ as the functor

$$S \longmapsto \left\{ f_1, f_2 \colon S \to X, P, \beta \right\}$$

where β is a trivialization of P on $X \times S - (\Gamma_{f_1} \cup \Gamma_{f_2})$. The functor $\mathrm{Gr}_{X \times X}$ comes with a natural projection to $X \times X$. The fiber over (x, y) when $x \neq y$ is $\mathrm{Gr}_G \times \mathrm{Gr}_G$, but it is Gr_G if $x = y$. In particular, $\mathrm{Gr}_{X \times X}$ is not a fiber bundle (as opposed to what happens with Gr_X).

Proposition 4.4. *The functor* $\mathrm{Gr}_{X \times X}$ *is representable by an* ind-*scheme. Let* $\Delta \subset X \times X$ *be the diagonal. We have the following isomorphisms*

$$\mathrm{Gr}_{X \times X}|_{X \times X - \Delta} \cong \mathrm{Gr}_X \times \mathrm{Gr}_X |_{X \times X - \Delta}$$
$$\mathrm{Gr}_{X \times X}|_{\Delta} \cong \mathrm{Gr}_X .$$

Let $\mathcal{S}$ be a $G(\widehat{\mathcal{O}})$-equivariant perverse sheaf on Gr_G. Since it is also $\mathrm{Aut}(\widehat{\mathcal{O}})$-equivariant (Lemma 4.2) we can associate with it a perverse sheaf on Gr_X as in (4.3)

$$\mathrm{Perv}_{G(\widehat{\mathcal{O}})}(\mathrm{Gr}_G) \longrightarrow \mathrm{Perv}(\mathrm{Gr}_X)$$
$$\mathcal{S} \longmapsto \mathcal{S}_X := \mathbb{C}_X \widetilde{\boxtimes} \mathcal{S}$$

where $\mathbb{C}_X$ is the constant sheaf on X.

Given a principal divisor Y_0 on a scheme Y (i.e., Y_0 is defined as the zero of a section of a line bundle on Y), we consider the nearby cycles functor

$$\Psi \colon \mathrm{Perv}(Y - Y_0) \longrightarrow \mathrm{Perv}(Y_0).$$

Taking $Y = \mathrm{Gr}_{X \times X}$, $Y_0 = \mathrm{Gr}_{X \times X}|_{\Delta}$ and using Proposition 4.4, we define

$$\mathrm{Perv}(\mathrm{Gr}_G) \times \mathrm{Perv}_{G(\widehat{\mathcal{O}})}(\mathrm{Gr}_G) \longrightarrow \mathrm{Perv}(\mathrm{Gr}_X)$$
$$(\mathcal{S}_1, \mathcal{S}_2) \longmapsto \Psi\left((\mathcal{S}_{1,X} \boxtimes \mathcal{S}_{2,X})|_{X \times X - \Delta} \right).$$

The following theorem results from [31, (5.10)].

Theorem 4.5. *We have*

$$(\mathcal{S}_1 * \mathcal{S}_2)_X \cong \Psi\left((\mathcal{S}_{1,X} \boxtimes \mathcal{S}_{2,X})|_{X \times X - \Delta} \right).$$

Therefore (cf. [31, (5.11)]):

$$(\mathcal{S}_1 * \mathcal{S}_2)_X \cong \Psi\big((\mathcal{S}_{1,X} \boxtimes \mathcal{S}_{2,X})|_{X \times X - \Delta}\big) \cong \Psi\big((\mathcal{S}_{2,X} \boxtimes \mathcal{S}_{1,X})|_{X \times X - \Delta}\big) \cong (\mathcal{S}_2 * \mathcal{S}_1)_X$$

and specializing to any point in X we get an isomorphism

$$\psi' \colon \mathcal{S}_1 * \mathcal{S}_2 \xrightarrow{\;\cong\;} \mathcal{S}_2 * \mathcal{S}_1.$$

This commutativity isomorphism provides Sph_G with the structure of a tensor category. We modify the commutativity isomorphisms ψ' with a sign in the following way. Given a connected component of Gr, all the $G(\widehat{\mathcal{O}})$-orbits have even (respectively odd) dimension ([6, Proposition 4.5.11]), and then we say that this component is even (respectively odd). Given an irreducible perverse sheaf $\mathcal{S}$ on Gr, define $p(\mathcal{S}) = 1$ if the support is even and $p(\mathcal{S}) = -1$ if it is odd. If $\mathcal{S}_1$ and $\mathcal{S}_2$ are irreducible, we modify the commutativity with the following sign: $\psi = (-1)^{p(\mathcal{S}_1)p(\mathcal{S}_2)}\psi'$.

Proposition 4.6. *The hyper-cohomology functor* $\mathbb{H}^* \colon \mathrm{Sph}_G \to \mathrm{Vect}_{\mathbb{C}}$, *sending a sheaf $\mathcal{S}$ to $\oplus \mathbb{H}^i(\mathcal{S})$, is a tensor functor with respect to the commutativity isomorphism ψ* [31, Proposition 6.3].

Theorem 4.7. *There is a canonical equivalence of tensor categories*

$$\mathrm{Rep}(^LG) \xrightarrow{\;\cong\;} \mathrm{Sph}_G,$$

where ^LG is the Langlands dual group to G. This equivalence sends the representation V^λ to the sheaf $\mathrm{IC}_{\overline{\mathrm{Gr}}^\lambda}$, where λ is a cocharacter of G, and hence a weight of ^LG.

This was proved by Ginzburg [19] for characteristic 0 and by Mirkovic and Vilonen [31, Theorems 7.3 and 12.1] in a more general setting.

4.2. Hecke stacks and Hecke functors

In this section we introduce the Hecke stacks. These give analogs for principal G-bundles of the Hecke transformation for vector bundles. There are two versions of this stack, depending on whether the point x is fixed or is allowed to move in X. Using these stacks, we define the Hecke functor (4.4). This is a product between objects of Sph_G and $\mathcal{D}$-modules on the moduli stack Bun_G of principal G-bundles on X. This product is used to define the notion of Hecke eigensheaf for $\mathcal{D}$-modules on Bun_G.

Recall (3.1) that $\mathrm{Bun}_{G,x}$ is a principal $G(\widehat{\mathcal{O}})$-bundle on Bun_G. Let $_x\mathcal{H}_G$ be the associated Gr_G-fiber bundle.

$$_x\mathcal{H}_G = \mathrm{Bun}_{G,x} \times_{G(\widehat{\mathcal{O}})} \mathrm{Gr}_G\,.$$

It represents the 2-functor

$$S \longmapsto \big(P, P', \beta \colon P|_{X_S^*} \longrightarrow P'|_{X_S^*}\big),$$

where P and P' are principal G-bundles on $X \times S$ and β is a trivialization on $(X - x) \times S$. We define morphisms of stacks

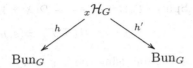

sending (P, β) to P and P'. We use this stack to define a product, called the Hecke functor

$$_xH(\cdot, \cdot)\colon \mathrm{Sph}_G \times \mathcal{D}(\mathrm{Bun}_G) \longrightarrow \mathcal{D}(\mathrm{Bun}_G)$$
$$(\mathcal{S}, \mathcal{F}) \longmapsto {_xH}(\mathcal{S}, \mathcal{F}) = h'_!\big(\mathcal{F} \widetilde{\boxtimes} \mathrm{DR}^{-1}(\mathcal{S})\big) \tag{4.4}$$

where DR is the de Rham functor, giving the Riemann–Hilbert correspondence between $\mathcal{D}$-modules and perverse sheaves, and $\mathcal{D}(\mathrm{Bun}_G)$ is the category of $\mathcal{D}$-modules on Bun_G (see [6, Section 7] for the definition of the category of $\mathcal{D}$-modules on a stack).

Lemma 4.8. *There is an isomorphism*

$$_xH\big(\mathcal{S}_1, {_xH}(\mathcal{S}_2, \mathcal{F})\big) \cong {_xH}(\mathcal{S}_1 * \mathcal{S}_2, \mathcal{F}).$$

This lemma follows from the associativity of the convolution. Now we define a global version of this stack and convolution product. We say "global" in the sense that now the point x will be allowed to move in X.

Recall (4.2) the definition of the canonical $\mathrm{Aut}(\widehat{\mathcal{O}})$-bundle $\pi\colon \mathfrak{X} \to X$. The group $\mathrm{Aut}(\widehat{\mathcal{O}})$ acts on $_x\mathcal{H}_G$, and we define $\mathcal{H}_G$ to be the associated fiber bundle on X

$$\mathcal{H}_G = \mathfrak{X} \times_{\mathrm{Aut}(\widehat{\mathcal{O}})} {_x\mathcal{H}_G}.$$

It represents the 2-functor

$$S \longmapsto \big(f\colon S \to X, P, P', \beta\colon P|_{X \times S - \Gamma_f} \xrightarrow{\cong} P'|_{X \times S - \Gamma_f}\big),$$

where Γ_f is the graph of f and β is an isomorphism on $X \times S - \Gamma_f$. There are morphisms

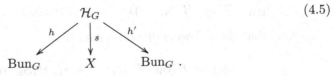

$$\tag{4.5}$$

Sending (f, P, P', β) to P, f and P'. Given $\mathcal{S} \in \mathrm{Sph}_G := \mathrm{Perv}_{G(\widehat{\mathcal{O}})}(\mathrm{Gr}_G)$ and $\mathcal{F} \in \mathcal{D}(\mathrm{Bun}_G)$, applying the construction of (4.1), we obtain $\mathcal{F} \widetilde{\boxtimes} \mathrm{DR}^{-1}(\mathcal{S}) \in \mathcal{D}(_x\mathcal{H}_G)$. Since $\mathcal{S}$ was $\mathrm{Aut}(\widehat{\mathcal{O}})$-equivariant (Lemma 4.2), the same holds for $\mathcal{F} \widetilde{\boxtimes} \mathrm{DR}^{-1}(\mathcal{S})$, hence we can define $(\mathcal{F} \widetilde{\boxtimes} \mathrm{DR}^{-1}(\mathcal{S}))_X \in \mathcal{D}(\mathcal{G}_G)$ as in (4.3). Hence we can define a product

$$H(\cdot, \cdot) \colon \mathrm{Sph}_G \times \mathcal{D}(\mathrm{Bun}_G) \longrightarrow \mathcal{D}(X \times \mathrm{Bun}_G)$$

$$(\mathcal{S}, \mathcal{F}) \longmapsto H(\mathcal{S}, \mathcal{F}) = (s, h')_!\big((\mathcal{F} \widetilde{\boxtimes} \mathrm{DR}^{-1}(\mathcal{S}))_X\big).$$

We can iterate this construction using the diagram

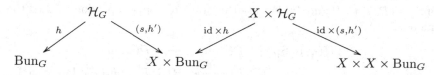

and we obtain a product

$$H(\cdot, \cdot) \colon \mathrm{Sph}_G \times \mathcal{D}(X \times \mathrm{Bun}_G) \longrightarrow \mathcal{D}(X \times X \times \mathrm{Bun}_G)$$

$$(\mathcal{S}, \mathcal{G}) \longmapsto H(\mathcal{S}, \mathcal{G}) = \big(\mathrm{id} \times (s, h')\big)_!\big((\mathcal{G} \widetilde{\boxtimes} \mathrm{DR}^{-1}(\mathcal{S}))_X\big).$$

Proposition 4.9. *Let* $\sigma_{12} \colon X \times X \to X \times X$ *be the morphism exchanging the factors. We have*

- $H(\mathcal{S}_1, H(\mathcal{S}_2, \mathcal{F}))|_{(X \times X - \Delta) \times \mathrm{Bun}_G} \cong \sigma_{12}^* H(\mathcal{S}_2, H(\mathcal{S}_1, \mathcal{F}))|_{(X \times X - \Delta) \times \mathrm{Bun}_G}.$
- $\Psi(H(\mathcal{S}_1, H(\mathcal{S}_2, \mathcal{F}))) \cong H(\mathcal{S}_1 * \mathcal{S}_2, \mathcal{F}).$

4.3. Statement of Hecke eigenproperty

Recall (Theorem 4.3) that there is an equivalence of categories between $\mathrm{Rep}(^L G)$ and Sph_G. We denote by $\mathcal{S}_V$ the sheaf corresponding to the representation V. Let $\sigma \colon \pi(X) \longrightarrow {}^L G$ be a representation of the fundamental group. Given a representation $V \in \mathrm{Rep}(^L G)$, we denote by V_σ the induced local system on X.

Let $\mathcal{F} \in \mathcal{D}(\mathrm{Bun}_G)$ be a $\mathcal{D}$-module on Bun_G. Assume that for all $V \in \mathrm{Rep}(^L G)$ we are given an isomorphism

$$\phi_V \colon H(\mathcal{S}_V, \mathcal{F}) \xrightarrow{\;\cong\;} V_\sigma \boxtimes \mathcal{F} \in \mathcal{D}(X \times \mathrm{Bun}_G).$$

Iterating this isomorphism, we obtain

$$\phi_{V,W} \colon H\big(\mathcal{S}_V, H(\mathcal{S}_W, \mathcal{F})\big) \xrightarrow{\;\cong\;} V_\sigma \boxtimes W_\sigma \boxtimes \mathcal{F} \in \mathcal{D}(X \times X \times \mathrm{Bun}_G).$$

Assume that the following diagrams commute

$$
\begin{array}{ccc}
H\big(\mathcal{S}_V, H(\mathcal{S}_W, \mathcal{F})\big)|_{\mathcal{U}} & \xrightarrow{\;\cong\;} & \sigma_{12}^* H\big(\mathcal{S}_V, H(\mathcal{S}_W, \mathcal{F})\big)|_{\mathcal{U}} \\
{\scriptstyle \phi_{V,W}} \downarrow {\scriptstyle \cong} & & {\scriptstyle \sigma_{12}^* \phi_{V,W}} \downarrow {\scriptstyle \cong} \\
V_\sigma \boxtimes W_\sigma \boxtimes \mathcal{F} & \xrightarrow{\;\cong\;} & \sigma_{12}^* W_\sigma \boxtimes V_\sigma \boxtimes \mathcal{F} \,,
\end{array}
$$

$$\Psi\Big(H\big(\mathcal{S}_V, H(\mathcal{S}_W, \mathcal{F})\big)\Big) \overset{\cong}{\longrightarrow} H(\mathcal{S}_V * \mathcal{S}_W, \mathcal{F})$$

$$\Psi(\phi_{V,W}) \Big\downarrow \cong \qquad\qquad \phi_{V\otimes W} \Big\downarrow \cong$$

$$\Psi\big(V_\sigma \boxtimes W_\sigma \boxtimes \mathcal{F}\big) \overset{\cong}{\longrightarrow} (V\otimes W)_\sigma \boxtimes \mathcal{F}$$

where $\mathcal{U} = (X \times X - \Delta) \times \mathrm{Bun}_G$. Then we say that $\mathcal{F}$ is a Hecke eigensheaf with "eigenvalue" V_σ.

It follows that if we restrict to a point x, we obtain the following commutative diagram

$$_xH\big(\mathcal{S}_V, {}_xH(\mathcal{S}_W, \mathcal{F})\big) \overset{\cong}{\longrightarrow} {}_xH(\mathcal{S}_V * \mathcal{S}_W, \mathcal{F})$$

$$_xH(\mathrm{id}, {}_x\phi_W) \Big\downarrow \cong \qquad\qquad {}_x\phi_{V\otimes W} \Big\downarrow \cong$$

$$_xH\big(\mathcal{S}_V, {}_xW_\sigma \otimes \mathcal{F}\big) \overset{\cong}{\underset{{}_x\phi_V}{\longrightarrow}} {}_xV_\sigma \otimes {}_xW_\sigma \otimes \mathcal{F} .$$

5. Opers

We are mainly interested in the case of a semisimple group of adjoint type, but some of the results will be given for an arbitrary connected reductive group which we will denote H. Let X be a smooth curve (or $\mathbb{D}$, or $\mathbb{D}^\times$).

Definition 5.1. A GL_n-oper on X is a triple

$$\big(E, \nabla\colon E \longrightarrow E \otimes \Omega_X, 0 = E_0 \subset E_1 \subset \cdots \subset E_{n-1} \subset E_n = E\big)$$

where E is a vector bundle on E, ∇ is a connection on E, and $M_\bullet$ is a filtration by vector bundles with $\mathrm{rk}\, M_i = i$, such that

1. $\nabla(E_i) \subset (E_{i+1} \otimes \Omega_X)$.
2. The following induced morphism is an isomorphism

$$(\nabla)_i \colon M_i/M_{i-1} \overset{\cong}{\longrightarrow} M_{i+1}/M_i \otimes \Omega_X.$$

Note that this morphism is $\mathcal{O}_X$-linear.

Now we will generalize this definition for a connected reductive group H with Lie algebra $\mathfrak{h}$. Fix a Borel subgroup B, $N = [B, B]$, so that $T = B/N$ is isomorphic to a Cartan subgroup. Denote by $\mathfrak{n} \subset \mathfrak{b} \subset \mathfrak{h}$, $\mathfrak{t}$ the corresponding Lie algebras. We have a filtration of B-modules (using the adjoint action)

$$\mathfrak{h}^0 = \mathfrak{b} \subset \mathfrak{h}^{-1} = \mathfrak{b} \oplus \bigoplus_{\alpha \in I} \mathfrak{h}^I,$$

where I is the set of negative simple roots, and a canonical isomorphism of B-modules.

$$\mathfrak{h}^{-1}/\mathfrak{b} \cong \bigoplus_{\alpha \in I} \mathfrak{h}^\alpha.$$

Note that the action of B on $\mathfrak{h}^\alpha$ factors through $B \to T$. Hence, for any principal B-bundle P_B, the associated bundle $P_B \times_B (\mathfrak{h}^{-1}/\mathfrak{b})$ splits as a direct sum of line bundles

$$P_B \times_B (\mathfrak{h}^{-1}/\mathfrak{b}) \cong \bigoplus_{\alpha \in I} \alpha(P_T),$$

where P_T is the principal T-bundle associated with P_B and the morphism $B \to T$, and $\alpha(P_T) = P_T \times_{T,\alpha} \mathbb{C}$ is the line bundle associated with P_T and the root α.

Given a connection ∇ on P_G and a reduction P_B of structure group to B, let

$$c(\nabla) \in \Gamma\big(X, \big(P_B \times_B (\mathfrak{h}/\mathfrak{b})\big) \otimes \Omega_X\big)$$

be the second fundamental form.

Definition 5.2. An H-oper on X is a triple

$$(P_H, \nabla, P_B)$$

where P_H is a principal H-bundle, ∇ is a connection on P_H, and P_B is a reduction of structure group to B, such that

1. $c(\nabla) \subset \Gamma(X, (P_H \times_H (\mathfrak{h}^{-1}/\mathfrak{b})) \otimes \Omega_X)$.
2. For all $\alpha \in I$, the component

$$c(\nabla)^\alpha \in H^0(X, \alpha(P_T) \otimes \Omega_X)$$

doesn't vanish at any point of X.

Definition 5.3. If $\mathfrak{h}$ is a semisimple Lie algebra, then we define an $\mathfrak{h}$-oper to be an H_{ad}-oper, where H_{ad} is the corresponding adjoint group.

Recall that giving a local system on a curve X with group H is equivalent to giving a pair (H, ∇), where P_H is a holomorphic principal H-bundle and ∇ a holomorphic connection. We can think of an H-oper as a local system (P_H, ∇) on X plus an oper structure: a reduction of P_H to a Borel subgroup satisfying Conditions 1 and 2.

Proposition 5.4. *Let (P_H, ∇, P_B) be an H-oper. If X is connected, then $\mathrm{Aut}(P_H, \nabla, P_B) = Z$, Z the center of the group H.*

Proposition 5.5. *Let X be a projective connected curve of genus $g > 1$. Let (P_H, ∇) be a local system that admits an oper structure. Then, the following holds:*

1. *The oper structure is unique. In fact, the reduction P_B is the Harder–Narasimhan reduction.*
2. *$\mathrm{Aut}(P_H, \nabla) = Z$.*
3. *The local system (P_H, ∇) is irreducible, i.e., the local system doesn't admit a reduction to a nontrivial parabolic subgroup.*
4. *If H is of adjoint type, the underlying principal H-bundle P_H is always the same (up to isomorphism) for all H-opers.*

Assume that X is a projective curve. Then H-opers form an algebraic stack $\mathrm{Op}_H(X)$. If H is semisimple, it is a Deligne–Mumford stack, and if H is of adjoint type, then it is an affine scheme. From now on we will assume that H is of adjoint type.

Lemma 5.6. *The space* $\mathrm{Op}_{\mathfrak{sl}_2}(X)$ *of* $\mathfrak{sl}_2$*-opers on* X (*cf. Definition 5.3*) *is a principal homogeneous space for* $\Gamma(X, \Omega_X^{\otimes 2})$.

Proof. The set of isomorphism classes of principal PGL_2-bundles on a curve X is equal to the set of isomorphism classes of vector vector bundles with $\det(E) = \Delta$, where Δ is either $\mathcal{O}_X$ or $\mathcal{O}_X(p)$ for some fixed point $p \in X$, modulo tensoring with a line bundle L with $L^{\otimes 2} \cong \mathcal{O}_X$. This vector bundle can be written as an extension

$$0 \longrightarrow M \longrightarrow E \longrightarrow M^{-1} \otimes \Delta \longrightarrow 0 \qquad (5.1)$$

for some line bundle M. Given a PGL_2-oper, the second fundamental form

$$c(\nabla) \in H^0(M^{-2} \otimes \Delta \otimes \Omega_X)$$

is required to be a non-vanishing section, so this forces $M^{\otimes 2} \otimes \Delta \cong \Omega_X$. Since $\deg \Omega_X$ is even, this forces $\Delta = \mathcal{O}_X$. Therefore, the PSL_2-bundle lifts to an SL_2-bundle E, which is an extension as in (5.1), and the holomorphic connection lifts to a holomorphic connection on E. A theorem of Weil says that a vector bundle on a curve admits a holomorphic connection if and only if each indecomposable summand is of degree 0. Therefore, the extension (5.1) is non-trivial.

Summing up, the set of isomorphism classes of PGL_2-opers is equal to the set of equivalence classes of pairs

$$\left(0 \longrightarrow \Omega_X^{1/2} \xrightarrow{\ i\ } E \xrightarrow{\ p\ } \Omega_X^{-1/2} \longrightarrow 0, \quad \nabla\right) \qquad (5.2)$$

where $\Omega_X^{1/2}$ is a square root of Ω_X, E is the unique non-trivial extension

$$0 \longrightarrow \Omega_X^{-1/2} \longrightarrow E \longrightarrow \Omega_X^{1/2} \longrightarrow 0 \,,$$

∇ is a connection on E inducing an isomorphism

$$\Omega_X^{1/2} \xrightarrow{\ i\ } E \xrightarrow{\ \nabla\ } E \otimes \Omega_X \xrightarrow{\ p \otimes \mathrm{id}\ } \Omega_X^{-1/2} \otimes \Omega_X,$$

and two pairs are equivalent if there exists a rank one local system (L, ∇_L) of order 2 such that $(E', \nabla') \cong (E, \nabla) \otimes (L, \nabla_L)$. Note that we obtain item (4) of Proposition 5.5, because if E and E' differ by a line bundle, then the associated PGL_2-bundles are isomorphic.

Let $\beta \in \Gamma(X, \Omega_X^{\otimes 2})$. Given an $\mathfrak{sl}_2$-oper, consider the corresponding extension as in (5.2). The action is defined by sending the connection ∇ to $\nabla + (i \otimes \mathrm{id}_{\Omega_X}) \circ \beta \circ p$. It is easy to check that this action is free and transitive. $\qquad \square$

Let B_0 be a Borel subgroup of PGL_2, $N_0 = [B_0, B_0]$, and let (e, f, h) be a standard basis of $\mathfrak{sl}_2$ with $e \in \mathfrak{n}_0$.

Let $e' \in \mathfrak{h}$ be a regular (i.e., its centralizer has minimal dimension equal to the rank of $\mathfrak{h}$) nilpotent element. There exists a group morphism $\iota \colon \mathrm{PGL}_2 \to H$ that, at the level of Lie algebras, sends $e \in \mathfrak{sl}_2$ to $d\iota(e) = e'$. Let $V = \ker([e', \cdot]) \subset \mathfrak{h}$. Define an action of $\mathbb{G}_m$ on V by

$$\mathbb{G}_m \times V \longrightarrow V$$
$$(t, v) \longmapsto a_t(v) := t \operatorname{Ad}(\varphi(t))v$$

where $\varphi \colon \mathbb{G}_m \longrightarrow B_0/N_0 = T$ is an isomorphism between $\mathbb{G}_m$ and the torus T of PGL_2.

Theorem 5.7 (Kostant). *Let $\mathfrak{h}_{\mathrm{nd}}^{-1} \subset \mathfrak{h}^{-1}$ be the subset where the projection to each root space $\mathfrak{h}^{-1} \twoheadrightarrow \mathfrak{h}^\alpha$ is nonzero for all $\alpha \in I$ ("nd" stands for non-degenerate). Consider the adjoint action of B on $\mathfrak{h}_{\mathrm{nd}}^{-1}$. The following morphism is an isomorphism*

$$\psi \colon V \longrightarrow \mathfrak{h}_{\mathrm{nd}}^{-1}/B$$
$$v \longmapsto \overline{v + f'}$$

where $f' = d\iota(f)$ is the image of the standard generator. Furthermore, this isomorphism is $\mathbb{G}_m$-equivariant, i.e.,

$$\psi(a_t(v)) = t\psi(v).$$

Considering Ω_X as a principal $\mathbb{G}_m$-bundle on X and using this action, the embedding $\mathbb{C}e' \hookrightarrow V$ produces a vector bundle embedding

$$\Omega_X^{\otimes 2} \cong (\Omega_X \times_{\mathbb{G}_m} \mathbb{C}e') \hookrightarrow (\Omega_X \times_{\mathbb{G}_m} V)$$

(the first isomorphism follows from $a_t(e) = t^2 e$). Hence

$$\Gamma(X, \Omega_X^2) \subset \Gamma(X, \Omega_X \times_{\mathbb{G}_m} V).$$

Since $\Gamma(X, \Omega_X^2)$ acts on $\mathrm{Op}_{\mathfrak{sl}_2}(X)$ (Lemma 5.6), we can consider the space

$$\mathrm{Op}_{\mathfrak{sl}_2}(X) \times_{\Gamma(X, \Omega_X^2)} \Gamma(X, \Omega_X \times_{\mathbb{G}_m} V). \tag{5.3}$$

Proposition 5.8. *The space (5.3) is canonically isomorphic to $\mathrm{Op}_H(X)$. In particular, $\mathrm{Op}_H(X)$ is a principal homogeneous space for $\Gamma(X, \Omega_X \times_{\mathbb{G}_m} V)$.*

Proof. We will define the map from (5.3) to $\mathrm{Op}_G(X)$.

Let $(P_{\mathrm{PGL}_2}, \nabla, P_{B_0})$ be an $\mathfrak{sl}_2$-oper. By Condition 2 in Definition 5.2, we have that $(P_{B_0} \times_{\mathrm{Ad}} \mathbb{C}f) \otimes \Omega_X$ is a trivial line bundle, and hence $(P_{B_0} \times_{\mathrm{Ad}} \mathbb{C}e) \cong \Omega_X$. It follows that

$$\Omega_X \times_{\mathbb{G}_m} V \cong P_{B_0} \times_{B_0, \varphi'} V$$

where φ' is the action

$$\varphi' \colon B_0 \times V \longrightarrow V$$
$$\left(\begin{pmatrix} a & b \\ 0 & a^{-1} \end{pmatrix}, v \right) \longmapsto a^2 v.$$

On the other hand

$$P_B \times_B \mathfrak{g} = P_{B_0} \times_{B_0} \mathfrak{g}$$

where $P_B = \iota_* P_{B_0}$ is the induced B-bundle, B has the adjoint action on $\mathfrak{h}$, and B_0 acts on $\mathfrak{h}$ via ι and the adjoint action. Furthermore, this last action, when restricted to $V \subset \mathfrak{h}$, is exactly the action φ', and hence

$$\Omega_X \times_{\mathbb{G}_m} V \subset P_B \times_B \mathfrak{g}. \tag{5.4}$$

Now let $\eta \in \Gamma(X, \Omega_X \times_{\mathbb{G}_m} V)$. Define the map from (5.3) to $\mathrm{Op}_H(X)$ by sending $(P_{\mathrm{PGL}_2}, \nabla, P_{B_0})$ to

$$(\iota_* P_{\mathrm{PGL}_2}, \nabla + \eta, P_B).$$

By (5.4), η can be considered as a section of $P_B \times_B \mathfrak{h}$, so this definition makes sense. It remains to show that this map is an isomorphism (Kostant's theorem is used here). $\qquad\square$

We define $\mathrm{Op}_H^{\mathrm{cl}}(X) := \Gamma(X, \Omega_X \times_{\mathbb{G}_m} V)$. Then we have $\mathrm{gr}\,\mathcal{O}(\mathrm{Op}_H(X)) = \mathcal{O}(\mathrm{Op}_H^{\mathrm{cl}}(X))$. Alternatively, we can introduce $\mathrm{Op}_H^{\mathrm{cl}}$ using λ-connections:

Definition 5.9 (λ-connection). Let M be a coherent sheaf on X. Let $\lambda \in \mathbb{C}$. A λ-connection on M is an operator

$$\nabla^\lambda : M \longrightarrow M \otimes \Omega_X$$

such that

$$\nabla^\lambda(fm) = \lambda df \otimes m + f\nabla^\lambda(m).$$

If $\lambda \neq 0$, then ∇^λ is a λ-connection if and only if $(1/\lambda)\nabla^\lambda$ is a usual connection. For $\lambda = 0$, a 0-connection is just an $\mathcal{O}_X$-linear homomorphism from M to $M \otimes \Omega_X$.

Definition 5.10 (λ-opers). An H-λ-oper is a triple $(P_H, \nabla^\lambda, P_B)$ where ∇^λ is a λ-connection, with the same properties as in Definition 5.2.

A classical oper is an H-λ-oper for $\lambda = 0$. Equivalently, a classical H-oper is a pair

$$\left(P_B, \nabla^{\mathrm{cl}} \in \Gamma(X, (P_B \times_B \mathfrak{h}) \otimes \Omega_X)\right).$$

Proposition 5.11. *Assume H is of adjoint type. There are canonical isomorphisms*

$$\mathrm{Op}_{\mathfrak{sl}_2}^{\mathrm{cl}}(X) \;\cong\; \Gamma(X, \Omega_X^{\otimes 2})$$
$$\mathrm{Op}_H^{\mathrm{cl}}(X) \;\cong\; \Gamma(X, \Omega_X \times_{\mathbb{G}_m} V).$$

6. Constructing $\mathcal{D}$-modules

If Y is a scheme and L a line bundle on Y, we can define $\mathcal{D}_Y^L = \mathrm{Diff}(L, L)$, the sheaf of differential operators in L. Let $\mathcal{D}_Y^L$-mod be the category of left $\mathcal{D}_Y^L$-modules on Y. It is equivalent to the category $\mathcal{D}_Y$-mod of left $\mathcal{D}_Y$-modules.

$$(\mathcal{D}_Y\text{-mod}) \longrightarrow (\mathcal{D}_Y^L\text{-mod}) \tag{6.1}$$
$$M \longmapsto L \otimes_{\mathcal{O}_Y} M$$

This is defined locally as follows: let U be an open set of Y with a trivialization of $L|_U$. This trivialization induces an isomorphism between $\mathcal{D}_U^L$ and $\mathcal{D}_U$ and an isomorphism between $L \otimes M|_U$ and $\mathcal{O}_U \otimes M_U$, and hence the natural $\mathcal{D}_U$-module structure of $\mathcal{O}_U \otimes M_U$ induces a $\mathcal{D}_U^L$-module structure. Then we check that this structure is independent of the trivialization used. As an example of this equivalence, if $M = \mathcal{D}_Y$, then $L \otimes \mathcal{D}_Y = \mathrm{Diff}(\mathcal{O}_Y, L)$.

Let $\mathcal{Y}$ be a DG-free algebraic stack (cf. Definition 1.2). Let L be a line bundle on $\mathcal{Y}$. We define the category of L-twisted $\mathcal{D}$-modules on $\mathcal{Y}$ as the category of sheaves on $\mathcal{Y}$ of the form $L \otimes_{\mathcal{O}_\mathcal{Y}} M$, where M is a $\mathcal{D}$-module on $\mathcal{Y}$.

In our application, $\mathcal{Y}$ will be Bun_G, and L will be the positive square root of the determinant bundle associated with the adjoint vector bundle on Bun_G. We assume that G is semisimple and simply connected (in particular, Bun_G is DG-free), and then this square root is uniquely defined.

For the construction of the determinant line bundle and its square root, see [34, Section 6]. Here we will give a brief sketch of the ingredients that are needed.

Recall that a family of vector bundles on a curve X parameterized by S is a vector bundle F on $X \times S$. This data produces a line bundle on S, called the determinant line bundle $\mathcal{D}_F$, whose fiber over a point s is canonically isomorphic to

$$\bigwedge^{\max} H^1(X, F_s) \otimes \Big(\bigwedge^{\max} H^0(X, F_s) \Big)^{-1}$$

([34, Section 6.1]). If we endow F with a nondegenerate quadratic form σ with values in the canonical line bundle Ω_X of the curve, then we can define a canonical square root of the determinant line bundle, called the Pfaffian line bundle $\mathcal{P}_{(F,\sigma)}$ ([34, Section 6.3]). In our case, we consider the universal adjoint bundle $\mathcal{E}(\mathfrak{g})$. The Cartan–Killing form gives a nondegenerate quadratic form with values in $\mathcal{O}_X$, so, tensoring with a theta characteristic (i.e., a square root of Ω_X), we obtain a nondegenerate quadratic form σ with values in Ω_X, and the Pfaffian $\mathcal{P}_{(\mathcal{E}(\mathfrak{g}),\sigma)}$ gives a square root of $\mathcal{D}_{\mathcal{E}(\mathfrak{g})}$. Note that, in general, this square root depends on the choice of a theta characteristic, but, when G is semisimple, $\mathrm{Pic}(\mathrm{Bun}_G) \cong \mathbb{Z}$ ([34, Corollary 10.3.4]), so there is a unique positive square root, and this is the line bundle we take as L.

The sheaf $\mathcal{D}_\mathcal{Y}^L$ is defined as follows: for any smooth morphism $\pi \colon Z \to \mathcal{Y}$ we set

$$\pi^*(\mathcal{D}_\mathcal{Y}^L) := \mathcal{D}_Z^L / (\mathcal{D}_Z^L \cdot T_{Z/\mathcal{Y}}).$$

The modules $\mathrm{End}_{\mathcal{D}^L\text{-mod}}(\mathcal{D}_\mathcal{Y}^L)$ and $\Gamma(\mathcal{Y}, \mathcal{D}_\mathcal{Y}^L)$ are isomorphic. Hence we give a ring structure to $\Gamma(\mathcal{Y}, \mathcal{D}_\mathcal{Y}^L)$ as follows

$$\Gamma(\mathcal{Y}, \mathcal{D}_\mathcal{Y}^L) = \mathrm{End}_{\mathcal{D}^L\text{-mod}}(\mathcal{D}_\mathcal{Y}^L)^{\mathrm{opp}}.$$

We take the opposite ring structure so that if $\mathcal{Y}$ is a scheme, we obtain the natural ring structure on $\Gamma(\mathcal{Y}, \mathcal{D}_\mathcal{Y}^L)$.

Definition 6.1. A quantum integrable system on $\mathcal{Y}$ is a ring homomorphism

$$h\colon A \longrightarrow \Gamma(\mathcal{Y}, \mathcal{D}_{\mathcal{Y}}^L)$$

where A is a commutative ring, L is a line bundle on $\mathcal{Y}$.

The tangent bundle $T_{\mathcal{Y}}$ has a Lie algebra structure, and this Lie bracket extends uniquely to a Poisson bracket $\{\cdot,\cdot\}$ on the sheaf of rings $\operatorname{Sym} T_{\mathcal{Y}}$ by the Leibniz rule: $\{f, gh\} = \{f, g\}h + \{f, h\}g$.

Definition 6.2. A classical integrable system on $\mathcal{Y}$ is a ring homomorphism

$$h^{\mathrm{cl}}\colon A^{\mathrm{cl}} \longrightarrow \Gamma(\mathcal{Y}, \operatorname{Sym} T_{\mathcal{Y}}),$$

where A^{cl} is a commutative ring,

$$\{h^{\mathrm{cl}}(a_1), h^{\mathrm{cl}}(a_2)\} = 0$$

for all a_1, a_2 in A^{cl}, and $\{,\}$ is the Poisson bracket.

We say that h is a quantization of h^{cl} if A is filtered, h is compatible with the filtration, $\operatorname{gr} A \cong A^{\mathrm{cl}}$, and the following diagram commutes

$$
\begin{array}{ccc}
A^{\mathrm{cl}} & \xrightarrow{\ h^{\mathrm{cl}}\ } & \Gamma(\mathcal{Y}, \operatorname{Sym} T_{\mathcal{Y}}) \\
\| & & \downarrow \\
\operatorname{gr} A & \xrightarrow{\ \operatorname{gr} h\ } & \operatorname{gr} \Gamma(\mathcal{Y}, \mathcal{D}_{\mathcal{Y}}^L) \ .
\end{array}
$$

Given a quantum integrable system, we associate with each closed point $\sigma \in \operatorname{Spec} A$ a $\mathcal{D}^L$-module on $\mathcal{Y}$

$$\mathcal{F}_\sigma^L = \mathcal{D}_{\mathcal{Y}}^L \otimes_A A/\mathfrak{m}_\sigma, \tag{6.2}$$

where $\mathfrak{m}_\sigma \subset A$ is the maximal ideal corresponding to σ, and $a \in A$ acts on $\mathcal{D}_{\mathcal{Y}}^L$ by sending $d \in \mathcal{D}_{\mathcal{Y}}^L$ to $d \cdot h(a)$. In our application, $\mathcal{D}_{\mathcal{Y}}^L$ will be flat over A (Lemma 11.2), and hence this tensor product is equivalent to the tensor product in the sense of derived categories. The fact that the tensor product coincides with the derived tensor product in the case at hand is used in the proof of the Hecke eigenproperty.

Untwisting with L we obtain a $\mathcal{D}$-module on $\mathcal{Y}$:

$$\mathcal{F}_\sigma = L^{-1} \otimes_{\mathcal{O}_Y} \mathcal{F}_\sigma^L. \tag{6.3}$$

Example. Let $\mathcal{Y} = V = \operatorname{Spec} \mathbb{C}[x_1, \ldots, x_n]$ be a vector space of dimension n considered as a scheme. Let $A = \mathbb{C}[\partial_1, \ldots, \partial_n]$, where $\partial_i = \partial/\partial x_i$. We take L to be the trivial bundle. We have

$$\Gamma(V, \mathcal{D}_V) = \mathbb{C}[x_1, \ldots, x_n, \partial_1, \ldots, \partial_n].$$

We can identify the dual vector space $V^* = \operatorname{Spec} A$. Consider the inclusion map

$$h\colon \mathbb{C}[\partial_1, \ldots, \partial_n] \longrightarrow \Gamma(V, \mathcal{D}_V).$$

Let $v^* \in V^*$. The induced $\mathcal{D}_V$-module is the pullback via $v^*\colon V \longrightarrow \mathbb{A}^1$ of the $\mathcal{D}_{\mathbb{A}^1}$-module generated by one generator ξ and with relation $\partial_t \xi = \xi$.

In other words, if $(\partial_1-a_1,\ldots,\partial_n-a_n)$, $a_i \in \mathbb{C}$, is a maximal ideal, the induced $\mathcal{D}_V$-module is the trivial bundle $\mathcal{O}_V$ with flat connection given by $d + \sum a_i dx_i$.

Example. Let X be a smooth projective curve. Let $\mathcal{Y} = J(X)$ be the Jacobian and $L = \mathcal{O}_{J(X)}$, the trivial line bundle on $J(X)$. Let $A = \operatorname{Sym} H^1(X, \mathcal{O}_X)$. We have

$$\Gamma(J(X), \mathcal{D}_{J(X)}) \cong \operatorname{Sym} H^1(X, \mathcal{O}_X).$$

Take $h\colon A \to \Gamma(J(X), \mathcal{D}_{J(X)})$ to be the identity.

A 1-dimensional local system on X is equivalent to a pair (M, ∇), where M is a holomorphic line bundle and ∇ is a holomorphic connection. If $M = \mathcal{O}_X$, then a holomorphic connection is written as $\nabla = d + \omega$ with $\omega \in H^0(X, \Omega_X)$, i.e., there is a bijection between $H^0(X, \Omega_X)$ and local systems whose holomorphic line bundle is trivial.

By Serre duality,

$$H^0(X, \Omega_X) \cong H^1(X, \mathcal{O}_X)^* \cong \operatorname{Spec} \operatorname{Sym} H^1(X, \mathcal{O}_X).$$

Therefore, a local system of the form $(\mathcal{O}_X, d + \omega)$ gives, by Serre duality, an element $\sigma \in H^1(X, \mathcal{O}_X)^*$, and hence a maximal ideal $\mathfrak{m}_\sigma$ of A, and the previous construction produces a $\mathcal{D}_{J(X)}$-module $\mathcal{F}_\sigma$, which is the trivial line bundle $\mathcal{O}_{J(X)}$ with connection $d + \sigma$, where $\sigma \in H^0(J(X), T^*_{J(X)}) = H^1(X, \mathcal{O}_X)^*$.

This example gives the geometric Langlands correspondence for $G = \mathbb{C}^*$, but only for those local systems on X whose holomorphic line bundle is trivial.

In our application, A will be the ring of functions $\mathcal{O}(\operatorname{Op}_{L_G}(X))$ of the affine scheme $\operatorname{Op}_{L_G}(X)$ of $^L G$-opers, and $Y = \operatorname{Bun}_G$. This system will be a quantization of a classical system where A^{cl} is $\mathcal{O}(\operatorname{Op}^{cl}_{L_G}(X))$, the ring of functions on the space of classical $^L G$-opers, and the map h^{cl} will be the Hitchin integrable system. This construction will give the Langlands correspondence, but only for those local systems on X which admit an oper structure, i.e., we get the "oper part" of the Langlands correspondence.

7. Hitchin integrable system I: definition

Recall the definition of the cotangent bundle for a DG-free stack (Definition 1.1). A point in the cotangent bundle $T^* \operatorname{Bun}_G$ is a pair (P, γ) where P is a principal G-bundle and

$$\gamma \in H^1(X, P \times_G \mathfrak{g})^* = H^0(X, \Omega_X \otimes P \times_G \mathfrak{g}^*). \qquad (7.1)$$

Consider the GIT quotient of the adjoint action of G on $\mathfrak{g}^*$, that is, the spectrum of the ring of polynomial invariants on $\mathfrak{g}^*$:

$$\mathfrak{g}^* \longrightarrow \mathfrak{g}^* / \operatorname{Ad} G := \operatorname{Spec}(\operatorname{Sym} \mathfrak{g})^G.$$

By a theorem of Chevalley, the ring $(\operatorname{Sym} \mathfrak{g})^G$ of invariant polynomials on the vector space $\mathfrak{g}^*$ is free, so this quotient is an affine space. Fixing the zero to be the image of the zero vector in $\mathfrak{g}^*$, this becomes a vector space. Let p_i, $i = 1, \ldots, k$, be a basis for the ring of invariants. The space (7.1) maps to

$$H^0\big(X, \Omega_X \otimes [P \times_G (\mathfrak{g}^*/\operatorname{Ad} G)]\big) = H^0\big(X, \Omega_X \times_{\mathbb{G}_m} (\mathfrak{g}^*/\operatorname{Ad} G)\big) = H^0\Big(X, \bigoplus_{i=1}^{k} \Omega_X^{d_i}\Big)$$

where the first equality follows from the fact that $P \times_G (\mathfrak{g}^*/\operatorname{Ad}(G))$ is the trivial vector bundle with fiber $\mathfrak{g}^*/\operatorname{Ad}(G)$, and $d_i = \deg p_i$. We denote by $\overline{\gamma}$ the image of γ. The Hitchin map is defined as the morphism

$$\begin{aligned} T^*\operatorname{Bun}_G &\longrightarrow \operatorname{Hitch}_G(X) := \Gamma(X, \Omega_X \times_{\mathbb{G}_m} \mathfrak{g}^*/\operatorname{Ad} G) \\ (P, \gamma) &\longmapsto \overline{\gamma}. \end{aligned}$$

Lemma 7.1. *The Hitchin map is flat, and moreover it induces an isomorphism at the level of global functions*

$$\mathcal{O}\big(\operatorname{Hitch}_G(X)\big) \xrightarrow{\cong} \mathcal{O}(T^*\operatorname{Bun}_G). \tag{7.2}$$

Proof. We start by showing that the Hitchin map is flat. Since $T^*\operatorname{Bun}_G$ is a complete intersection, it is enough to prove that the fibers are equidimensional. Both sides are cones, i.e., they have $\mathbb{C}^*$-actions, and the map is equivariant with respect to this action, so it is enough to show that the dimension doesn't jump at the origin. The fiber over the origin is the nilpotent cone. It is a Lagrangian substack [20], hence its dimension is equal to the dimension of Bun_G, and we conclude that the map is flat.

Now we look at the morphism at the level of global sections. It is injective because it is flat (hence dominant, because the base is irreducible), and it is surjective because the generic fiber is projective. $\qquad\square$

Lemma 7.2. *There is a natural isomorphism*

$$\operatorname{Hitch}_G(X) \cong \operatorname{Op}^{\mathrm{cl}}_{{}^L G}(X). \tag{7.3}$$

Proof. Recall (cf. Proposition 5.11) that

$$\operatorname{Op}^{\mathrm{cl}}_{{}^L G}(X) \cong \Gamma(X, \Omega_X \times_{\mathbb{G}_m} {}^L V),$$

where ${}^L V = \ker([e', \cdot]) \subset {}^L\mathfrak{g}$ and ${}^L\mathfrak{g}$ is the Lie algebra of the Langlands dual group ${}^L G$. Then we have to show that ${}^L V \cong \mathfrak{g}^*/\operatorname{Ad}_G$. Using Kostant's isomorphism (Theorem 5.7) we have

$$ {}^L V \cong {}^L\mathfrak{g}^{-1}_{\mathrm{nd}}/{}^L B \cong {}^L\mathfrak{g}/\operatorname{Ad}({}^L G) \cong {}^L\mathfrak{t}/{}^L W$$

where ${}^L\mathfrak{t}$ is a Cartan subalgebra of ${}^L\mathfrak{g}$, and ${}^L W$ is the Weyl group. We have ${}^L\mathfrak{t} = \mathfrak{t}^*$ (and ${}^L W = W$), hence

$$ {}^L\mathfrak{t}/{}^L W \cong \mathfrak{t}^*/W \cong \mathfrak{g}^*/\operatorname{Ad} G. \qquad\square$$

Remark 7.3. Note that in (7.3) we have the group G on the left-hand side, but the Langlands dual group ${}^L G$ on the right-hand side.

Using $\Gamma(T^* \operatorname{Bun}_G, \mathcal{O}_{T^* \operatorname{Bun}_G}) = \Gamma(\operatorname{Bun}_G, \operatorname{Sym} T_{\operatorname{Bun}_G})$, (7.2) and (7.3) gives an isomorphism

$$h^{\mathrm{cl}} \colon \mathcal{O}\big(\operatorname{Op}^{\mathrm{cl}}_{{}^L G}(X)\big) \longrightarrow \Gamma(\operatorname{Bun}_G, \operatorname{Sym} T_{\operatorname{Bun}_G}) \tag{7.4}$$

with $A^{\mathrm{cl}} = \mathcal{O}(\operatorname{Op}^{\mathrm{cl}}_{{}^L G}(X))$, the ring of functions on the affine space $\operatorname{Op}^{\mathrm{cl}}_{{}^L G}(X)$. We will show later that the image of h^{cl} consists of Poisson-commuting functions, hence h^{cl} is a classical integrable system.

8. Localization functor

We start by defining the "untwisted" localization functor, which takes a Harish-Chandra module acting on a scheme and produces a $\mathcal{D}$-module on the quotient of that scheme. This functor is called "localization" because it is adjoint to taking global sections (see [5]). Then we define a "twisted" version, giving $\mathcal{D}^L$-modules. It is this twisted version that will be used to give the quantization of the Hitchin system. As a reference for this section, see [14, Section 7.4].

Definition 8.1. A Harish-Chandra pair $(\mathfrak{l}, K)$ consists of a Lie algebra $\mathfrak{l}$ and a Lie group K, together with an inclusion $i \colon \mathfrak{k} = \operatorname{Lie}(K) \hookrightarrow \mathfrak{l}$ and an action of K on $\mathfrak{l}$ compatible with the adjoint action coming from the inclusion.

An $(\mathfrak{l}, K)$-module is a vector space V with a group representation $f \colon K \longrightarrow \operatorname{GL}(V)$ and a Lie algebra representation $\alpha \colon \mathfrak{l} \longrightarrow \mathfrak{gl}(V)$ such that the following diagram commutes

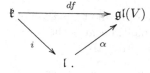

We say that a Harish-Chandra pair acts on Z if K acts on Z and there is a Lie algebra homomorphism

$$\mathfrak{l} \longrightarrow \Gamma(Z, T_Z), \tag{8.1}$$

extending the map $\mathfrak{k} \longrightarrow \Gamma(Z, T_Z)$ induced by the action. Let $\mathcal{Y} = [Z/K]$ be the quotient stack, and assume that $\mathcal{Y}$ is DG-free (Definition 1.2). The (untwisted) localization functor is defined as follows

$$\operatorname{Loc} \colon \big((\mathfrak{l}, K)\text{-mod}\big) \longrightarrow (\mathcal{D}\text{-mod on } \mathcal{Y})$$
$$M \longmapsto \mathcal{D}_Z \otimes_{U(\mathfrak{l})} M$$

where $U(\mathfrak{l})$ is the universal enveloping algebra. Note that on the right we have a K-equivariant $\mathcal{D}$-module on Z, hence it gives a $\mathcal{D}$-module on $\mathcal{Y}$.

For example, consider the trivial Lie representation of $\mathfrak{k}$ on $\mathbb{C}$, and let

$$\mathbb{V} = \operatorname{Ind}^{\mathfrak{l}}_{\mathfrak{k}}(\mathbb{C}) := U(\mathfrak{l}) \otimes_{U(\mathfrak{k})} \mathbb{C}. \tag{8.2}$$

Clearly, $\mathbb{V}$ is an $(\mathfrak{l}, K)$-module. We have $\operatorname{Loc} \mathbb{V} \cong \mathcal{D}_{\mathcal{Y}}$. This follows from:

$$\pi^*(\operatorname{Loc} \mathbb{V}) \cong \mathcal{D}_Z \otimes_{U(\mathfrak{l})} U(\mathfrak{l}) \otimes_{U(\mathfrak{k})} \mathbb{C} \cong \mathcal{D}_Z \otimes_{U(\mathfrak{k})} \mathbb{C} \cong \mathcal{D}_Z/(\mathcal{D}_Z \cdot \mathfrak{k}).$$

Remark 8.2. For an arbitrary smooth algebraic stack $\mathcal{Y}$ the localization functor is defined using the derived tensor product $\mathcal{D}_Z \otimes^L_{U(\mathfrak{l})} M$, and therefore we obtain an object in the derived category of $\mathcal{D}$-modules. If we take $M = \mathbb{V}$ (cf. (8.2)), we obtain that $\mathcal{D}_Z \otimes^L_{U(\mathfrak{l})} \mathbb{V}$ is precisely the complex (1.1), but, if $\mathcal{Y}$ is DG-free, this complex is exact except in degree 0, so we can take $\mathcal{D}_Z \otimes_{U(\mathfrak{l})} \mathbb{V}$.

Now, $\mathrm{End}(\mathcal{D}_\mathcal{Y}) \cong \Gamma(\mathcal{Y}, \mathcal{D}_\mathcal{Y})^{\mathrm{opp}}$, and on the other hand, for any Harish-Chandra module V we have $V^K = \mathrm{Hom}(\mathbb{V}, V)$, the K-invariant vectors of V, so there is a bijection $\mathbb{V}^K = \mathrm{End}(\mathbb{V})$ that is actually an anti-isomorphism of algebras. Since Loc is a functor, we obtain an algebra homomorphism

$$\mathbb{V}^K = \mathrm{End}(\mathbb{V})^{\mathrm{opp}} \xrightarrow{\mathrm{Loc}} \mathrm{End}(\mathcal{D}_\mathcal{Y})^{\mathrm{opp}} = \Gamma(\mathcal{Y}, \mathcal{D}_\mathcal{Y}). \tag{8.3}$$

Now we will study the graded of this morphism. We have $\mathrm{gr}\, \mathbb{V} = \mathrm{Sym}(\mathfrak{l}/\mathfrak{k})$, and this induces an inclusion of rings

$$\sigma_{(\mathfrak{l}, K)} \colon \mathrm{gr}(\mathbb{V}^K) \hookrightarrow \mathrm{Sym}(\mathfrak{l}/\mathfrak{k})^K.$$

Proposition 8.3. *If $\mathcal{Y}$ is DG-free, the following diagram is commutative*

$$
\begin{array}{ccc}
\mathrm{Sym}(\mathfrak{l}/\mathfrak{k})^K & \xrightarrow{\;f\;} & \Gamma(\mathcal{Y}, p_*\mathcal{O}_{T^*\mathcal{Y}}) \\
\sigma_{(\mathfrak{l},K)} \Big\uparrow & & \Big\uparrow \sigma_\mathcal{Y} \\
\mathrm{gr}(\mathbb{V}^K) & \xrightarrow{\mathrm{gr\,Loc}} & \mathrm{gr}\,\Gamma(\mathcal{Y}, \mathcal{D}_\mathcal{Y})
\end{array}
$$

where the morphism f is induced from the action of $\mathfrak{l}$ (8.1), and $\sigma_\mathcal{Y}$ was defined in Lemma 1.4.

Now we will introduce a twisted version of the localization functor. Let the Harish-Chandra pair $(\widetilde{\mathfrak{l}}, K)$ act on Z, and let L be a line bundle on $\mathcal{Y}$. We assume that the Lie algebra $\widetilde{\mathfrak{l}}$ is a central extension of a Lie algebra $\mathfrak{l}$, $\mathfrak{k} \subset \mathfrak{l}$, and we assume that this is split on $\mathfrak{k}$, i.e., there is a commutative diagram

Let $\mathrm{Diff}^n(A, B)$ denote the sheaf of differential operators of order n between the vector bundles A and B. We say that a central extension $\widetilde{\mathfrak{l}}$ acts on L if there is a map

$$\widetilde{\mathfrak{l}} \longrightarrow \mathrm{Diff}^1(\pi^*L, \pi^*L), \tag{8.4}$$

where $\pi\colon Z \longrightarrow [Z/K]$, and such that the restriction of this map to the central element $\mathbf{1}$ acts by scalar multiplication, i.e., there is a commutative diagram

$$
\begin{array}{ccc}
\widetilde{\mathfrak{l}} & \longrightarrow & \mathrm{Diff}^1(\pi^*L, \pi^*L) \\
\uparrow & & \uparrow \\
\mathbb{C}\mathbf{1} \xrightarrow{\;j\;} \mathbb{C} \subset \mathcal{O}_Z & = & \mathrm{Diff}^0(\pi^*L, \pi^*L)
\end{array}
$$

where the map j sends $\mathbf{1}$ to $1 \in \mathbb{C}$.

Let $(\widetilde{\mathfrak{l}}, K)$-mod$'$ be the category of Harish-Chandra modules such that the central element $\mathbf{1}$ acts as multiplication by $1 \in \mathbb{C}$. We define the twisted localization functor

$$
\begin{aligned}
\mathrm{Loc}'\colon \big((\widetilde{\mathfrak{l}}, K)\text{-mod}'\big) &\longrightarrow (\mathcal{D}^L\text{-mod on } \mathcal{Y}) \\
M &\longmapsto \mathcal{D}_Z^{\pi^*L} \otimes_{U(\widetilde{\mathfrak{l}})} M.
\end{aligned}
$$

For example, we can take the twisted vacuum Harish-Chandra module

$$
\mathbb{V}' = U'(\widetilde{\mathfrak{l}}) \otimes_{U(\mathfrak{k})} \mathbb{C}
$$

where $U'(\widetilde{\mathfrak{l}}) = U(\widetilde{\mathfrak{l}})/(\mathbf{1}-1)$ and $\mathfrak{k}$ acts on $\mathbb{C}$ as the trivial Lie algebra representation (multiplication by zero). Alternatively, we can define $\mathbb{V}'$ as

$$
\mathbb{V}' = \mathrm{Ind}_{\mathfrak{k}\oplus\mathbb{C}\mathbf{1}}^{\widetilde{\mathfrak{l}}} \mathbb{C}
$$

where $\mathbf{1}$ acts on $\mathbb{C}$ as multiplication by $1 \in \mathbb{C}$ and $\mathfrak{k}$ acts trivially on $\mathbb{C}$. We have

$$
\mathrm{Loc}'(\mathbb{V}') = \mathcal{D}_{\mathcal{Y}}^L, \tag{8.5}
$$

$\mathrm{End}(\mathcal{D}_{\mathcal{Y}}^L) \cong \Gamma(\mathcal{Y}, \mathcal{D}_{\mathcal{Y}}^L)^{\mathrm{opp}}$, $\mathbb{V}'^K = \mathrm{End}(\mathbb{V}')^{\mathrm{opp}}$, and Loc' gives an algebra homomorphism

$$
\mathbb{V}'^K = \mathrm{End}(\mathbb{V}')^{\mathrm{opp}} \xrightarrow{\;\mathrm{Loc}'\;} \mathrm{End}(\mathcal{D}_{\mathcal{Y}}^L)^{\mathrm{opp}} = \Gamma(\mathcal{Y}, \mathcal{D}_{\mathcal{Y}}^L).
$$

There is an inclusion

$$
\sigma_{(\widetilde{\mathfrak{l}}, K)}\colon \mathrm{gr}\,\mathrm{End}(\mathbb{V}') \hookrightarrow \mathrm{Sym}(\mathfrak{l}/\mathfrak{k})^K.
$$

Proposition 8.4. *If Y is DG-free, the following diagram is commutative*

$$
\begin{array}{ccc}
\mathrm{Sym}(\mathfrak{l}/\mathfrak{k})^K & \xrightarrow{\;f\;} & \Gamma(\mathcal{Y}, p_*\mathcal{O}_{T^*\mathcal{Y}}) \\
\sigma_{(\widetilde{\mathfrak{l}}, K)} \uparrow & & \uparrow \sigma_{\mathcal{Y}} \\
\mathrm{gr}(\mathbb{V}'^K) & \xrightarrow{\;\mathrm{gr}\,\mathrm{Loc}'\;} & \mathrm{gr}\,\Gamma(\mathcal{Y}, \mathcal{D}_{\mathcal{Y}}^L)
\end{array}
$$

where the morphism f is induced from the action of $\mathfrak{l}$ (cf. (8.1)). The morphism $\sigma_{\mathcal{Y}}$ was defined for $\mathcal{D}$-modules on $\mathcal{Y}$ in Lemma 1.4, but it can also be defined for twisted differentials.

9. Quantum integrable system h

Now we will apply the twisted localization functor to define a quantum integrable system h. In our application we will take $Z = \mathrm{Bun}_{G,x}$ (cf. (3.1)) and $(\mathfrak{l}, K) = (\mathfrak{g} \otimes \widehat{\mathcal{K}}_x, G(\widehat{\mathcal{O}}_x))$, and hence $Y = \mathrm{Bun}_G$ (cf. (3.2)).

Recall that a point of $\mathrm{Bun}_{G,x}$ corresponds to a pair (P, α), where P is a principal G-bundle on X and α is a trivialization of P on the disk $\mathbb{D}_x$ (in this picture, $G(\widehat{\mathcal{O}}_x)$ acts by change of trivialization). The tangent space at (P, α) is

$$T_{(P,\alpha)} \mathrm{Bun}_{G,x} \cong \Gamma(X - x, P \times_G \mathfrak{g}) \backslash ((P \times_G \mathfrak{g}) \otimes_{\mathcal{O}_X} \widehat{\mathcal{K}}_x) \cong \Gamma(X - x, P \times_G \mathfrak{g}) \backslash \mathfrak{g} \otimes \widehat{\mathcal{K}}_x,$$

where $(P \times_G \mathfrak{g}) \otimes_{\mathcal{O}_X} \widehat{\mathcal{K}}_x = (P \times_G \mathfrak{g})|_{\mathbb{D}_x^\times}$ is the associated bundle on the punctured disk, and the second isomorphism is obtained using α. Then we have a map

$$\mathfrak{g} \otimes \widehat{\mathcal{K}}_x \longrightarrow \Gamma(\mathrm{Bun}_{G,x}, T_{\mathrm{Bun}_{G,x}}), \tag{9.1}$$

compatible with the action of $\mathfrak{g} \otimes \widehat{\mathcal{O}}_x$, and hence the Harish-Chandra pair $(\mathfrak{g} \otimes \widehat{\mathcal{K}}_x, G(\widehat{\mathcal{O}}_x))$ acts on $Z = \mathrm{Bun}_{G,x}$.

Proposition 9.1. *Let* $q \colon \mathfrak{g} \otimes \mathfrak{g} \longrightarrow \mathfrak{g}$ *be an invariant quadratic form, and let* $\widehat{\mathfrak{g}}_q$ *be the central extension*

$$0 \longrightarrow \mathbb{C}\mathbf{1} \longrightarrow \widehat{\mathfrak{g}}_q \longrightarrow \mathfrak{g} \otimes \widehat{\mathcal{K}}_x \longrightarrow 0$$

with Lie algebra structure given by

$$[g_1 \otimes f_1, g_2 \otimes g_2] = [g_1, g_2] \otimes f_1 f_2 + q(g_1, g_2)\mathrm{Res}((df_1)f_2)\mathbf{1}.$$

Take $q = -q_0$, *where* q_0 *is the Killing form. Then* $\widehat{\mathfrak{g}}_{-q_0}$ *acts on* $p^* \mathcal{L}_{\det}$, *where* $p \colon \mathrm{Bun}_{G,x} \longrightarrow \mathrm{Bun}_G$ *is the projection and* $\mathcal{L}_{\det}$ *is the determinant line bundle on* Bun_G.

Let $c \in \mathbb{C}$. If we take $q = -cq_0$, then $\widehat{\mathfrak{g}}_{-cq_0}$ acts on $\mathcal{L}_{\det}^c$, provided that this line bundle exists (for an arbitrary $c \in \mathbb{C}$, $\mathcal{L}_{\det}^c \in \mathrm{Pic}(\mathrm{Bun}_G) \otimes_{\mathbb{Z}} \mathbb{C}$). We will be interested in $q = -\frac{1}{2}q_0$, the so-called critical level. In the case $c = 1/2$ the line bundle exists because $\mathcal{L}_{\det}$ admits a square root, using the Pfaffian construction (see Section 6).

Corollary 9.2. *Let* L *be the square root of the determinant bundle* $\mathcal{L}_{\det}$ *on* Bun_G, *i.e.,* $L^{\otimes 2} = \mathcal{L}_{\det}$ *(it is unique if* G *is simply connected). The central extension for the critical level* $\widehat{\mathfrak{g}}_{\mathrm{crit}} := \widehat{\mathfrak{g}}_{-\frac{1}{2}q_0}$ *acts on* L.

Define the vacuum Harish-Chandra module for $\widehat{\mathfrak{g}}_q$

$$\mathbb{V}_q = \mathrm{Ind}_{\mathfrak{g} \otimes \widehat{\mathcal{O}}_x \oplus \mathbb{C}\mathbf{1}}^{\widehat{\mathfrak{g}}_q} \mathbb{C}. \tag{9.2}$$

It is a filtered module. The endomorphism algebra $\mathrm{End}(\mathbb{V}_q)$ of the vacuum is called the chiral center. The following proposition shows that it is non-trivial if we take the critical level.

Theorem 9.3 (Feigin–Frenkel). *If $q \neq -\frac{1}{2}q_0$, then*

$$\operatorname{End} \mathbb{V}_q = \mathbb{C}.$$

For the critical level $q = -\frac{1}{2}q_0$, the inclusion $\sigma_{(\widehat{\mathfrak{g}}_{\mathrm{crit}}, G(\widehat{\mathcal{O}}_x))}$ is an isomorphism

$$\sigma_{(\widehat{\mathfrak{g}}_{\mathrm{crit}}, G(\widehat{\mathcal{O}}_x))} \colon \operatorname{gr} \operatorname{End} \mathbb{V}_{\mathrm{crit}} \xrightarrow{\cong} \operatorname{Sym}(\mathfrak{g} \otimes \widehat{\mathcal{K}}_x / \mathfrak{g} \otimes \widehat{\mathcal{O}}_x)^{G(\widehat{\mathcal{O}}_x)}. \qquad (9.3)$$

The twisted localization functor gives a functor

$$\operatorname{Loc}' \colon ((\widehat{\mathfrak{g}}_{\mathrm{crit}}, G(\widehat{\mathcal{O}}_x))\text{-mod}') \longrightarrow (\mathcal{D}^L\text{-mod on } \operatorname{Bun}_G),$$

and hence a morphism

$$\mathbb{V}_{\mathrm{crit}}^{G(\widehat{\mathcal{O}}_x)} = \operatorname{End}(\mathbb{V}_{\mathrm{crit}}) \longrightarrow \Gamma(\operatorname{Bun}_G, \mathcal{D}^L_{\operatorname{Bun}_G}). \qquad (9.4)$$

This morphism is the main ingredient in the construction of the quantization of the Hitchin integrable system.

Recall that when $x \in X$ varies, the modules $\mathbb{V}_q$ give a $\mathcal{D}$-module on X. More precisely, consider the Kac–Moody Lie* algebra $\mathfrak{g} \otimes \mathcal{D}_X \oplus \Omega_X$ and define

$$\mathcal{B}(\mathfrak{g}, q) := \mathcal{U}(\mathfrak{g} \otimes \mathcal{D}_X \oplus \Omega_X)/(u(\Omega_X) - \Omega_X),$$

where $\mathcal{U}(\mathfrak{g} \otimes \mathcal{D}_X \oplus \Omega_X)$ is the chiral enveloping algebra, and

$$u \colon \Omega_X \longrightarrow \mathcal{U}(\mathfrak{g} \otimes \mathcal{D}_X \oplus \Omega_X)$$

is the unit (cf. (2.2)). Note that this is a filtered chiral algebra. The fiber over $x \in X$ is canonically isomorphic to the vacuum module

$$\mathcal{B}(\mathfrak{g}, q)_x \cong \mathbb{V}_q.$$

Define the center of the chiral algebra $\mathcal{B}(\mathfrak{g}, q)$

$$\mathfrak{Z} = \mathcal{Z}(\mathcal{B}(\mathfrak{g}, q)) = \{b \in \mathcal{B}(\mathfrak{g}, q) : [b \boxtimes b'] = 0 \quad \forall\, b' \in \mathcal{B}(\mathfrak{g}, q)\},$$

where [ ] is the Lie* algebra bracket. It is a filtered chiral algebra. Since it is defined as a center, it is a commutative chiral algebra, i.e., there is a morphism $\mathfrak{Z} \otimes \mathfrak{Z} \longrightarrow \mathfrak{Z}$ giving the corresponding $\mathcal{D}$-module the structure of a $\mathcal{D}$-algebra. Then we can consider the algebra of coinvariants $H_\nabla(X, \mathfrak{Z})$.

Let the fiber on $x \in X$ be denoted by

$$\mathfrak{Z}_x = \mathfrak{Z} \otimes \mathcal{O}_X/\mathfrak{m}_x.$$

It has an induced filtration. There is a canonical surjection

$$\mathfrak{Z}_x \twoheadrightarrow H_\nabla(X, \mathfrak{Z}). \qquad (9.5)$$

Using this surjection and the filtration in $\mathfrak{Z}_x$, the algebra of coinvariants $H_\nabla(X, \mathfrak{Z})$ becomes a filtered algebra.

Theorem 9.4 (Feigin–Frenkel, first iteration). *There is an isomorphism of filtered algebras*

$$H_\nabla(X, \mathfrak{Z}) \cong \mathcal{O}(\operatorname{Op}_{{}^L G}(X))$$

between the coinvariants and the ring of functions on the affine space of ${}^L G$-opers.

Proposition 9.5. *There is an isomorphism* $\mathfrak{Z}_x \cong \mathbb{V}_q^{G(\widehat{\mathcal{O}}_x)}$, *and we have a commutative diagram*

$$
\begin{array}{ccc}
\mathfrak{Z}_x & \xrightarrow{\ \cong\ } & \mathbb{V}_q^{G(\widehat{\mathcal{O}}_x)} \\
\downarrow & & \cup \\
\mathcal{B}(\mathfrak{g}, q)_x & \xrightarrow{\ \cong\ } & \mathbb{V}_q.
\end{array}
$$

Furthermore, the ring structure on $\mathbb{V}_q^{G(\widehat{\mathcal{O}}_x)} \cong \operatorname{End}(\mathbb{V}_q)$ *coincides with the algebra structure on* $\mathfrak{Z}_x$. *In particular,* $\operatorname{End}(\mathbb{V}_q)$ *is commutative.*

Using the isomorphism of Proposition 9.5 and the morphism (9.4), we have a morphism

$$
h_x \colon \mathfrak{Z}_x \longrightarrow \Gamma(\operatorname{Bun}_G, \mathcal{D}_{\operatorname{Bun}_G}^L). \tag{9.6}
$$

Using Feigin–Frenkel's isomorphism (Theorem 9.4) and the surjection (9.5), we have a morphism

$$
f \colon \mathfrak{Z}_x \longrightarrow \mathcal{O}(\operatorname{Op}_{{}^L G}(X)).
$$

The following theorem defines the quantum integrable system h.

Theorem 9.6. *The morphism* h_x *factors through* f

$$
\begin{array}{ccc}
\mathfrak{Z}_x & \xrightarrow{\quad h_x \quad} & \Gamma(\operatorname{Bun}_G, \mathcal{D}_{\operatorname{Bun}_G}^L) \\
& \searrow{\scriptstyle f} \qquad \nearrow{\scriptstyle h} & \\
& \mathcal{O}(\operatorname{Op}_{{}^L G}(X)). &
\end{array}
$$

Sketch of proof. As $x \in X$ varies, the morphism h_x in (9.6) defines a morphism of $\mathcal{O}_X$-algebras

$$
\mathfrak{Z} \longrightarrow \Gamma(\operatorname{Bun}_G, \mathcal{D}_{\operatorname{Bun}_G}^L) \otimes \mathcal{O}_X.
$$

It can be shown that this is in fact a morphism of $\mathcal{D}_X$-algebras [6, 2.8], and then the left adjoint property of $H_\nabla(X, \cdot)$ gives a morphism of algebras

$$
H_\nabla(X, \mathfrak{Z}) \longrightarrow \Gamma(\operatorname{Bun}_G, \mathcal{D}_{\operatorname{Bun}_G}^L).
$$

Using the Feigin–Frenkel isomorphism (Theorem 9.4) this map defines h, and one checks that $h \circ f = h_x$. $\qquad\square$

In Section 11 we will show that this morphism h is a quantization of the Hitchin integrable system.

10. Hitchin integrable system II: $\mathcal{D}$-algebras

In this section we give an alternative description of the Hitchin integrable system, using chiral algebras. This will be used to quantize this system.

Think of the cotangent bundle Ω_X as a principal $\mathbb{G}_m$-bundle on X. Recall (from Section 7) that $\mathfrak{g}^* / \operatorname{Ad} G$ is a vector space. Consider the associated vector bundle defined as $\Omega_X \times_{\mathbb{G}_m} (\mathfrak{g}^*/\operatorname{Ad} G)$ and define the sheaf of algebras

$$C := \operatorname{Sym}\left(\Omega_X \times_{\mathbb{G}_m} (\mathfrak{g}/\operatorname{Ad} G)\right), \tag{10.1}$$

i.e., $\underline{\operatorname{Spec}} C \longrightarrow X$ is the total space of the vector bundle.

Definition 10.1. Let

$$\mathfrak{z}^{\mathrm{cl}} := J(C)$$

be the jet construction of C (cf. (2.1)) and

$$\mathfrak{z}^{\mathrm{cl}}_x := \mathfrak{z}^{\mathrm{cl}} \otimes_{\mathcal{O}_X} \mathcal{O}_X/\mathfrak{m}_x,$$

the fiber of at x.

Lemma 10.2. *We have*

$$\mathcal{O}\left(\operatorname{Hitch}^{\mathrm{cl}}_G(X)\right) \cong H_\nabla(X, \mathfrak{z}^{\mathrm{cl}}),$$

the algebra of coinvariants.

Proof. We will check the lemma for $\mathbb{C}$-valued points (the general case is analogous). Recalling the definition of $\operatorname{Hitch}(X)$, this isomorphism follows from an easy calculation, using the definitions of the jet construction and the algebra of coinvariants as left adjoint functors

$$\operatorname{Hitch}(X) := \Gamma(X, \Omega_X \times_{\mathbb{G}_m} \mathfrak{g}^*/\operatorname{Ad} G) = \operatorname{Hom}_{\mathcal{O}_X\text{-alg}}(C, \mathcal{O}_X) =$$
$$= \operatorname{Hom}_{\mathcal{D}_X\text{-alg}}(J(C), \mathcal{O}_X) = \operatorname{Hom}_{\mathbb{C}\text{-alg}}(H_\nabla(X, \mathfrak{z}^{\mathrm{cl}}), \mathbb{C}) = \operatorname{Spec} H_\nabla(X, \mathfrak{z}^{\mathrm{cl}}),$$

where C is the sheaf of algebras defined in (10.1). $\qquad\square$

Recall that the Harish-Chandra pair $(\mathfrak{g} \otimes \widehat{\mathcal{K}}_x, G(\widehat{\mathcal{O}}_x))$ acts on $Z = \operatorname{Bun}_{G,x}$, and $[Z/G(\widehat{\mathcal{O}}_x)] = \operatorname{Bun}_G$. In particular, there is a map $\mathfrak{g} \otimes \widehat{\mathcal{K}}_x \longrightarrow \Gamma(Z, T_Z)$ (cf. 9.1), and this map induces the following morphism

$$\operatorname{Sym}(\mathfrak{g} \otimes \widehat{\mathcal{K}}_x/\mathfrak{g} \otimes \widehat{\mathcal{O}}_x)^{G(\widehat{\mathcal{O}}_x)} \longrightarrow \Gamma(\operatorname{Bun}_G, \operatorname{Sym} T_{\operatorname{Bun}_G}). \tag{10.2}$$

Lemma 10.3. *There is an isomorphism*

$$\mathfrak{z}^{\mathrm{cl}}_x \cong \operatorname{Sym}(\mathfrak{g} \otimes \widehat{\mathcal{K}}_x/\mathfrak{g} \otimes \widehat{\mathcal{O}}_x)^{G(\widehat{\mathcal{O}}_x)}.$$

This isomorphism together with morphism (10.2) give a morphism

$$h^{\mathrm{cl}}_x : \mathfrak{z}^{\mathrm{cl}}_x \longrightarrow \Gamma(\operatorname{Bun}_G, \operatorname{Sym} T_{\operatorname{Bun}_G}).$$

Lemma 10.4. *The following diagram is commutative*

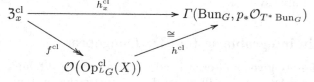

where f^{cl} comes from the canonical surjection $\mathfrak{z}^{\mathrm{cl}}_x \twoheadrightarrow H_\nabla(X, \mathfrak{z}^{\mathrm{cl}})$, Lemma 10.2, and the isomorphism (7.3), and h^{cl} is the isomorphism (7.4).

11. Quantization condition

In this section we will show that the quantum integrable system h is a quantization of the Hitchin integrable system.

For any filtered chiral algebra, in particular for 3, there is a commutative diagram

$$\operatorname{gr} H_{\nabla}(X, 3) \longleftarrow \operatorname{gr} 3_x \tag{11.1}$$

$$H_{\nabla}(X, \operatorname{gr} 3) \; .$$

Putting together Proposition 9.5, Proposition 9.3, and Lemma 10.3, we have an isomorphism $\operatorname{gr} 3_x \cong 3_x^{\mathrm{cl}}$. This globalizes to give an isomorphism

$$\operatorname{gr} 3 \cong 3^{\mathrm{cl}}. \tag{11.2}$$

Proposition 11.1. *There is a commutative diagram*

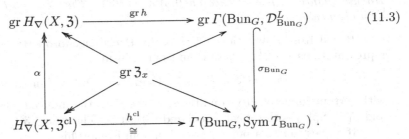

$$(11.3)$$

Proof. The left triangle is commutative by (11.1) and (11.2). The top triangle is commutative by Theorem 9.6. The right triangle is commutative by Proposition 8.4 and Proposition 9.5, and the bottom triangle is commutative by Lemma 10.4. It follows that the outer square is also commutative. $\qquad\square$

Since h^{cl} is an isomorphism, it follows from the commutativity of (11.3) that the other three maps α, σ_{Bun_G}, and $\operatorname{gr} h$ are also isomorphisms. Hence, using the Feigin–Frenkel isomorphism (Theorem 9.4) and the isomorphism of Lemma 10.2, the diagram (11.3) becomes

$$\operatorname{gr} \mathcal{O}\big(\operatorname{Op}_{L_G}(X)\big) \xrightarrow{\;\operatorname{gr} h\;} \operatorname{gr} \Gamma(\operatorname{Bun}_G, \mathcal{D}^L_{\operatorname{Bun}_G})$$

$$\Big\downarrow{\scriptstyle \alpha} \qquad\qquad\qquad \cong \Big\downarrow{\scriptstyle \sigma_{\operatorname{Bun}_G}}$$

$$\mathcal{O}\big(\operatorname{Op}^{\mathrm{cl}}_{L_G}(X)\big) \xrightarrow{\;h^{\mathrm{cl}}\;} \Gamma(\operatorname{Bun}_G, \operatorname{Sym} T_{\operatorname{Bun}_G}) \; .$$

In other words, h is the quantization of the Hitchin system h^{cl}.

Lemma 11.2. *Let $\pi\colon Z \longrightarrow \operatorname{Bun}_G$ be an affine cover. Then $\Gamma(Z, \pi^* \mathcal{D}_{\operatorname{Bun}_G})$ is a flat $\mathcal{O}(\operatorname{Op}_{L_G}(X))$-module.*

Proof. Both objects are filtered (in a way compatible with the module structure). We have

$$\operatorname{gr}\Gamma(Z, \pi^* \mathcal{D}_{\mathrm{Bun}_G}) = \Gamma(Z, \pi^* \operatorname{Sym} T_{\mathrm{Bun}_G})$$

$$\operatorname{gr}\mathcal{O}(\operatorname{Op}_{^L G}(X)) \overset{\alpha}{\cong} \mathcal{O}(\operatorname{Op}_{^L G}^{\mathrm{cl}}(X)).$$

Lemma 7.1 implies that $\Gamma(Z, \pi^* \operatorname{Sym} T_{\mathrm{Bun}_G})$ is a flat $\mathcal{O}(\operatorname{Op}_{^L G}^{\mathrm{cl}}(X))$-module, hence $\operatorname{gr}\Gamma(Z, \pi^* \mathcal{D}_{\mathrm{Bun}_G})$ is a flat $\operatorname{gr}\mathcal{O}(\operatorname{Op}_{^L G}(X))$-module, and the result follows. $\square$

12. Proof of the Hecke eigenproperty

Recall that if a local system σ admits an oper structure, this oper structure is unique, so we can also denote the oper by σ.

Theorem 12.1 (Beilinson–Drinfeld). *Let σ be an irreducible $^L G$-local system on X that admits an oper structure. Let $\mathcal{F}_\sigma$ be the $\mathcal{D}$-module on Bun_G obtained from the Hitchin quantum system (Theorem 9.6) as in (6.3). Then $\mathcal{F}_\sigma$ is a Hecke eigensheaf with eigenvalue σ.*

Recall from Subsection 4.3 that the Hecke eigenproperty says that for all representations V of $^L G$ there is an isomorphism

$$\phi_V : H(\mathcal{S}_V, \mathcal{F}_\sigma) \overset{\cong}{\longrightarrow} V_\sigma \boxtimes \mathcal{F}_\sigma \in \mathcal{D}(X \times \mathrm{Bun}_G)$$

with certain compatibility conditions, where V_σ is the induced local system on X and $\mathcal{S}_V \in \mathrm{Sph}_G$ is the sheaf associated with it by Theorem 4.3.

If we restrict to a point x, we obtain an isomorphism

$$_x\phi_V : {}_xH(\mathcal{S}_V, \mathcal{F}_\sigma) \overset{\cong}{\longrightarrow} {}_xV_\sigma \otimes \mathcal{F}_\sigma \in \mathcal{D}(\mathrm{Bun}_G). \qquad (12.1)$$

In this section we will only explain how to prove this local version.

Before we continue, we need to introduce some notions from $\mathcal{D}_X$-schemes. An affine $\mathcal{D}_X$-scheme is a pair

$$\left(\pi: Z \to X, \psi: \mathcal{O}_Z \to \mathcal{O}_Z \otimes \pi^* \Omega_X\right)$$

where Z is an affine X-scheme, ψ is $\mathcal{O}_X$-linear, and there is a $\mathcal{D}_X$-algebra $\mathcal{A}$ (cf. Section 2) such that

$$Z = \underline{\operatorname{Spec}}(\mathcal{A})$$

(here we only use the $\mathcal{O}_X$-algebra structure of $\mathcal{A}$) and ψ is induced by the $\mathcal{D}_X$-algebra structure $\mathcal{A} \to \mathcal{A} \otimes \Omega_X$.

An arbitrary $\mathcal{D}_X$-scheme is a pair (Z, ψ) such that Z can be covered by affine X-schemes U_i and $(U_i, \psi|_{U_i})$ is an affine $\mathcal{D}_X$-scheme.

Now we will define the $\mathcal{D}_X$-scheme $\mathfrak{Op}_{^L G}$. We do this by describing the functor that it represents

$$(\mathcal{D}_X\text{-}\mathrm{Sch}) \longrightarrow (\mathrm{Sets})$$
$$(Z \overset{\pi}{\to} X) \longmapsto \{^L G\text{-opers on } Z \text{ relative to } X\}$$

where an oper on Z relative to X is an ^LG-bundle on Z, a reduction to a Borel subgroup of ^LG, and a connection along X satisfying the usual oper condition. A connection along X is a map

$$\mathcal{O}_Z \longrightarrow {}^L\mathfrak{g} \otimes \pi^* \Omega_X$$

that satisfies the Leibniz rule with respect to the map

$$\mathcal{O}_Z \longrightarrow \mathcal{O}_Z \otimes \pi^* \Omega_X$$

coming from the $\mathcal{D}_X$-scheme structure of Z. In particular, $\mathfrak{Op}_{LG}(X) = \mathrm{Op}_{LG}(X)$, the scheme of usual opers. Note that $\mathfrak{Op}$ is affine over X. We denote $\mathcal{O}(\mathfrak{Op})$ the corresponding $\mathcal{D}_X$-algebra.

Theorem 12.2 (Feigin–Frenkel, second iteration). *There is an isomorphism of $\mathcal{D}_X$-algebras*

$$\mathfrak{Z} \cong \mathcal{O}(\mathfrak{Op}).$$

Note that the 'first iteration' (Theorem 9.4) follows from this. Indeed, we have

$$\mathrm{Hom}_{\mathrm{alg}}\big(H_\nabla(X, \mathcal{O}(\mathfrak{Op})), \mathbb{C}\big) = \mathrm{Hom}_{\mathcal{D}_X\text{-alg}}(\mathcal{O}(\mathfrak{Op}), \mathcal{O}_X) = \mathfrak{Op}(X).$$

Hence $H_\nabla(X, \mathfrak{Op}) = \mathcal{O}(\mathfrak{Op}(X))$, and Theorem 12.2 implies Theorem 9.4.

The $\mathcal{D}_X$-scheme $\mathfrak{Op} \longrightarrow X$ has a universal ^LG-bundle. Hence, giving a representation $V \in \mathrm{Rep}(^LG)$, we obtain a vector bundle $\mathcal{V}$ on $\mathfrak{Op}$. Using Theorem 12.2, the fiber of $\mathfrak{Op}$ over $x \in X$ is

$$_x\mathfrak{Op} = \mathrm{Spec}(\mathfrak{Z}_x) = \mathrm{Op}_{LG}(\mathbb{D}_x)$$

and the restriction $_x\mathcal{V}$ of $\mathcal{V}$ to $_x\mathfrak{Op}$ is a $\mathfrak{Z}_x$-module.

Recall that $\mathcal{F}_\sigma$ is defined as a twist of $\mathcal{F}_\sigma^L$ (cf. 6.3) and that $\mathcal{F}_\sigma^L$ is defined as a quotient of $\mathcal{D}_{\mathrm{Bun}_G}^L$ (cf. 6.2). Therefore it is enough to construct an isomorphism

$$_x\phi_V : {}_xH(\mathcal{S}_V, \mathcal{D}_{\mathrm{Bun}_G}^L) \overset{\cong}{\longrightarrow} {}_x\mathcal{V} \otimes_{\mathfrak{Z}_x} \mathcal{D}_{\mathrm{Bun}_G}^L. \tag{12.2}$$

Lemma 12.3. *We have*

$$\mathrm{Loc}'\big(\Gamma(G(\widehat{\mathcal{K}})/G(\widehat{\mathcal{O}}), \mathcal{S}_V)\big) \cong \mathcal{S}_V * \mathcal{D}_{\mathrm{Bun}_G}.$$

Let $\mathbb{V} = \mathbb{V}_{-\frac{1}{2}q_0}$ be the vacuum Harish-Chandra module for the critical level (cf. 9.2).

Theorem 12.4 (Feigin–Frenkel, third iteration).

1. *We have a noncanonical isomorphism of Harish-Chandra modules*

$$\Gamma(G(\widehat{\mathcal{K}})/G(\widehat{\mathcal{O}}), \mathcal{S}_V^L) \cong \mathbb{V}^{\oplus n} \quad \text{for some } n.$$

2. *We have a canonical isomorphism of modules*

$$\mathrm{Hom}_{\widehat{\mathfrak{g}}_{\mathrm{crit}}}\big(\mathbb{V}, \Gamma(G(\widehat{\mathcal{K}})/G(\widehat{\mathcal{O}}), \mathcal{S}_V^L)\big) \cong {}_x\mathcal{V}.$$

In item 2, the left-hand side is an $\mathrm{End}(\mathbb{V})$-module, and the right-hand side is a $\mathfrak{Z}_x$-module. The statement means that the module structures coincide, under the identification of these two rings given in Proposition 9.5.

Then, using item 1, we obtain a canonical isomorphism

$$\Gamma(G(\widehat{\mathcal{K}})/G(\widehat{\mathcal{O}}), \mathcal{S}_V) \cong \mathrm{Hom}\big(\mathbb{V}, \Gamma(G(\widehat{\mathcal{K}})/G(\widehat{\mathcal{O}}), \mathcal{S}_V)\big) \otimes_{\mathrm{End}(\mathbb{V})} \mathbb{V}. \tag{12.3}$$

Using item 2, this is isomorphic to

$$_x\mathcal{V} \otimes_{\mathrm{End}(\mathbb{V})} \mathbb{V} = {}_x\mathcal{V} \otimes_{\mathfrak{Z}_x} \mathbb{V}. \tag{12.4}$$

Hence

$$\begin{aligned}
_xH(\mathcal{S}_V, \mathcal{D}^L_{\mathrm{Bun}_G}) &= \mathcal{S}_V * \mathcal{D}^L_{\mathrm{Bun}_G} = \mathrm{Loc}'\big(\Gamma(G(\widehat{\mathcal{K}})/G(\widehat{\mathcal{O}}), \mathcal{S}^L_V)\big) = \\
&= \mathrm{Loc}'({}_x\mathcal{V} \otimes_{\mathfrak{Z}_x} \mathbb{V}) = {}_x\mathcal{V} \otimes_{\mathfrak{Z}_x} \mathcal{D}^L_{\mathrm{Bun}_G}
\end{aligned}$$

where the first equality is by definition, the second equality is Lemma 12.3, the third follows from applying the functor Loc to (12.3) and (12.4), and the fourth follows from (8.5). Then we have proved (12.2), and hence also (12.1).

Acknowledgement

This article is based on the lectures of Prof. Gaitsgory at the Tata Institute of Fundamental Research (Mumbai, India) in February 2001, during the special year on the Geometric Langlands Program. I am very grateful to him for the lectures, his patience answering my questions when preparing this article and his permission for publishing it. Of course, the responsibility for any mistakes is only mine. In February 2007 I gave a lecture based on this material in the German–Spanish workshop on moduli spaces of vector bundles. I am very grateful to the organizers for the invitation to give the lecture. I thank Indranil Biswas for discussions and also the anonymous referee, whose comments helped improve the article.

References

[1] S. Arkhipov and D. Gaitsgory, *Differential operators on the loop group via chiral algebras.* Int. Math. Res. Not. 2002, no. 4, 165–210.

[2] A. Beauville and Y. Laszlo, *Un lemme de descente.* C. R. Acad. Sci. Paris Sér. I Math. **320** (1995), no. 3, 335–340.

[3] K. Behrend, *Differential Graded Schemes II: The 2-category of Differential Graded Schemes.* Preprint. `arXiv:math/0212226`.

[4] A. Beilinson, J. Bernstein and P. Deligne, *Faisceaux pervers,* Astérisque **100** (1982).

[5] A. Beilinson, and J. Bernstein, *A proof of Jantzen conjectures.* I.M. Gelfand Seminar, 1–50, Adv. Soviet Math. **16**, Part 1, Amer. Math. Soc. Providence, RI, 1993.

[6] A. Beilinson and V. Drinfeld, *Quantization of Hitchin's integrable system and Hecke eigensheaves,* Preprint.
Available at `http://www.math.uchicago.edu/~mitya/langlands.html`.

[7] A. Beilinson and V. Drinfeld, *Chiral algebras,* American Mathematical Society Colloquium Publications, 51. AMS, Providence, RI, 2004. vi+375 pp.

[8] A.A. Belavin, A.M. Polyakov and A.B. Zamolodchikov, *Infinite conformal symmetry in two-dimensional quantum field theory*. Nuclear Phys. B **241** (1984), no. 2, 333–380.

[9] R.E. Borcherds, *Vertex algebras, Kac–Moody algebras, and the Monster*. Proc. Nat. Acad. Sci. U.S.A. **83** (1986), no. 10, 3068–3071.

[10] A. Borel, P.-P. Grivel, B. Kaup, A. Haefliger, B. Malgrange, and F. Ehlers, Algebraic *D*-modules. Perspectives in Mathematics, 2. Academic Press, Inc., Boston, MA, 1987. xii+355 pp.

[11] R. Donagi and T. Pantev, *Langlands duality for Hitchin systems*, `arXiv:math/0604617`

[12] V. Drinfeld, *Two-dimensional l-adic representations of the fundamental group of a curve over a finite field and automorphic forms on* GL(2). Amer. J. Math. **105** (1983), no. 1, 85–114.

[13] V.G. Drinfeld and C. Simpson, *B-structures on G-bundles and local triviality*. Math. Res. Lett. **2** (1995), no. 6, 823–829.

[14] E. Frenkel, Lectures on the Langlands program and conformal field theory. Frontiers in number theory, physics, and geometry. II, 387–533, Springer, Berlin, 2007. `arXiv:hep-th/0512172`

[15] E. Frenkel, D. Gaitsgory and K. Vilonen, *On the geometric Langlands conjecture*. J. Amer. Math. Soc. **15** (2002), no. 2, 367–417.

[16] E. Frenkel and D. Ben-Zvi, *Vertex algebras and algebraic curves*. Second edition. Mathematical Surveys and Monographs, 88. American Mathematical Society, Providence, RI, 2004. xiv+400 pp.

[17] D. Gaitsgory, Notes on 2D conformal field theory and string theory. *Quantum fields and strings: a course for mathematicians*, Vol. 1, 2 (Princeton, NJ, 1996/1997), 1017–1089, Amer. Math. Soc. Providence, RI, 1999. `arXiv:math/9811061v2`

[18] D. Gaitsgory, *On a vanishing conjecture appearing in the geometric Langlands correspondence*. Ann. of Math. (2) **160** (2004), no. 2, 617–682.

[19] V. Ginzburg, *Perverse sheaves on a loop group and Langlands' duality*, `arXiv:alg-geom/9511007`

[20] V. Ginzburg, *The global nilpotent variety is Lagrangian*. Duke Math. J. **109** (2001), no. 3, 511–519.

[21] N. Hitchin, Langlands duality and G_2 spectral curves. Q. J. Math. **58** (2007), no. 3, 319–344. `arXiv:math/0611524`

[22] L. Ilusie, Complexe cotangent et déformations. I. Lecture Notes in Mathematics, Vol. 239. Springer-Verlag, Berlin–New York, 1971. xv+355 pp.

[23] L. Ilusie, Complexe cotangent et déformations. II. Lecture Notes in Mathematics, Vol. 283. Springer-Verlag, Berlin–New York, 1972. vii+304 pp.

[24] A. Kapustin and E. Witten, *Electric-magnetic duality and the geometric Langlands program*. Commun. Number Theory Phys. **1** (2007), no. 1, 1–236. `arXiv:hep-th/0604151`

[25] L. Lafforgue, *Chtoucas de Drinfeld et correspondance de Langlands*. Invent. Math. **147** (2002), no. 1, 1–241.

[26] G. Laumon, *Correspondance de Langlands géométrique pour les corps de fonctions,* Duke Math J. **54** (1987) 309–359.

[27] G. Laumon, L. Moret-Bailly, Champs algébriques. Ergebnisse der Mathematik und ihrer Grenzgebiete. 3. Folge. A Series of Modern Surveys in Mathematics, 39. Springer-Verlag, Berlin, 2000. xii+208 pp.

[28] Y. Laszlo, M. Olsson, *The six operations for sheaves on Artin stacks. I. Finite coefficients.* Publ. Math. Inst. Hautes Études Sci. **107** (2008), 109–168.

[29] Y. Laszlo, M. Olsson, *The six operations for sheaves on Artin stacks. II. Adic coefficients.* Publ. Math. Inst. Hautes Études Sci. **107** (2008), 169–210.

[30] Y. Laszlo, M. Olsson, *Perverse t-structure on Artin stacks.* Math. Z. **261** (2009), no. 4, 737–748.

[31] I. Mirković and K. Vilonen, *Geometric Langlands duality and representations of algebraic groups over commutative rings,* Ann. of Math (2) **166** (2007), 95–143.

[32] L. Moret-Bailly, *Un problème de descente.* Bull. Soc. Math. France **124** (1996), no. 4, 559–585.

[33] M. Olsson, *Sheaves on Artin stacks.* J. reine angew. Math. **603** (2007), 55–112.

[34] C. Sorger, *Lectures on moduli of principal G-bundles over algebraic curves.* School on Algebraic Geometry (Trieste, 1999), 1–57. ICTP Lect. Notes, 1, Abdus Salam Int. Cent. Theoret. Phys., Trieste, 2000. (Available at http://publications.ictp.it).

Tomás L. Gómez
Instituto de Ciencias Matemáticas
(CSIC-UAM-UC3M-UCM)
Serrano 113bis
E-28006 Madrid, Spain

and

Facultad de Ciencias Matemáticas
Universidad Complutense de Madrid
E-28040 Madrid, Spain
e-mail: tomas.gomez@icmat.es

Affine Flag Manifolds and Principal Bundles
Trends in Mathematics, 91–122
© 2010 Springer Basel AG

Faltings' Construction of the Moduli Space of Vector Bundles on a Smooth Projective Curve

Georg Hein

Abstract. In 1993 Faltings gave a construction of the moduli space of semistable vector bundles on a smooth projective curve X over an algebraically closed field k. This construction was presented by the author at the *German–Spanish Workshop on Moduli Spaces of Vector Bundles* at the University of Essen in February 2007.

    To ease notation and to simplify the necessary proofs only the case of rank two vector bundles with determinant isomorphic to ω_X is considered.

    These notes give a self-contained introduction to the moduli spaces of vector bundles and the generalized Θ-divisor.

Mathematics Subject Classification (2000). 14H60, 14D20, 14F05.

Keywords. Moduli space, vector bundles on a curve, generalized Theta divisor.

1. Outline of the construction

In his paper [8] Faltings gives a "GIT-free" construction of the moduli space M_X of semistable vector bundles on a smooth projective curve X. This moduli space was constructed using Mumford's *Geometric Invariant Theory* (see [22]). By construction M_X is a projective variety when we fix the rank and the degree of the vector bundles. Thus, we have an ample line bundle $\mathcal{L}$ on M_X.

    We next present the central idea of Faltings' construction. The usual GIT construction of M_X yields the following data:

- (i) a quasi-projective scheme Q
- (ii) a surjection $\pi \colon Q \to M_X$ with connected fibers
- (iii) a ample line bundle $\mathcal{L}$ on M_X
- (iv) global sections $\{s_i \in H^0(M_X, \mathcal{L})\}_{i=0,\ldots,N}$ defining a finite morphism
  $s \colon M_X \to \mathbb{P}^N$.

The idea is to recover these data without assuming a priori that M_X exists.

We proceed as follows:

(1) We give a vector bundle $\mathcal{E}$ on $\mathbb{P}(V) \times X$ parameterizing all semistable vector bundles E of rank two and determinant ω_X on the curve X.

(2) We construct a line bundle $\mathcal{O}_{\mathbb{P}(V)}(\Theta)$ on $\mathbb{P}(V)$ and global sections s_L of this line bundle.

(3) We identify the open subset $Q \subset \mathbb{P}(V)$ where the global sections $\{s_L\}$ generate $\mathcal{O}_{\mathbb{P}(V)}(\Theta)$ with the locus parameterizing semistable vector bundles.

(4) We show that the image of $s \colon Q \to \mathbb{P}^N$ defined by the global sections $\{s_L\}$ has proper image.

(5) We analyze the connected fibers of $s \colon Q \to \mathbb{P}^N$.

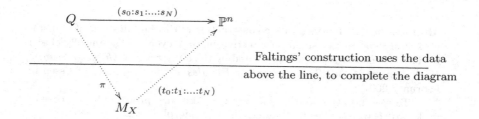

FIGURE 1. A sketch of the construction.

To simplify the presentation, we discuss this construction only for vector bundles of rank two and determinant ω_X. We indicate in 8.3 the generalization to arbitrary ranks and degrees.

2. Background and notation

2.1. Notation

We fix a smooth projective curve X defined over an algebraically closed field k. As usual the genus of X is denoted by g. For coherent sheaves F and E on X we use the following notations.

k	algebraically closed field
X	smooth projective curve over k
g	the genus of X
ω_X	the dualizing sheaf $\omega_X = \Omega_{X/k}$
$\mathrm{tors}(E)$	the torsion subsheaf of E
$E^\vee$	the dual sheaf $E^\vee = \mathcal{H}om(E, \mathcal{O}_X)$
$\mathrm{rk}(E)$	the rank of E
$\deg(E)$	the degree of E
$\mu(E) = \frac{\deg(E)}{\mathrm{rk}(E)}$	the slope of E
$h^i(E)$	the dimension of the k-vector space $H^i(E) = H^i(X, E)$

$\chi(E)$ the Euler characteristic of E

$\chi(E) = h^0(E) - h^1(E)$, and by the Riemann–Roch theorem

$\chi(E) = \deg(E) - \mathrm{rk}(E)(g-1)$

$\hom(E, F)$ the dimension of the k-vector space $\mathrm{Hom}(E, F)$

$\mathrm{ext}^1(E, F)$ the dimension of the k-vector space $\mathrm{Ext}^1(E, F)$

$\chi(E, F)$ the difference $\hom(E, F) - \mathrm{ext}^1(E, F)$

If $\mathrm{tors}(E) = 0$ we say that E is torsion free, or a vector bundle. In general a torsion free sheaf is not a vector bundle. This is a special feature for coherent sheaves on smooth curves.

When classifying vector bundles we have two numerical invariants, the rank and the degree. These invariants do not vary in flat families. Thus, it is natural to classify vector bundles with given rank and degree. Classifying means to answer the following questions:

1. Do vector bundles of given rank r and degree d exist?
2. Do they form a family?
3. What is the dimension of this family?
4. Is this family irreducible?

The answer to question 1 is yes. To understand and answer question 2 for the case of rank two bundles with determinant ω_X, we consider the following functor from the category ($\underline{k\text{-schemes}}$) of schemes over k to the category ($\underline{\text{sets}}$) of sets

$$\mathrm{SU}_X(2, \omega_X)\colon (\underline{k\text{-schemes}}) \to (\underline{\text{sets}}), \qquad S \mapsto \mathrm{SU}_X(2, \omega_X)(S).$$

Here $\mathrm{SU}_X(2, \omega_X)(S)$ stands for the set

$$\mathrm{SU}_X(2, \omega_X)(S) := \left\{ \begin{array}{l} \text{rank two vector bundles } \mathcal{E} \text{ on } S \times X \\ \text{such that for each } s \in S(k) \text{ the bundle} \\ \mathcal{E}_s = \mathcal{E}_S|_{\{s\} \times X} \text{ is semistable and we} \\ \text{have an isomorphism} \\ \det(\mathcal{E}) \cong \mathrm{pr}_S^* L_S \otimes \mathrm{pr}_X^* \omega_X \\ \text{for some line bundle } L_S \in \mathrm{Pic}(S) \end{array} \right\} \Big/ \sim \,.$$

Here $\sim$ denotes an equivalence relation we discuss in 2.2. It turns out that $\sim$ should be chosen suitably, and we should restrict to S-equivalence classes of (semi)stable vector bundles. Now this functor cannot be represented by a scheme. That means the answer to the second question is no. However there exists a coarse moduli space which is also called SU_X which is an irreducible projective variety of dimension $3g-3$. We interpret this coarse moduli space as an affirmative answer to the second question. This also answers questions 3 and 4.

2.2. The Picard torus and the Poincaré line bundle

Fix an integer $d \in \mathbb{Z}$. We consider the Picard functor

$$\mathrm{Pic}_X^d\colon (\underline{k\text{-schemes}}) \to (\underline{\text{sets}}), \qquad S \mapsto \mathrm{Pic}_X^d(S),$$

where $\operatorname{Pic}_X^d(S)$ is the set of all line bundles $\mathcal{L}$ on $S \times X$ which have degree d on each fiber over S modulo a certain equivalence relation. Let us explain next what the correct equivalence relation between two such line bundles $\mathcal{L}_1$ and $\mathcal{L}_2$ is.

symbol	name	explanation
$\sim_1$	global equivalence	there exists an isomorphism $\mathcal{L}_1 \overset{\psi}{\to} \mathcal{L}_2$.
$\sim_2$	twist equivalence	there exist a line bundle M on S and an isomorphism $\mathcal{L}_1 \overset{\psi}{\to} \mathcal{L}_2 \otimes \operatorname{pr}_S^* M$.
$\sim_3$	local equivalence	there exist an open covering $S = \bigcup_i S_i$ and isomorphisms $\mathcal{L}_1|_{S_i \times X} \overset{\psi_i}{\to} \mathcal{L}_2|_{S_i \times X}$.
$\sim_4$	fiberwise equivalence	for all points $s \in S(k)$ we have an isomorphism $\psi_s \colon \mathcal{L}_1|_{\{s\} \times X} \to \mathcal{L}_2|_{\{s\} \times X}$.

There are some implications among these relations, which also hold when we consider vector bundles on $S \times X$:

$\sim_1 \Longrightarrow \sim_2$ The strongest equivalence relation is $\sim_1$ and it implies the others.

$\sim_2 \Longrightarrow \sim_3$ Since for any line bundle M on S we have a covering $S = \bigcup_i S_i$ such that $M|_{S_i} \cong \mathcal{O}_{S_i}$, we deduce that $\mathcal{L}_1 \sim_2 \mathcal{L}_2$ implies $\mathcal{L}_1 \sim_3 \mathcal{L}_2$.

$\sim_3 \Longrightarrow \sim_4$ This implication is obvious.

If there are nontrivial line bundles M on S, then we conclude that $\sim_2$ does not imply $\sim_1$.

To see that $\sim_4$ does not imply $\sim_3$, take $S = \operatorname{Spec}(k[\varepsilon]/\varepsilon^2)$ and $\mathcal{L}$ a nontrivial deformation. If we were to restrict ourselves to reduced schemes (which we don't do) then $\sim_4$ would imply $\sim_3$.

Since the local isomorphisms of line bundles form the gluing data for a line bundle on S we can conclude that $\sim_3$ implies $\sim_2$. It is this equivalence which we use for the definition of Pic_X^d and which guarantees that Pic_X^d is a scheme (or, more precisely, is a functor which is represented by a scheme which is (usual abuse of notation) denoted by Pic_X^d).

The equivalence of the relations $\sim_2$ and $\sim_3$ comes from the following facts.

1. If L_1 and L_2 are two line bundles of the same degree on X, then we have

$$L_1 \cong L_2 \iff \operatorname{Hom}(L_1, L_2) \neq 0.$$

2. For any line bundle L on X we have $\operatorname{Hom}(L, L) = k$ and $\operatorname{Isom}(L, L) = k^*$.

These properties do not hold for vector bundles of rank two or greater. However, restricting to stable bundles we regain these two properties.

To guarantee that the reader is really used to the concept of stability we repeat the basic properties around this concept in the next subsection.

2.3. Stability

For a sheaf E we define its slope $\mu(E)$ to be the quotient $\mu(E) = \deg(E)/\mathrm{rk}(E)$ of degree and rank. For a nonzero torsion sheaf we define its slope to be ∞, and for the zero sheaf we define $\mu(0) = -\infty$. Now we come to the definition.

Definition 2.1. A sheaf E on X is *stable* if for all proper subsheaves $E' \subsetneq E$ we have the inequality $\mu(E') < \mu(E)$.
A sheaf E on X is *semistable* if for all subsheaves $E' \subset E$ we have the inequality $\mu(E') \leq \mu(E)$.

In their book [17] Huybrechts and Lehn introduced the following convention, which has become standard by now: if in a statement (*semi*)*stable* and ($\leq$) both appear, then this statement stands for two, one with *stable* and strict inequality $<$ and the other with *semistable* and the mild inequality $\leq$. As a final test for the reader, and to show the use of this convention, we repeat the above

Definition 2.2. A sheaf E on X is (*semi*)*stable* if for all proper subsheaves $E' \subsetneq E$ we have the inequality $\mu(E') \,(\leq)\, \mu(E)$.

Remark 2.3. A torsion sheaf T is by definition always semistable. It is stable, if the only proper subsheaf is the zero sheaf, that is when T is of length one which means (remember that we work over an algebraically closed field) T is isomorphic to the skyscraper sheaf $k(P)$ for a closed point $P \in X(k)$.
If E is a sheaf of positive rank, then semistability implies that the torsion subsheaf of E is 0, which means E is a vector bundle.
If E is not semistable, then there exists a subsheaf $E' \subset E$ with $\mu(E') > \mu(E)$. Such a sheaf E' is called a *destabilizing* sheaf.

From the Riemann–Roch theorem for curves we deduce the inequality

$$\mu(E) = (g-1) + \frac{\chi(E)}{\mathrm{rk}(E)} \leq (g-1) + \frac{h^0(E)}{\mathrm{rk}(E)} \leq (g-1) + h^0(E)$$

for sheaves E of rank $\mathrm{rk}(E) \geq 1$. This inequality gives also an upper bound for the slope of all subsheaves of E provided that E is torsion free. Thus, there is a maximal slope for all subsheaves. A subsheaf $E_1 \subset E$ of this maximal slope with the maximal possible rank is called the *maximal destabilizing subsheaf* of E. It is easy to check that E_1 is semistable and unique. Considering the maximal destabilizing subsheaf of E/E_1 we derive inductively the existence of the

Harder–Narasimhan filtration. For any sheaf E there exists a unique filtration

$$0 \subset \mathrm{tors}(E) = E_0 \subset E_1 \subset E_2 \subset \cdots \subset E_l = E$$

with the property that E_k/E_{k-1} is semistable and $\mu(E_{k+1}/E_k) < \mu(E_k/E_{k-1})$ (see [27]).

For semistable sheaves which are not stable we can refine the above. This is the

Jordan–Hölder filtration. If E is a semistable bundle then there exists a filtration

$$0 = E_0 \subset E_1 \subset \cdots \subset E_l = E$$

where E_k/E_{k-1} is stable of slope $\mu(E_k/E_{k-1}) = \mu(E)$. Here we have no uniqueness (think of the direct sum of two stable objects of the same slope). However, the graded object $\mathrm{gr}(E) := \bigoplus_{k=1}^{l} E_k/E_{k-1}$ is unique.

S-equivalence. Two semistable sheaves E_1 and E_2 are defined to be S-equivalent when their graded objects $\mathrm{gr}(E_1)$ and $\mathrm{gr}(E_2)$ are isomorphic.

2.4. Properties of vector bundles on algebraic curves

Here we repeat basic properties of vector bundles on a smooth projective curve X of genus g over a field k. We sketch the proofs or give a reference.

Theorem 2.4 (The Riemann–Roch theorem and Serre duality on curves). *We have for the Euler characteristic $\chi(E)$ of the bundle E*

$$\chi(E) = h^0(E) - h^1(E) = \deg(E) - \mathrm{rk}(E)(g-1).$$

The $h^1(E)$-dimensional vector space $H^1(E)$ is dual to $H^0(\omega_X \otimes E^\vee)$.

For a proof see §IV.1 in Hartshorne's book [10]. We deduce the next result.

Corollary 2.5. *For a vector bundle E we have the implications*

$$\begin{aligned}
\chi(E) > 0 &\implies h^0(E) \geq \chi(E) > 0 \\
\chi(E) < 0 &\implies h^1(E) \geq -\chi(E) > 0 \\
\chi(E) < 0 &\implies h^0(\omega_X \otimes E^\vee) \geq -\chi(E) > 0.
\end{aligned}$$

Proposition 2.6 (Some basic properties of vector bundles). *Let E and F be two vector bundles on a smooth projective curve X of genus g defined over an algebraically closed field k.*

(i) $\mu(E \otimes F) = \mu(E) + \mu(F)$.

(ii) *E is (semi)stable if and only if, for all surjections $E \to E''$, we have $\mu(E)(\leq)\mu(E'')$.*

(iii) *E is (semi)stable if and only if $E^\vee$ is (semi)stable.*

(iv) *If E is semistable of slope $\mu(E) < 0$, then $h^0(E) = 0$.*

(v) *If M is a line bundle on X, then E is (semi)stable if and only if $E \otimes M$ is (semi)stable.*

(vi) *If E is semistable and $\mu(E) > 2g - 2$, then $h^1(E) = 0$.*

(vii) *If E is semistable and $\mu(E) > 2g - 1$, then E is globally generated.*

(viii) *If E is globally generated, then there exists a short exact sequence*

$$0 \to \det(E)^{-1} \to \mathcal{O}_X^{\mathrm{rk}(E)+1} \to E \to 0.$$

(ix) *If E is globally generated, then there exists a short exact sequence*

$$0 \to \mathcal{O}_X^{\mathrm{rk}(E)-1} \to E \to \det(E) \to 0.$$

(x) *If E and F are semistable with $\mu(E) > \mu(F)$, then $\mathrm{Hom}(E, F) = 0$.*

(xi) *If E and F are stable with $\mu(E) = \mu(F)$, then $\mathrm{Hom}(E, F) = 0$ or $E \cong F$.*

(xii) *For E stable we have $\mathrm{Hom}(E, E) = \{\lambda \cdot \mathrm{id}_E \mid \lambda \in k\}$.*

Proof. (i) Computing the rank and the degree of the tensor product yields this formula.

(ii) This is just the observation that we can reconstruct an epimorphism $\pi \colon E \to E''$ by knowing its kernel, and an injection $\iota \colon E' \to E$ can be reconstructed from the cokernel. Elementary manipulations of inequalities give for a short exact sequence

$$0 \to E' \to E \to E'' \to 0$$

that $\mu(E')(\leq)\mu(E)$ is equivalent to $\mu(E)(\leq)\mu(E'')$.

(iii) Passing to the dual interchanges injections to E with surjections from E. Since both can be used to express (semi)stability the statement follows.

(iv) Indeed, any section $s \in H^0(E)$ with $s \neq 0$ defines an injection $\mathcal{O}_X \xrightarrow{s} E$.

(v) Twisting with M is an equivalence on the category of coherent sheaves.

(vi) Semistability of E is equivalent to the semistability of $\omega_X \otimes E^\vee$ by (iii) and (v). Now the statement follows from $\mu(\omega_X \otimes E^\vee) = 2g - 2 - \mu(E)$, the equality $h^1(E) = h^0(\omega_X \otimes E^\vee)$ and (iv).

(vii) For any point $P \in X(k)$ we have $\mu(E(-P)) = \mu(E) - 1 > 2g - 2$. Thus, $H^1(E(-P)) = 0$ by (vi). Thus, applying the functor H^0 to the short exact sequence

$$0 \to E(-P) \to E \to E \otimes k(P) \to 0$$

yields that E is generated by global sections at the point P.

(viii) If E is globally generated, then $H^0(E) \otimes \mathcal{O}_X \xrightarrow{\alpha} E$ is surjective. If $h^0(E) = \mathrm{rk}(E) + 1$, then we are done. If $h^0(E) = \mathrm{rk}(E)$, then α is an isomorphism. We consider the morphism $\mathcal{O}_X \oplus H^0(E) \otimes \mathcal{O}_X \xrightarrow{0 \oplus \alpha} E$; this has kernel $\mathcal{O}_X$ which is the determinant of E. Thus, we have to discuss only the case when $h^0(E) > \mathrm{rk}(E) + 1$. The idea is to consider the Grassmannian G of $(\mathrm{rk}(E) + 1)$-dimensional subspaces of $H^0(E)$. For any point $P \in X(k)$, the $(\mathrm{rk}(E) + 1)$-dimensional subspaces of G which do not surject to $E \otimes k(P)$ form a codimension two subvariety of G. Since a one-dimensional family of codimension two subvarieties cannot cover G, a general subspace $V \subset H^0(E)$ of dimension $\mathrm{rk}(E) + 1$ gives a surjection $V \otimes \mathcal{O}_X \xrightarrow{\beta} E$. The kernel of β is locally free of rank one, a line bundle. Considering the determinant we obtain $\ker \beta = \det(\ker \beta) = \det(V \otimes \mathcal{O}_X) \otimes \det(E)^{-1}$.

(ix) Here the proof works similarly to (viii). We show that the $(\mathrm{rk}(E) - 1)$-dimensional subspaces V of $H^0(E)$ which do not give a $(\mathrm{rk}(E) - 1)$-dimensional subspace of $E \otimes k(P)$ under the composition $V \otimes \mathcal{O}_X \to H^0(E) \otimes \mathcal{O}_X \to E \to E \otimes k(P)$ form a codimension two subvariety in the Grassmannian of $(\mathrm{rk}(E) - 1)$-dimensional subspaces of $H^0(E)$. Thus, for a general $V \subset H^0(E)$ of dimension $\mathrm{rk}(E) - 1$ we obtain a subbundle $V \otimes \mathcal{O}_X \to E$. The cokernel is a line bundle. Considering the determinant of the cokernel and $\det(V \otimes \mathcal{O}_X) \cong \mathcal{O}_X$ gives the asserted short exact sequence.

(x) For a nontrivial morphism $\alpha \colon E \to F$ we consider its image. Semistability of F implies $\mu(\mathrm{im}\,\alpha) \leq \mu(F)$. Semistability of E implies by (ii) that $\mu(E) \leq \mu(\mathrm{im}\,\alpha)$.

(xi) Suppose $\mathrm{Hom}(E, F) \neq 0$. As in the proof of (x) we consider the image $\mathrm{im}\alpha$ of a nontrivial $\alpha\colon E \to F$. We deduce $\mu(E) \leq \mu(\mathrm{im}\alpha) \leq \mu(F)$. Since $\mu(E) = \mu(F)$ and equality holds only when $E \cong \mathrm{im}\alpha \cong F$ is fulfilled, we conclude the statement.

(xii) Suppose that there exists an endomorphism $\beta \in \mathrm{End}(E)$ which is not a scalar multiple of the identity. Then for some $P \in X(k)$ we have that the restricted morphism $\beta_P \in \mathrm{End}(E \otimes k(P))$ is not a scalar multiple of the identity. Take an eigenvalue $\lambda \in k$ of β_P. Then $\beta - \lambda \cdot \mathrm{id} \in \mathrm{End}(E) \setminus \{0\}$ is not surjective. However by the proof of (xi) this endomorphism is an isomorphism, a contradiction. $\qquad\square$

The next proposition is as elementary as each statement in the preceding one. However, here a new criterion for semistability pops up for the first time which we will investigate later on.

Proposition 2.7. *Let X be a smooth projective curve and E and F two coherent sheaves with $F \neq 0$. If $H^*(E \otimes F) = 0$, i.e., $h^0(E \otimes F) = 0 = h^1(E \otimes F)$, then E is semistable.*

Proof. Assume we have two sheaves E and F with $H^*(E \otimes F) = 0$. We may assume that $E \neq 0$. Further we remark that $\mathrm{tors}(F) \neq 0$ implies E and F are torsion sheaves with disjoint support. Hence we assume that F is torsion free. This implies F is a vector bundle and from $H^0(E \otimes F) = 0$ we conclude that E is torsion free too. From $H^*(E \otimes F) = 0$ we deduce that $\chi(E \otimes F) = 0$, or $\mu(E) + \mu(F) = g - 1$. Suppose $E' \subset E$ is a subsheaf with $\mu(E') > \mu(E)$. From $\mu(E') + \mu(F) > g - 1$ we deduce that $h^0(E' \otimes F) > 0$. However, $H^0(E' \otimes F)$ is a subspace of $H^0(E \otimes F) = 0$, a contradiction. $\qquad\square$

3. A nice over-parameterizing family

Our aim is to find a nice family parameterizing all semistable vector bundles on a smooth projective curve X of genus g. We concentrate on the case of rank two vector bundles of determinant isomorphic to ω_X. Fix a line bundle L of degree $g + 1$.

Proposition 3.1. *For a semistable vector bundle E with $\mathrm{rk}(E) = 2$ and $\det(E) \cong \omega_X$ we have a short exact sequence*

$$0 \to L^{-1} \to E \to L \otimes \omega_X \to 0\,.$$

Proof. The vector bundle $E \otimes L$ is semistable by Proposition 2.6 (v) of slope $\mu(E \otimes L) = 2g$. By Proposition 2.6 (vii) $E \otimes L$ is globally generated and appears by 2.6 (ix) in a short exact sequence

$$0 \to \mathcal{O}_X \to E \otimes L \to \det(E \otimes L) \to 0\,.$$

Twisting this short exact sequence with L^{-1} we obtain the result. $\qquad\square$

Thus, all semistable vector bundles E of rank two with $\det(E) \cong \omega_X$ can be parameterized by extensions in the vector space $\mathrm{Ext}^1(L \otimes \omega_X, L^{-1})$. Clearly not all these extensions parameterize semistable vector bundles. To construct a universal family over $(V \setminus \{0\})/k^*$ we consider the vector space

$$V := H^1(L^{-2} \otimes \omega_X^{-1})^\vee = \mathrm{Ext}^1(L \otimes \omega_X, L^{-1})^\vee.$$

When writing $\mathbb{P}(V)$ we follow Grothendieck's definition that $\mathbb{P}(V)$ is the space of linear hyperplanes in V. With our settings a k-valued point in $\mathbb{P}(V)$ corresponds to a one-dimensional linear subspace in $\mathrm{Ext}^1(L \otimes \omega_X, L^{-1})$. We consider the product space $\mathbb{P}(V) \times X$ with the projections

$$\mathbb{P}(V) \xleftarrow{\ \ p\ \ } \mathbb{P}(V) \times X \xrightarrow{\ \ q\ \ } X \ .$$

On $\mathbb{P}(V) \times X$ we have a canonical extension class in

$$
\begin{aligned}
\mathrm{Ext}^1\big(p^*\mathcal{O}_{\mathbb{P}(V)}(-1) \otimes q^*(L \otimes \omega_X), q^*L^{-1}\big) &= \\
&= H^1\big(\mathbb{P}(V) \times X, p^*\mathcal{O}_{\mathbb{P}(V)}(1) \otimes q^*(L^{-2} \otimes \omega_X^{-1})\big) \\
&= H^0\big(\mathbb{P}(V), \mathcal{O}_{\mathbb{P}(V)}(1)\big) \otimes H^1(X, L^{-2} \otimes \omega_X^{-1}) \\
&= V \otimes H^1(L^{-2} \otimes \omega_X^{-1}) \\
&= \mathrm{End}(V)
\end{aligned}
$$

corresponding to the identity $\mathrm{id}_V \in \mathrm{End}(V)$. This way we obtain the

Lemma 3.2. *There exists a universal extension sequence on $\mathbb{P}(V) \times X$, namely*

$$0 \to q^*L^{-1} \to \mathcal{E} \to p^*\mathcal{O}_{\mathbb{P}(V)}(-1) \otimes q^*(L \otimes \omega_X) \to 0\ .$$

Next we want to investigate how over-parameterizing the family $\mathcal{E}$ is. To do so, we introduce the following notation. For a point $e \in \mathbb{P}(V)(k)$ we denote by $\mathcal{E}_e$ the sheaf $\mathcal{E}|_{\{e\} \times X}$ considered as a sheaf on X. To understand how many extensions parameterize a sheaf isomorphic to a given sheaf E (of rank two and determinant ω_X) we define the two sets

$$\mathrm{Hom}^{\max}(L^{-1}, E) := \{\varphi \colon L^{-1} \to E \text{ which have a line bundle as cokernel}\}$$

$$\mathbb{P}(V)_E := \{e \in \mathbb{P}(V)(k) \mid e \leftrightarrow (L^{-1} \to \mathcal{E}_e \to L \otimes \omega_X) \text{ with } E \cong \mathcal{E}_e\}.$$

Proposition 3.3. *For a stable bundle E of rank two and determinant $\det(E) \cong \omega_X$ we we have a 1-1 correspondence*

$$\mathrm{Hom}^{\max}(L^{-1}, E)/k^* \xleftarrow{\quad 1\text{-}1 \quad} \mathbb{P}(V)_E\ .$$

In particular, we see that $\mathbb{P}(V)_E$ is an irreducible quasi-projective variety.

Proof. Suppose $e \in \mathbb{P}(V)_E$. Then there is, up to scalar multiple, only one isomorphism $\mathcal{E}_e \cong E$ by Proposition 2.6 (xii). Thus, the map $L^{-1} \to \mathcal{E}_e$ determines a unique subbundle isomorphic to L^{-1}. On the other hand for a subbundle $L^{-1} \to E$ considering the determinant gives that the cokernel is isomorphic to $L \otimes \omega_X$. The rest is Exercise A3.26 (pages 645–647) in Eisenbud's book [7]. $\qquad\square$

Our family $\mathcal{E}$ parameterized by $\mathbb{P}(V)$ induces locally any other family. The precise statement for this is the content of the following

Proposition 3.4. *Let $\mathcal{E}_S$ be a family of semistable bundles on X of rank two and determinant ω_X on $S \times X$. Then there exist a covering $S = \bigcup_i S_i$ of S and morphisms $\varphi_i \colon S_i \to \mathbb{P}(V)$ such that we have isomorphisms*

$$\mathcal{E}_{S_i} = \mathcal{E}|_{S_i \times X} \cong (\varphi_i \times \mathrm{id}_X)^* \mathcal{E}\,.$$

Proof. We may assume that S is affine. We consider the morphisms

$$S \xleftarrow{\;p_S\;} S \times X \xrightarrow{\;q_S\;} X \,.$$

Since $\mathcal{E}_s$ is semistable of slope $g - 1$ for all points $s \in S$, the sheaf $\mathcal{E}_s \otimes L$ is semistable of slope $2g$ for all $s \in S$. Therefore the vector bundle $\mathcal{E}_S \otimes q_S^* L$ has no higher direct image with respect to p_S. Take a closed point $s \in S(k)$. Having in mind that S was affine we obtain a surjection

$$(p_S)_*(\mathcal{E}_S \otimes q_S^* L) \xrightarrow{\;\beta\;} (p_S)_*(\mathcal{E}_S \otimes q_S^* L) \otimes k(s) = H^0(\mathcal{E}_s \otimes L) \,.$$

By Proposition 2.6 (vii) and (ix) we may find a section $m_s \in H^0(\mathcal{E}_s \otimes L)$ such that $\mathcal{O}_X \xrightarrow{m_s} \mathcal{E}_s \otimes L$ defines a subbundle. Since β is surjective, this section m_s can be lifted to a section $m_S \colon \mathcal{O}_{S \times X} \to \mathcal{E}_S \otimes q_S^* L$. After passing to a smaller neighborhood of s in S we may assume m_S defines a line subbundle. The cokernel is fiberwise (with respect to p_S) isomorphic to $L^2 \otimes \omega_X$. Thus, after twisting with $q_S^* L^{-1}$ we obtain a short exact sequence

$$0 \to q_S^* L^{-1} \to \mathcal{E}_S \to p_S^* M \otimes q_S^*(L \otimes \omega_X) \to 0\,,$$

where M is the line bundle $p_{S*}\mathcal{H}om(q_S^*(L^2 \otimes \omega_X), \mathrm{coker}(m_S))$. Thus, there exists a morphism $S \to \mathbb{P}(V)$ as indicated. $\qquad\square$

Remark 3.5. So far we have only seen that $\mathbb{P}(V)$ parameterizes all semistable bundles of rank two with determinant ω_X. It parameterizes also some unstable bundles. However, these unstable bundles are contained in a closed subscheme of positive codimension. Or formulated differently: the points $s \in \mathbb{P}(V)$ which correspond to semistable bundles form a dense Zariski open subset. If follows from GIT that for any parameter space the set of points corresponding to stable (or semistable) objects forms a Zariski open subset (see Remark 4.5).

4. The generalized Θ-divisor

We consider our family $\mathcal{E}$ of vector bundles on X parameterized by $\mathbb{P}(V)$ as obtained in the preceding section. We construct the generalized Θ-line bundle on $\mathbb{P}(V)$ next. If we proceeded formally, then everything would be very easy. Since $\mathrm{Pic}(\mathbb{P}(V)) \cong \mathbb{Z}$ we just have to tell which $a \in \mathbb{Z}$ corresponds to the generalized Θ-divisor. We are not only giving the construction of $\mathcal{O}_{\mathbb{P}(V)}(\Theta)$ but we give invariant sections in this line bundle. However, we will finally compute the number $a \in \mathbb{Z}$ for the sake of completeness.

The construction of the generalized Θ-divisor follows Drezet and Narasimhan's work [6].

4.1. The line bundle $\mathcal{O}_{\mathbb{P}(V)}(R \cdot \Theta)$

We consider the morphisms

$$\mathbb{P}(V) \xleftarrow{\ p\ } \mathbb{P}(V) \times X \xrightarrow{\ q\ } X$$

and the universal vector bundle $\mathcal{E}$ on $\mathbb{P}(V) \times X$ appearing in the short exact sequence of Lemma 3.2.

We fix a line bundle $A \in \mathrm{Pic}^{-2g}(X)$ of degree $-2g$. Since $\deg(A \otimes L \otimes \omega_X) = g - 1$ we may assume that $h^0(A \otimes L \otimes \omega_X) = 0$. Now we define two vector bundles

$$A_{R,0} := A^{\oplus(R+1)} \quad \text{and} \quad A_{R,1} := A^{\otimes(R+1)}.$$

Computing the ranks and determinants of these two bundles we obtain the following equality in the Grothendieck group $K(X)$ of the curve X:

$$R[\mathcal{O}_X] = [A_{R,0}] - [A_{R,1}].$$

The short exact sequence of Lemma 3.2 and $h^0(A \otimes L \otimes \omega_X) = 0$ gives that for $i \in \{0,1\}$ we have $p_*(\mathcal{E} \otimes q^* A_{R,i}) = 0$. Thus, we have two vector bundles

$$B_{R,0} := R^1 p_*(\mathcal{E} \otimes q^* A_{R,0}) \quad \text{and} \quad B_{R,1} := R^1 p_*(\mathcal{E} \otimes q^* A_{R,1})$$

on $\mathbb{P}(V)$ of rank $4g(R+1)$. Now we have everything at our disposal to define the generalized Θ-divisor as the Cartier divisor associated with the Θ-line bundle

$$\mathcal{O}_{\mathbb{P}(V)}(R \cdot \Theta) := \det(B_{R,1})^{-1} \otimes \det(B_{R,0}) = \left(\bigwedge^{4g(R+1)} B_{R,1} \right)^{-1} \otimes \left(\bigwedge^{4g(R+1)} B_{R,0} \right).$$

4.2. The invariant sections

Consider a nontrivial morphism $\alpha \in \mathrm{Hom}(A_{R,1}, A_{R,0})$. This morphism must be injective and defines a short exact sequence

$$0 \to A_{R,1} \xrightarrow{\alpha} A_{R,0} \to F(\alpha) \to 0,$$

where $F(\alpha)$ is the cokernel. In $K(X)$ the sheaf $F(\alpha)$ equals $\mathcal{O}_X^{\oplus R}$. In particular $\mu(F(\alpha)) = 0$, and for any sheaf $\mathcal{E}_s$ parameterized by $s \in \mathbb{P}(V)$ we have $H^*(X, \mathcal{E}_s \otimes F(\alpha)) = 0$ if and only if $H^1(X, \mathcal{E}_s \otimes F(\alpha)) = 0$. Applying the functor $R^* p_*(\mathcal{E} \otimes q^* \_)$ to the previous short exact sequence we obtain the long cohomology sequence

$$0 \longrightarrow p_*(\mathcal{E} \otimes q^* F(\alpha)) \longrightarrow B_{R,1} \xrightarrow{R^1 \alpha} B_{R,0} \longrightarrow R^1 p_*(\mathcal{E} \otimes q^* F(\alpha)) \longrightarrow 0.$$

We obtain a section $\theta_\alpha \in \mathcal{O}_{\mathbb{P}(V)}(R \cdot \Theta)$ by passing to the top exterior power of $R^1\alpha$, in short: $\theta_\alpha := \bigwedge^{4g(R+1)}(R^1\alpha) \in \mathrm{Hom}(\bigwedge^{4g(R+1)} B_{R,1}, \bigwedge^{4g(R+1)} B_{R,0}) = \Gamma(\mathcal{O}_{\mathbb{P}(V)}(R \cdot \Theta))$.

Proposition 4.1. *The global section* $\theta_\alpha \in \Gamma(\mathcal{O}_{\mathbb{P}(V)}(R \cdot \Theta))$ *vanishes in*

$$V(\theta_\alpha) = \{\, s \in \mathbb{P}(V) \,|\, H^*(\mathcal{E}_s \otimes F(\alpha)) \neq 0 \,\}.$$

Proof. We remark that the morphism $R^1\alpha$ is surjective at the point $s \in \mathbb{P}(V)$ if and only if $H^1(\mathcal{E}_s \otimes F(\alpha)) = 0$. The last equality implies $H^0(\mathcal{E}_s \otimes F(\alpha)) = 0$ because $\chi(\mathcal{E}_s \otimes F(\alpha)) = 0$. Since $R^1\alpha \otimes k(s)$ is a morphism of vector spaces of the same dimension we have that $R^1\alpha$ is surjective if and only if $\det(R^1(\alpha)) \neq 0$. $\square$

Now let us investigate which cokernels $F(\alpha)$ yield nontrivial sections of the generalized Θ-bundle. Suppose that $F(\alpha)$ is not semistable. If $H^*(\mathcal{E}_s \otimes F) = 0$ for some $s \in \mathbb{P}(V)(k)$, then F is semistable by Proposition 2.7, a contradiction. Hence, we are interested only in those α with semistable cokernel $F(\alpha)$. On the other hand, let F be a semistable rank R vector bundle of determinant $\mathcal{O}_X$. Then $F \otimes A^{-1}$ is globally generated by Proposition 2.6 (vii). So there exists a short exact sequence

$$0 \to \det(F \otimes A^{-1})^{-1} \to \mathcal{O}_X^{\oplus(R+1)} \to F \otimes A^{-1} \to 0$$

by (viii) of 2.6. Twisting this short exact sequence with A yields the next result.

Lemma 4.2. *If* F *is semistable of rank* R *and* $\det(F) \cong \mathcal{O}_X$, *then there exists a morphism* $A_{R,1} \overset{\alpha}{\to} A_{R,0}$ *with* $\operatorname{coker}(\alpha) = F(\alpha) \cong F$. *For* $F(\alpha)$ *not semistable we have* $\theta_\alpha = 0$.

Remark 4.3. It is very tempting to write θ_F instead of $\theta_{F(\alpha)}$ or θ_α. Doing this, the formula of Proposition 4.1 reads

$$V(\theta_F) = \{\, s \in \mathbb{P}(V) \,|\, H^*(\mathcal{E}_s \otimes F) \neq 0 \,\}.$$

The author yielded to this temptation. However, we have in mind that we have to choose a framing of F (that is a short exact sequence $0 \to A_{R,1} \to A_{R,0} \to F \to 0$) to obtain the section θ_F. The vanishing locus is by 4.1 not affected by this choice.

Remark 4.4. On the reduced variety $\mathbb{P}(V)$ a reduced divisor D is given by the set of underlying points. Therefore, very sloppily one could define

$$\Theta_R := \{\, s \in \mathbb{P}(V) \,|\, H^*(\mathcal{E}_s \otimes F_R) \neq 0 \,\}$$

for a fixed vector bundle F_R of rank R and determinant $\mathcal{O}_X$. If we had done so, then we could define $\mathcal{O}_{\mathbb{P}(V)}(R \cdot \Theta) := \mathcal{O}_{\mathbb{P}(V)}(\Theta_{F_R})$. However, it is not clear why F_R gives a reduced divisor of $\mathcal{O}_{\mathbb{P}(V)}(R \cdot \Theta)$.

Remark 4.5. If $R = 1$ then there exists only one semistable sheaf of rank R of determinant $\mathcal{O}_X$, namely $\mathcal{O}_X$. So the definition of Θ is independent of F_1 and reads

$$\Theta := \{\, s \in \mathbb{P}(V) \,|\, H^*(\mathcal{E}_s) \neq 0 \,\}.$$

Since there are line bundles M of degree $g - 1$ with $h^0(M) = 0$, we deduce that $E := M \oplus (\omega_X \otimes M^{-1})$ is a rank two vector bundle of determinant ω_X which satisfies $H^*(E) = 0$. So not all points $s \in \mathbb{P}(V)$ are in Θ. By Proposition 2.7 the complement of Θ is a dense open subset which parameterizes semistable vector bundles on X.

Remark 4.6. If $\mathcal{E}_S$ is a family of semistable sheaves of rank two and determinant ω_X on $S \times X$, then we can copy the above definition to define the generalized Θ divisor $\mathcal{O}_S(R \cdot \Theta)$ on S. The same holds for the section θ_α.

4.3. The multiplicative structure

So far we have defined $\mathcal{O}(R \cdot \Theta)$ for each $R \in \mathbb{N}$ individually. Next we show the expected equality that $\mathcal{O}(R \cdot \Theta) \cong \mathcal{O}(\Theta)^{\otimes R}$. We start with two nontrivial morphisms $A_{R,1} \xrightarrow{\alpha} A_{R,0}$ and $A_{R',1} \xrightarrow{\alpha'} A_{R',0}$ defining two sections $\theta_\alpha \in \Gamma(\mathcal{O}(R \cdot \Theta))$ and $\theta_{\alpha'} \in \Gamma(\mathcal{O}(R' \cdot \Theta))$ respectively.

If we choose a general surjection $A_{R,0} \oplus A_{R',0} \xrightarrow{\pi} A$, then the composition $A_{R,1} \oplus A_{R',1} \to A_{R,0} \oplus A_{R',0} \to A$ will also be a surjection. To see this, we remark that $A_{R,1} \oplus A_{R',1} \to A_{R,0} \oplus A_{R',0}$ is a rank two vector subbundle, and we have an isomorphism $\operatorname{Hom}(A_{R,0} \oplus A_{R',0}, A) \cong \operatorname{Hom}(k^{R+R'+2}, k)$. Now for any point $P \in X(k)$ the morphisms $A_{R,0} \oplus A_{R',0} \to A$ for which the induced morphism $A_{R,1} \oplus A_{R',1} \to A$ is not surjective at P form a linear subspace V_P of codimension two in $\operatorname{Hom}(k^{R+R'+2}, k)$. The one-dimensional family $\{V_P\}_{P \in X(k)}$ of those linear subspaces is contained in a divisor in $\operatorname{Hom}(k^{R+R'+2}, k)$. This proves the above claim. Now we take π outside this divisor.

The kernel of the surjection $\pi|_{A_{R,1} \oplus A_{R',1}} : A_{R,1} \oplus A_{R',1} \to A$ is a vector bundle of rank one and determinant $A^{\otimes(R+R'+1)}$. Thus, it is isomorphic to $A_{R+R',1}$. From the two surjections we obtain therefore the commutative diagram

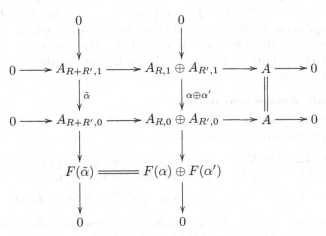

with exact rows and columns. Applying the functor $R^1 p_*(\mathcal{E} \otimes q^* \_)$ to this diagram yields an isomorphism $\mathcal{O}((R+R') \cdot \Theta) \cong \mathcal{O}(R \cdot \Theta) \otimes \mathcal{O}(R' \cdot \Theta)$ identifying the section $\theta_\alpha \cdot \theta_{\alpha'}$ with $\theta_{\tilde{\alpha}}$. Thus we obtain

Lemma 4.7. *We have isomorphisms* $\mathcal{O}((R+R') \cdot \Theta) \cong \mathcal{O}(R \cdot \Theta) \otimes \mathcal{O}(R' \cdot \Theta)$. *Under these isomorphisms the global sections* $\theta_\alpha \cdot \theta_{\alpha'}$ *and* $\theta_{\tilde{\alpha}}$ *are identified. This also reads* $\theta_{F \oplus F'} = \theta_F \cdot \theta_{F'}$.

We finish this section with an aside which gives a useless description of the line bundles $\mathcal{O}_{\mathbb{P}(V)}(R \cdot \Theta)$ on $\mathbb{P}(V)$.

Proposition 4.8. *On $\mathbb{P}(V)$ with ample generator $\mathcal{O}_{\mathbb{P}(V)}(1)$ of the Picard group $\mathrm{Pic}(\mathbb{P}(V))$ we have an isomorphism $\mathcal{O}(R \cdot \Theta) \cong \mathcal{O}_{\mathbb{P}(V)}(2gR)$.*

Proof. Consider the push forward Rp_* of the universal short exact sequence of Lemma 3.2 to obtain the long exact cohomology sequence

$$0 \to p_*\mathcal{E} \to \mathcal{O}_{\mathbb{P}(V)}(-1) \otimes H^0(X, L \otimes \omega_X) \xrightarrow{\gamma} \mathcal{O}_{\mathbb{P}(V)} \otimes H^1(X, L^{-1}) \to R^1 p_*\mathcal{E} \to 0$$

on $\mathbb{P}(V)$. The complex $A_{1,1} \xrightarrow{\alpha} A_{1,0}$ equals $\mathcal{O}_X$ in the Grothendieck group $K(X)$. Thus we can replace $R^i p_*\mathcal{E}$ in the above exact sequence by $R^i p_*(\mathcal{E} \otimes q^* F(\alpha))$. Thus this exact sequence and the one before Proposition 4.1 yield $\mathcal{O}_{\mathbb{P}(V)}(\Theta) = \mathcal{O}(1)^{\otimes h^0(X, L \otimes \omega_X)}$. Since $h^0(X, L \otimes \omega_X) = 2g$ we are done. $\square$

5. Raynaud's vanishing result for rank two bundles

In 1982 Raynaud published in [26] the following result:

Theorem 5.1 (Proposition 1.6.2 in [26]). *Let E be a rank two vector bundle of degree $2g - 2$ on a smooth projective curve X. We have an equivalence:*

$$E \text{ is semistable} \iff \text{there exists } L \in \mathrm{Pic}^0(X) \text{ with } H^*(E \otimes L) = 0.$$

We remark that $H^*(E \otimes L) = 0$ implies the semistability of E, as we have seen in Proposition 2.7. Thus, we only have to show the implication $\implies$. We give different proofs of this result which also catch a glimpse of the generalizations of Le Potier (see [20]) and Popa (see [25]) of this result to the case of higher ranks. But first things first.

5.1. The case of genus zero and one

For the case of genus zero and one we can without further difficulties prove the following generalization of Theorem 5.1.

Proposition 5.2. *Let X be a smooth projective curve of genus $g \in \{0,1\}$. For a rank r vector bundle of slope $g - 1$ we have the equivalence:*

$$E \text{ is semistable} \iff \text{there exists } L \in \mathrm{Pic}^0(X) \text{ with } H^*(E \otimes L) = 0.$$

Proof for genus zero. If E is semistable, then there exists no global section of E because the slope of the semistable bundle $\mathcal{O}_X$ is greater than the slope of E. We have $\chi(E) = 0$ and we conclude $h^0(E) = 0 = h^1(E)$. In short, we have $H^*(E \otimes \mathcal{O}_X) = 0$ as claimed. $\square$

Proof for genus one. Here we proceed inductively by proving the statement:

For E semistable of rank r and slope zero, there are at most r line bundles $L \in \mathrm{Pic}^0(X)$ with $H^*(E \otimes L) \neq 0$.

For rank $r = 1$, E is a line bundle of degree zero. Thus for all $L \not\cong E^\vee$ we have $H^*(E \otimes L) = 0$.

Now we assume the statement holds for $r - 1$ and take a semistable E of slope zero and rank r. Suppose we have $H^*(E \otimes L_r) \neq 0$ for a line bundle L_r of degree zero. We obtain a nontrivial morphism $L_r^\vee \to E$. We denote the cokernel by E''. It is a sheaf of rank $r - 1$ and degree zero. If E'' were not semistable, then we would have a surjection $E'' \to F$ to a sheaf of negative degree. However, the composition $E \to E'' \to F$ contradicts the semistability of E. Thus, E'' is semistable too and we have a short exact sequence

$$0 \to L_r^\vee \to E \to E'' \to 0$$

of semistable sheaves of slope zero. If a line bundle L of degree zero is not isomorphic to L_r, then we have $H^*(E \otimes L) = H^*(E'' \otimes L)$ and conclude by induction. $\square$

5.2. Preparations for the proof of 5.1

We collect here some facts which will be used in the sequel. The first fact is just an observation which allows us to concentrate on stable bundles for the proof of 5.1. The next is a detail we will need later on.

Lemma 5.3. *If E is a semistable but not stable vector bundle of rank two and degree $2g - 2$, then there exist line bundles $L \in \mathrm{Pic}^0(X)$ with $H^*(E \otimes L) = 0$.*

Proof. Since E is not stable we have a short exact sequence

$$0 \to L_1 \to E \to L_2 \to 0 \,,$$

where the L_i are line bundles of degree $g - 1$. Thus, they define two Θ-divisors in $\mathrm{Pic}^0(X)$, namely the divisors

$$\Theta_i = \left\{ L \in \mathrm{Pic}^0(X) \,|\, H^*(L \otimes L_i) \neq 0 \right\} .$$

If L is any line bundle of degree zero not contained in $\Theta_1 \cup \Theta_2$, then we have $H^*(L \otimes L_i) = 0$ and the above short exact sequence yields $H^*(L \otimes E) = 0$. $\square$

Lemma 5.4. *Let E be a rank two vector bundle and L a line bundle of degree d. Suppose that $\hom(L, E) \geq 2$ and $\deg(E) \neq 2d$ holds. Under these assumptions there exists $\alpha \colon L \to E$ such that the cokernel $\mathrm{coker}(\alpha)$ has torsion different from zero.*

Proof. Take two linearly independent morphisms $\beta_1, \beta_2 \in \mathrm{Hom}(L, E)$. We consider the resulting morphism $\beta = \beta_1 \oplus \beta_2 \colon L \oplus L \to E$. The morphism β cannot be surjective, because we assumed $\deg(E) \neq \deg(L \oplus L)$. Thus, we may take a geometric point P in the support of $\mathrm{coker}(\beta)$. It follows that the composite morphism $L \oplus L \xrightarrow{\beta} E \to E \otimes k(P) \cong k(P) \oplus k(P)$ is not surjective. Thus, for a suitable nontrivial linear combination $\alpha = \lambda_1 \beta_1 + \lambda_2 \beta_2$ we have $\alpha \otimes k(P) = 0$. $\square$

5.3. A proof for genus two using the rigidity theorem

Let us start with the following result about proper morphisms (see [5]).

Theorem 5.5 (Rigidity theorem). *We consider morphisms of varieties*

$$Y \xleftarrow{\;f\;} X \xrightarrow{\;g\;} Z$$

where f is proper with connected fibers. If for one point $y_0 \in Y$ the inverse image $f^{-1}(y_0)$ is mapped to a point under g, then there exists an open neighborhood U of y_0 in Y such that for all $y \in U$ the image $g(f^{-1}(y))$ is a point.

Proof. Take an affine open subset $\mathrm{Spec}(A) \subset Z$ containing the point $g(f^{-1}(y_0))$. Then $X' = g^{-1}\mathrm{Spec}(A)$ is open and contains the fiber over y_0. Thus the complement $X'' = X \setminus X'$ is closed and maps under the proper morphism f to some closed set $Y'' = f(X'')$. By construction y_0 is contained in the open subset $U = Y \setminus Y''$. For $y \in U$ we have by construction that $f^{-1}(y)$ is contained in X' and by assumption $f^{-1}(y)$ is proper and connected. Thus $g(f^{-1}(y)) \subset \mathrm{Spec}(A)$ is proper and connected. Hence, $g(f^{-1}(y))$ is a point. $\qquad\square$

Proof of Theorem 5.1 for curves of genus two. Let E be a semistable vector bundle of rank two and degree two on a smooth projective curve X of genus two. We want to show that there exist line bundles $L \in \mathrm{Pic}^0(X)$ with $H^*(E \otimes L) = 0$. We may assume that E is stable by Lemma 5.3. If for some $L \in \mathrm{Pic}^0(X)$ we have $\mathrm{Hom}(L, E) = 0$, then $H^*(E \otimes L^\vee) = 0$ and we are done. Thus, we may assume by Lemma 5.4 that $\hom(L, E) = 1$ for all $L \in \mathrm{Pic}^0(X)$. Indeed, if $\alpha \colon L \to E$ had cokernel with torsion supported at P, then α would give rise to some $\tilde{\alpha} \colon L(P) \to E$. This contradicts the stability of E. Thus, we have morphisms $\alpha_L \colon L \to E$. They are unique up to scalar multiplication, and their images are line subbundles.

This way we obtain for a point $P \in X(k)$ the morphism

$$\beta_P \colon \mathrm{Pic}^0(X) \to \mathbb{P}\big(E \otimes k(P)\big) \cong \mathbb{P}^1, \qquad L \mapsto \Big(L \otimes k(P) \xrightarrow{\;\alpha_L \otimes k(P)\;} E \otimes k(P) \Big).$$

This morphism is just the specialization of the X-morphism β

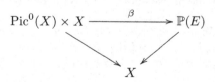

over $P \in X$. We take a proper curve $C \subset \mathrm{Pic}^0(X)$ which is contracted to a point under β_P which exists for dimensional reasons. Now we apply the rigidity theorem to the morphism

$$X \xleftarrow{\;\mathrm{pr}_2\;} C \times X \xrightarrow{\;\beta\;} \mathbb{P}(E) \ .$$

By definition the fiber of pr_2 over P is contracted to a point by β. Thus, by the rigidity theorem almost all fibers are contracted to a point by β. Thus, all line bundles parameterized by C describe the same line subbundle of E which is absurd. $\qquad\qquad\qquad\qquad\qquad\qquad\qquad\qquad\qquad\qquad\qquad\qquad\qquad\qquad\square$

5.4. A proof based on Clifford's theorem

For the sake of completeness we repeat here Clifford's theorem which will be the main ingredient in our proof of Theorem 5.1. For a proof we refer to Hartshorne's book [10, Theorem IV.5.4].

Theorem 5.6 (Clifford's theorem). *Let L be a line bundle on a smooth projective curve X with $0 \le \deg(L) \le 2g - 2$. Then we have the estimate*

$$h^0(L) \le \frac{\deg(L)}{2} + 1,$$

and equality holds only for a finite number of line bundles.

Proof of Theorem 5.1 for curves of genus $g \ge 2$. We fix a stable vector bundle E of rank two and degree $2g - 2$. Let us assume that for all line bundles L of degree zero we have $\mathrm{Hom}(L, E) \ne 0$. We consider the following schemes $\{B^d\}_{d\in\mathbb{N}}$ with the reduced subscheme structure. Indeed, we are computing dimensions in this proof only, so the underlying scheme structure is of no interest for us.

$$B^d := \left\{ (L \xrightarrow{\alpha} M) \text{ with } \begin{array}{l} L \in \mathrm{Pic}^0(X) \text{ and } M \in \mathrm{Pic}^d(X) \\ \alpha \in (\mathrm{Hom}(L, M) \setminus \{0\})/k^* \\ \text{and } \mathrm{Hom}(M, E) \ne 0 \end{array} \right\}.$$

Since E is stable we have $B^d = \varnothing$ for $d \ge g-1$. We consider the natural projection $B^d \xrightarrow{\beta_d} \mathrm{Pic}^0(X)$ which maps a triple $(L \xrightarrow{\alpha} M)$ to L. Suppose $d > 0$ and $(L \xrightarrow{\alpha} M) \in B^d$. For $P \in \mathrm{supp}(\mathrm{coker}(\alpha))$, we have that $(L \xrightarrow{\alpha} M(-P)) \in B^{d-1}$. Thus, we have inclusions $\mathrm{im}(\beta_d) \subseteq \mathrm{im}(\beta_{d-1}) \subseteq \cdots \subseteq \mathrm{im}(\beta_0)$. If $\mathrm{im}(\beta_0) \subsetneq \mathrm{Pic}^0(X)$, then we have for any line bundle L not contained in $\mathrm{im}(\beta_0)$ that $\mathrm{Hom}(L, E) = 0$ which contradicts our assumption. Let now d be the maximal integer, such that $\mathrm{im}(\beta_d) = \mathrm{Pic}^0(X)$. We consider the open set $U_d = \mathrm{Pic}^0(X)\backslash\mathrm{im}(\beta^{d+1})$. Take a triple $(L \xrightarrow{\alpha} M)$ with $L \in U_d$. Next we show that each nontrivial morphism $M \to E$ has torsion free cokernel. If $M \to E$ has cokernel with torsion supported in P, then we obtain a nonzero morphism $M(P) \to E$. However, from the composition $L \to M \to M(P)$ we deduce that $L \in \mathrm{im}(\beta_{d+1})$, so M is a line subbundle and $\mathrm{hom}(M, E) = 1$ by Lemma 5.4.

Now we consider the projection $B^d \xrightarrow{\alpha_d} \mathrm{Pic}^d(X)$ assigning M to $(L \xrightarrow{\alpha} M)$. Since α_d is an $X^{(d)}$-bundle, we obtain that $\alpha_d(\beta_d^{-1}(U_d))$ is a family of line bundles M of degree d of dimension at least $g - d$. We have an inclusion

$$\alpha_d(\beta_d^{-1}(U_d)) \subseteq \{ M \in \mathrm{Pic}^d(X) \mid \mathrm{hom}(M, E) = 1 \,\&\, M \to E \text{ a line subbundle } \}.$$

It follows that the Quot scheme of quotients of E of degree $2g - 2 - d$ is at least $(g-d)$-dimensional in a neighborhood of the point $[E \to E/M]$. Thus, the tangent space $\mathrm{Hom}(M, E/M)$ of the Quot scheme is at least of that dimension. However

$$\mathrm{Hom}(M, E/M) = H^0((E/M) \otimes M^\vee) = H^0(\det(E) \otimes M^{\otimes -2}).$$

By Clifford's theorem the dimension is at most $g - d$ and this equality can hold only for finitely many M. Since $g - d$ is positive, this contradicts our assumption that $\text{Hom}(L, E) \neq 0$ for all $L \in \text{Pic}^0(X)$, and we are done. $\qquad\square$

5.5. Generalizations and consequences

First of all we deduce two consequences from Theorem 5.1.

Corollary 5.7. *Let $r \geq 2$ be an integer. We consider a family $\mathcal{E}$ on $S \times X$ of rank two vector bundles on X of degree $2g-2$. The base points of the linear subsystem of $\mathcal{O}_S(r \cdot \Theta)$ given by the sections θ_F where F runs through all rank r vector bundles of trivial determinant on X is the set of all points $s \in S$ such that $\mathcal{E}_s$ is not semistable.*

Proof. We know by Theorem 5.1 that points $s \in S$ which are not base points parameterize semistable bundles $\mathcal{E}_s$. Suppose now that $E = \mathcal{E}_s$ is semistable. We have to find a rank r bundle F with trivial determinant such that $H^*(E \otimes F) = 0$. We find this by setting $F := L_1 \oplus L_2 \oplus \cdots \oplus L_r$. $\qquad\square$

Proposition 5.8. *Let X be a smooth projective curve. For a family $\mathcal{E}$ over $S \times X$ of rank two vector bundles of degree $2g-2$ on X the set $S^{\text{ss}} := \{s \in S \mid \mathcal{E}_s \text{ semistable}\}$ is an open subset of X.*

Proof. We consider the morphisms $S \xleftarrow{\ p\ } S \times X \xrightarrow{\ q\ } X$. Suppose $s \in S^{\text{ss}}$, then there exists by Theorem 5.1 a line bundle L such that $H^*(X, \mathcal{E}_s \otimes L) = 0$. Thus, the coherent sheaf $R^1 p_*(\mathcal{E} \otimes q^*L)$ is zero at s. So there exists an open $U \subset S$ containing s such that $R^1 p_*(\mathcal{E} \otimes q^*L)$ is zero on U. By base change we have $H^1(X, \mathcal{E}_t \otimes L) = 0$ for all $t \in U$. From Riemann–Roch we deduce $H^*(X, \mathcal{E}_t \otimes L) = 0$ for $t \in U$. Theorem 5.1 tells us that all these $\mathcal{E}_t$ are semistable. $\qquad\square$

The next results give us further equivalent conditions for semistability and show that these results generalize to sheaves E of arbitrary rank and degree. However, these results will not be used in the sequel. So the reader may skip to the next section.

Theorem 5.9. *Let X be a smooth projective curve of genus g. Then there exists a vector bundle P_{2g+1} on X such that for sheaves E of rank two and degree $2g - 2$ on X the following conditions are equivalent:*

(i) *E is semistable.*
(ii) *$\text{Hom}(L, E) = 0$ for a line bundle L of degree zero.*
(iii) *$H^*(E \otimes M) = 0$ for a line bundle M of degree zero.*
(iv) *$\text{Hom}(P_{2g+1}, E) = 0$.*

We have proved the equivalence of (i)–(iii) in Theorem 5.1. Considering the Fourier–Mukai transform on the pair Jacobian and Picard torus of X the condition (iv) is deduced (see [12] for details and a proof).

Theorem 5.10. *Let X be a smooth projective curve of genus g. Furthermore two integers $r \geq 1$ and d are given. Then there exist integers R and D and a vector bundle $P_{r,d}$ depending on r, d and g, such that for all vector bundles E of rank r and degree d the following are equivalent:*

(i) *E is semistable.*

(ii) *$H^*(E \otimes F) = 0$ for a sheaf F of rank R and degree D.*

(iii) *$\mathrm{Hom}(P_{r,d}, E) = 0$.*

The equivalence of (i) and (ii) is shown in [25]. The equivalence of (ii) and (iii) is essentially linear algebra and carried out in [13].

Remark 5.11. It follows that the construction, which we present here for the rank two case, allows an obvious generalization to the case of arbitrary rank and degree. However, in our case we can take line bundles as the parameters for our generalized Θ-divisors which is very convenient.

6. Semistable limits

6.1. Limits of vector bundles

We repeat here the definition of separatedness for noetherian schemes over an algebraically closed field. The concept of a limit is formalized in algebraic geometry by the concept of a discrete valuation domain R. We have $\mathrm{Spec}(R) = \{\eta, 0\}$ where η denotes the generic point and 0 denotes the closed point. The standard example is of course the localization of a curve at a smooth point. The closed point 0 is the limit of the generic point η.

The Hausdorff separation axiom from topology is too strong for the Zariski topology and is replaced by the concept of separatedness:

X is separated if for all discrete valuation domains R and all maps $\psi \colon \{\eta\} \to X$ there exists at most one extension $\tilde{\psi} \colon \mathrm{Spec}(R) \to X$.

If there exists a unique extension $\tilde{\psi} \colon \mathrm{Spec}(R) \to X$, then X is proper. The usual picture is the following:

——— —— ——	————•————	————•——
Separated, not proper	Separated and proper	Neither separated nor proper

Morphisms $S \to M_X$ to the (potential) moduli space M_X of vector bundles on X should correspond to families $\mathcal{E}_S$ of vector bundles on $S \times X$. Thus, we take a discrete valuation ring R with $\mathrm{Spec}(R) = \{\eta, 0\}$ and a vector bundle $\mathcal{E}_\eta$ on $\{\eta\} \times X$. The first thing we need is the following lemma which does not generalize to higher-dimensional varieties or singular curves.

Lemma 6.1. *Let X be a smooth projective curve. Any family $\mathcal{E}_\eta$ of vector bundles on $\{\eta\} \times X$ can be extended to a family $\mathcal{E}_R$ on $\mathrm{Spec}(R) \times X$.*

Proof. Set $r := \mathrm{rk}(\mathcal{E}_\eta)$. We consider the morphisms

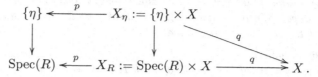

We may replace $\mathcal{E}_\eta$ by $\mathcal{E}_\eta \otimes q^*L$. Thus, we may assume $\mathcal{E}_\eta$ is globally generated and its determinant is of degree at least $2g$. On $\{\eta\} \times X$ we have by Proposition 2.6 a short exact sequence

$$0 \to \mathcal{O}_{X_\eta}^{\oplus r-1} \to \mathcal{E}_\eta \to \det(\mathcal{E}_\eta) \to 0 \,.$$

The properness of the Picard functor $\mathrm{Pic}(X)$ guarantees that there exists a(n) (unique) extension $\mathcal{L}_R$ of the line bundle $\det(\mathcal{E}_\eta)$. On the other hand $\mathcal{O}_{X_R}^{\oplus r-1}$ is an extension of $\mathcal{O}_{X_\eta}^{\oplus r-1}$ to X_R.

The extension $\mathcal{E}_\eta$ is given by a $k(\eta)$-valued section α_η of the coherent sheaf $R^1 p_*(\mathcal{H}om(\mathcal{L}_R, \mathcal{O}_{X_R}^{\oplus r-1}))$. Changing α_η by an element of $k(\eta)^*$ does not affect the isomorphism class of $\mathcal{E}_\eta$. This way we can obtain an R-valued section α_R of this sheaf which gives α_η at the point η. Now since

$$H^0\Big(R^1 p_*(\mathcal{H}om(\mathcal{L}_R, \mathcal{O}_{X_R}^{\oplus r-1}))\Big) = \mathrm{Ext}^1(\mathcal{L}_R, \mathcal{O}_{X_R}^{\oplus r-1})$$

we are done.   □

6.2. Changing limits – elementary transformations

Now let $\mathcal{E}_R$ be a vector bundle on X_R. The restriction $\mathcal{E}_0$ of $\mathcal{E}_R$ to the closed fiber $X_0 = p^{-1}(0)$ is sometimes called the limit of $\mathcal{E}_\eta = \mathcal{E}_R|_{X_\eta}$. Since X_0 is a Cartier divisor on X_R the structure sheaf $\mathcal{O}_{X_0}$ considered as an $\mathcal{O}_{X_R}$-module is of projective dimension one. Thus, every X_0-vector bundle F_0 is as a sheaf on X_R of projective dimension one. This observation allows the following construction.

Construction: Elementary transformation

Let $\mathcal{E}_0$ be the restriction of the vector bundle $\mathcal{E}_R$ to X_0. Furthermore, let $\mathcal{E}_0 \xrightarrow{\pi_0} F_0$ be a surjection of vector bundles on X_0. Composing we obtain a surjection $\mathcal{E}_R \xrightarrow{\pi} F_0$ of sheaves on X_R. The kernel of π is of projective dimension zero. Thus, $\ker(\pi)$ is a vector bundle. This vector bundle $\mathcal{E}'_R$ is called the elementary transformation of $\mathcal{E}_R$ along F_0. The bundle $\mathcal{E}'_R$ appears in a short exact sequence

$$0 \to \mathcal{E}'_R \to \mathcal{E}_R \xrightarrow{\pi} F_0 \to 0 \,.$$

When restricting this short exact sequence to X_0 and denoting the restriction of $\mathcal{E}'_R$ to X_0 by $\mathcal{E}'_0$ we obtain the exact sequence

$$0 \to \Big(\mathrm{Tor}_1^{\mathcal{O}_{X_R}}(\mathcal{O}_{X_0}, F_0) = F_0\Big) \to \mathcal{E}'_0 \to \mathcal{E}_0 \xrightarrow{\pi_0} F_0 \to 0 \,.$$

Obviously the sheaves $\mathcal{E}_\eta$ and $\mathcal{E}'_\eta$ coincide. Summing up we have the next result.

Proposition 6.2. *Let $\mathcal{E}'_R$ be the elementary transformation of $\mathcal{E}_R$ along F_0 then we have an isomorphism over X_η: $\mathcal{E}'_\eta \cong \mathcal{E}_\eta$.*
Over the special fiber X_0 we have two short exact sequences

$$0 \to F_0 \to \mathcal{E}'_0 \to \ker(\pi_0) \to 0 \quad and \quad 0 \to \ker(\pi_0) \to \mathcal{E}_0 \to F_0 \to 0\,.$$

In short we may say: Elementary transformations along F_0 do not change the bundle over the generic point, they transform the quotient F_0 of $\mathcal{E}_0$ to a subsheaf of $\mathcal{E}'_0$.

6.3. Example: Limits are not uniquely determined

The following example shows that there are infinitely many possible limits for a vector bundle $\mathcal{E}_\eta$. The example is intended to show that it makes sense to restrict to a nice class of bundles to avoid this plenitude of limits.

Proposition 6.3. *Let $\mathcal{E}_R$ be a rank two vector bundle on X_R such that $\det(\mathcal{E}_0) \cong \mathcal{O}_{X_R}$ and $\mathcal{E}_0$ is semistable. Then for any X-line bundle L of degree $\deg(L) \geq 2g$ there exists an elementary transformation $\mathcal{E}'_R$ of $\mathcal{E}_R$ such that $\mathcal{E}'_R \cong L \oplus L^{-1}$.*

Proof. By Proposition 2.6, $\mathcal{E}_0 \otimes L$ is globally generated, and by part (ix) there exists a surjection $\mathcal{E}_0 \otimes L \to L^{\otimes 2}$. Twisting with L^{-1} we obtain a short exact sequence

$$0 \to L^{-1} \to \mathcal{E}_0 \xrightarrow{\pi_0} L \to 0\,.$$

Applying an elementary transformation of $\mathcal{E}_R$ along L we obtain by Proposition 6.2 a vector bundle $\mathcal{E}'_R$ with $\mathcal{E}'_\eta \cong \mathcal{E}_\eta$ and a short exact sequence

$$0 \to L \to \mathcal{E}'_0 \to L^{-1} \to 0\,.$$

However, $\mathrm{Ext}^1(L^{-1}, L) = H^1(L^{\otimes 2}) = 0$ which implies that $\mathcal{E}'_0 \cong L \oplus L^{-1}$. $\qquad\square$

6.4. Semistable limits exist

The next result of Langton (see [18]) tells us that a semistable vector bundle $\mathcal{E}_\eta$ on X_η can be extended to a bundle $\mathcal{E}_R$ on X_R such that the restriction $\mathcal{E}_0$ of $\mathcal{E}_R$ to the special fiber X_0 is also semistable. The proof uses that we can *stabilize* a given extension by elementary transformations. The new idea is that we can control the number of these elementary transformations by the badness, a number introduced in [11] by the author. Please remember that Langton's theorem (as well as the proof below) holds for arbitrary rank and determinant.

Theorem 6.4 (Langton's theorem on the existence of semistable limits). *Let $\mathcal{E}_\eta$ be a semistable rank two vector bundle with determinant $q^* \omega_X$. Then there exists a vector bundle $\mathcal{E}_R$ on X_R such that $\mathcal{E}_R|_{X_\eta} \cong \mathcal{E}_\eta$ and $\mathcal{E}_0 = \mathcal{E}_R|_{X_0}$ is semistable.*

Proof. By Lemma 6.1 there exist extensions $\mathcal{E} = \mathcal{E}_R$ of $\mathcal{E}_\eta$ to X_R. We take such an extension $\mathcal{E}$ and assign it an integer $\mathrm{bad}(\mathcal{E})$ – the badness of $\mathcal{E}$.

$$\mathrm{bad}(\mathcal{E}) := \min_{L \in \mathrm{Pic}^0(X)} \left\{ \mathrm{length}\big(R^1 p_*(\mathcal{E} \otimes q^* L)\big) \right\}.$$

By Theorem 5.1 the semistability of $\mathcal{E}_\eta$ implies that there exist $L \in \text{Pic}^0(X)$ such that the sheaf $R^1 p_*(\mathcal{E} \otimes q^* L)$ is zero at η. We conclude that the badness is well defined. By base change and Theorem 5.1 we deduce the equivalence

$$\mathcal{E}_0 \text{ is semistable } \iff \text{bad}(\mathcal{E}) = 0.$$

It is enough to show the following statement. If $\mathcal{E}$ is an extension with positive badness, then there exists an elementary transformation $\mathcal{E}'$ of $\mathcal{E}$ such that $\text{bad}(\mathcal{E}') < \text{bad}(\mathcal{E})$. Let us show this. Since $\mathcal{E}_0$ is not semistable, there exists a line bundle quotient $\mathcal{E}_0 \to F$ on X_0 with $\deg(F) < g - 1$. This implies $\chi(F \otimes L) = \chi(F) < 0$ for all $L \in \text{Pic}^0(X)$. We consider the elementary transformation $\mathcal{E}'$ of $\mathcal{E}$ along F:

$$0 \to \mathcal{E}' \to \mathcal{E} \to F \to 0.$$

Next we choose a line bundle $L \in \text{Pic}^0(X)$ such that $\text{bad}(\mathcal{E}) = \text{length}(R^1 p_*(\mathcal{E} \otimes q^* L))$. This implies that $p_*(\mathcal{E} \otimes q^* L) = 0$ because it is torsion free and zero at η. From the above short exact sequence we obtain

$$0 \to p_*(F \otimes q^* L) \to R^1 p_*(\mathcal{E}' \otimes q^* L) \to R^1 p_*(\mathcal{E} \otimes q^* L) \to R^1 p_*(F \otimes q^* L) \to 0.$$

We have $p_*(F \otimes q^* L) = H^0(F \otimes L)$ and $R^1 p_*(F \otimes q^* L) = H^1(F \otimes L)$. Thus, both are sheaves of finite length on R. We deduce that $R^1 p_*(\mathcal{E}' \otimes q^* L)$ is also a sheaf of finite length which we compute to be

$$\text{length}\big(R^1 p_*(\mathcal{E}' \otimes q^* L)\big) = \text{length}\big(R^1 p_*(\mathcal{E} \otimes q^* L)\big) + h^0(F \otimes L) - h^1(F \otimes L).$$

Having in mind that $\chi(F \otimes L)$ is negative, we are done. $\qquad \square$

6.5. Semistable limits are almost uniquely determined

We have seen in Theorem 6.4 that semistable limits exist. However the example of Proposition 6.3 shows that we should not hope for a unique limit. One might hope that semistable limits are unique. Unfortunately this is not the case. Let us illustrate it.

Let L be a line bundle of degree $g - 1$ which is not isomorphic to $\omega_X \otimes L^{-1}$. The vector space $\text{Ext}^1(L, \omega_X \otimes L^{-1})$ is of dimension $g - 1$. If $g \geq 2$, then there exist nontrivial extensions fitting in short exact sequences

$$0 \to \omega_X \otimes L^{-1} \to E \to L \to 0.$$

Since $\deg(\omega_X \otimes L^{-1}) = \deg(L) = g - 1$, the vector bundle E is semistable but not stable. Taking the pull back $\mathcal{E} = q^* E$ we obtain a sheaf on X_R which gives E when restricted to X_0. The short exact sequence above allows an elementary transformation of $\mathcal{E}$ along L. This way we obtain (by Proposition 6.2) a sheaf $\mathcal{E}'$ on X_R with restriction $E' = \mathcal{E}'|_{X_0}$ fitting into an exact sequence

$$0 \to L \to E' \to \omega_X \otimes L^{-1} \to 0.$$

Since the extension was not trivial, E' and E are both semistable but not isomorphic. However, both appear as a limit of $\mathcal{E}_\eta = \mathcal{E}|_{X_\eta} \cong \mathcal{E}'|_{X_\eta}$.

We note that both sheaves are S-equivalent because

$$\text{gr}(E) = \text{gr}(E') = L \oplus \big(\omega_X \otimes L^{-1}\big).$$

Thus, we can only hope that the S-equivalence class of a limit is unique. This is generally the case and is shown in our case now.

Proposition 6.5. *Let $\mathcal{E}$ and $\mathcal{E}'$ be two rank two vector bundles on X_R such that $\mathcal{E}_\eta \cong \mathcal{E}'_\eta$. If the restrictions $\mathcal{E}_0$ and $\mathcal{E}'_0$ to X_0 are semistable, then they are S-equivalent.*

Proof. We consider the vector bundle $\mathcal{E}' \otimes \mathcal{E}^\vee \otimes q^* \omega_X$. When applying $R^1 p_*$ to it, we obtain by base change a sheaf whose restriction to the generic point η equals $H^1(X_\eta, \mathcal{E}'_\eta \otimes \mathcal{E}^\vee_\eta \otimes q^* \omega_X)$. This is the Serre dual of $\mathrm{Hom}(\mathcal{E}'_\eta, \mathcal{E}_\eta)$. Since both sheaves are isomorphic there exist nontrivial homomorphisms and we conclude that $R^1 p_*(\mathcal{E}' \otimes \mathcal{E}^\vee \otimes q^* \omega_X) \neq 0$. So the specialization of this sheaf to the special point $0 \in \mathrm{Spec}(R)$ is not zero. Again by base change and Serre duality we conclude $\mathrm{Hom}(\mathcal{E}'_0, \mathcal{E}) \neq 0$. We obtain a morphism $\varphi \colon \mathcal{E}'_0 \to \mathcal{E}$ between two semistable rank two vector bundles of the same determinant. If φ is an isomorphism, then we are done. Thus, we may assume $\mathrm{rk}(\ker(\varphi)) = 1$. The image $M = \mathrm{im}(\varphi)$ of φ is at the same time quotient and subbundle of semistable bundles with slope $g - 1$. Thus, M is a degree $g - 1$ line bundle. Considering the determinant we conclude $\ker(\varphi) \cong \mathrm{coker}(\varphi)$. This finishes the proof because the graded objects are given by $\mathrm{gr}(\mathcal{E}'_0) = \ker(\varphi) \oplus M$ and $\mathrm{gr}(\mathcal{E}_0) = M \oplus \mathrm{coker}(\varphi)$. $\square$

S-equivalence prevents us from non-separated moduli functors

So by passing to the moduli functor of S-equivalence classes, we obtain a proper moduli functor. That is there exist semistable extensions of $\mathcal{E}_\eta$ and all those extensions are in the same S-equivalence class. Note that S-equivalence classes are required to circumvent a non-separated moduli functor. It causes several problems because we don't have a universal object.

However, for a stable vector bundle E the S-equivalence class of E is the class of vector bundles isomorphic to E. Thus, on a dense open subset of the moduli space there is no difference between S-equivalence and isomorphism, whereas for strictly semistable bundles it makes a difference.

7. Positivity

7.1. Notation and preliminaries

For this section we fix the following objects:

X	our smooth projective curve of genus $g_X \geq 2$ over $k = \bar{k}$
C	a smooth projective curve over k of genus g_C
p, q	the natural projections $C \xleftarrow{\ p\ } C \times X \xrightarrow{\ q\ } X$
$A \boxtimes B$	a shorthand for the tensor product $p^*A \otimes q^*B$
$\mathcal{E}$	a rank two vector bundle on $C \times X$
$\mathcal{E}_c$	for a point $c \in C(k)$ the X-vector bundle $q_*(\mathcal{E} \otimes p^*k(c))$, that is the vector bundle parameterized by the point c
$\mathcal{E}_x$	for $x \in X(k)$ the vector bundle $p_*(\mathcal{E} \otimes q^*k(x))$ on C

As a warm up we consider a well-known result (see Corollary 7.3). Apart from the obvious proof via Proposition 7.1, we deduce this result from Proposition 7.2 which allows a generalization to vector bundles.

Proposition 7.1. *Let $\mathcal{L}$ be a line bundle on $C \times X$ with $\mathcal{L}_c \cong \mathcal{L}_{c'}$ for all points $c, c' \in C(k)$. Then there exist line bundles M and N on C and X respectively, such that $\mathcal{L} \cong M \boxtimes N$.*

Proof. Let $N = \mathcal{L}_c$ for a point $c \in C$. Since $R^1 p_*(\mathcal{L} \otimes q^* N^{-1})$ is a vector bundle, it follows that $p_*(\mathcal{L} \otimes q^* N^{-1})$ also commutes with base change (see [23, page 50, Corollary 2]). Thus, $M := p_*(\mathcal{L} \otimes q^* N^{-1})$ is a line bundle on C. The composition morphism $p^* M \to p^*(p_*(\mathcal{L} \otimes q^* N^{-1})) \to \mathcal{L} \otimes q^* N^{-1}$ is an isomorphism on all fibers of p. Thus, it is an isomorphism on $C \times X$. $\square$

Proposition 7.2. *Suppose $\mathcal{L}$ is a line bundle on $C \times X$. The degrees $d_1 := \deg(\mathcal{L}_x)$ and $d_2 := \deg(\mathcal{L}_c)$ of the line bundles parameterized by X and C do not depend on the choice of the points $x \in X(k)$ and $c \in C(k)$. We obtain two morphisms*

$$\varphi_1 \colon X \xrightarrow{\; x \mapsto \mathcal{L}_x \;} \operatorname{Pic}^{d_1}(C) \quad and \quad \varphi_2 \colon C \xrightarrow{\; c \mapsto \mathcal{L}_c \;} \operatorname{Pic}^{d_2}(X).$$

For the principal polarizations Θ_1 and Θ_2 on $\operatorname{Pic}^{d_1}(C)$ and $\operatorname{Pic}^{d_2}(X)$, the degrees of X in $\operatorname{Pic}^{d_1}(C)$ and C in $\operatorname{Pic}^{d_2}(X)$ coincide, that is

$$\deg_X\big(\varphi_1^* \mathcal{O}_{\operatorname{Pic}^{d_1}(C)}(\Theta_1)\big) = \deg_C\big(\varphi_2^* \mathcal{O}_{\operatorname{Pic}^{d_2}(X)}(\Theta_2)\big).$$

Proof. First we remark that the theorem is invariant under twisting L with line bundles of type $p^* M$ or $p^* N$. Therefore, we may assume $d_1 = g_C - 1$ and $d_2 = g_X - 1$. On $\operatorname{Pic}^{g_C - 1}(C)$ the Θ-line bundle is the determinant of cohomology, that is:

$$\deg_X\big(\varphi_1^* \mathcal{O}_{\operatorname{Pic}^{d_1}(C)}(\Theta_1)\big) = -\deg(R^* p_*(\mathcal{L}))$$

this we can rewrite as

$$= -\int_X \operatorname{ch}(R^* p_*(\mathcal{L}))$$

using Grothendieck–Riemann–Roch we obtain

$$= -\int_{C \times X} (\operatorname{ch}(\mathcal{L}) \cdot p^* \operatorname{Td}(C))$$
$$= -\int_{C \times X} (1 + c_1(\mathcal{L}) + \tfrac{c_1^2(\mathcal{L})}{2})(1 - \tfrac{p^* \omega_C}{2})$$
$$= \tfrac{1}{2} \int_{C \times X} \big(p^* \omega_C . c_1(\mathcal{L}) - c_1^2(\mathcal{L})\big).$$

Since $p^* \omega_C$ is numerically equivalent to $2g_C - 2$ fibers of p and the intersection of $c_1(\mathcal{L})$ with a fiber of p is the degree $d_2 = g_X - 1$, we conclude that

$$\deg_X\big(\varphi_1^* \mathcal{O}_{\operatorname{Pic}^{d_1}(C)}(\Theta_1)\big) = (g_C - 1)(g_X - 1) - \frac{1}{2} \int_{C \times X} c_1^2(\mathcal{L}).$$

By symmetry the right-hand side is also the degree of $\varphi_2^* \mathcal{O}_{\operatorname{Pic}^{d_2}(X)}(\Theta_2)$. $\square$

Corollary 7.3. *Let $\mathcal{L}$ be a line bundle on $C \times X$. If for any two points $c, c' \in C(k)$ the line bundles $\mathcal{L}_c$ and $\mathcal{L}_{c'}$ are isomorphic, then for all points $x, x' \in X(k)$ the line bundles $\mathcal{L}_x$ and $\mathcal{L}_{x'}$ are isomorphic.*

Proof. Indeed, this result can be obviously deduced from Proposition 7.1. In order to deduce it from Proposition 7.2, we consider the maps $\varphi_1 \colon X \to \mathrm{Pic}^{d_1}(C)$ and $\varphi_2 \colon C \to \mathrm{Pic}^{d_2}(X)$ from Proposition 7.2. By assumption C is mapped to a point. Hence $\deg_C(\varphi_2^*\Theta_2) = 0$. By Proposition 7.2 this implies $\deg_X(\varphi_1^*\Theta_1) = 0$. Since the Theta divisor Θ_1 is ample, X is also mapped to a point. $\qquad\square$

We assume $\det(\mathcal{E}_c) \cong \omega_X$ for all $c \in C(k)$. This implies that $\det(\mathcal{E}) = M \boxtimes \omega_X$ by Proposition 7.1. Suppose that $\mathcal{E}_c$ is semistable. By Theorem 5.1 there exists an $L \in \mathrm{Pic}^0(X)(k)$ such that $H^*(X, \mathcal{E}_c \otimes L) = 0$. By base change this implies that $R^1 p_*(\mathcal{E} \otimes q^*L)$ is not supported at c. Thus, it is a torsion sheaf of finite length. The bundle $\mathcal{E}_c$ is therefore semistable for all $c \notin \mathrm{supp} R^1 p_*(\mathcal{E} \otimes q^*L)$. As seen in Theorem 6.4 we can perform elementary transformations, such that $\mathcal{E}$ becomes semistable in the remaining points.

We will see C as a parameter curve mapping into the moduli space M_X (which we have not constructed so far). The aim of this section is to show that the intersection number of C with the Θ-divisor is not negative, and that this intersection number is zero if and only if all bundles parameterized by C are S-equivalent.

It will be convenient for us that the vector bundles $\mathcal{E}_x$ on C will have degree $2g_C - 2$. If the degree of $\mathcal{E}_x$ is an even number, then this is obtained by twisting $\mathcal{E}$ with a line bundle p^*M where M is of degree

$$g_C - 1 - \frac{\deg(\mathcal{E}_x)}{2}.$$

In case $\deg(\mathcal{E}_x)$ is odd, we replace C by an irreducible curve $\psi \colon C' \xrightarrow{2:1} C$ and $\mathcal{E}$ by $\mathcal{E}' := (\psi \times \mathrm{id}_X)^*\mathcal{E}$. This way we obtain the same bundles on X doubly parameterized by C' and an even degree of $\mathcal{E}'_x$.

Summing up we have the following data

$$\begin{array}{ll} \det(\mathcal{E}_c) \cong \omega_X & \text{for all } c \in C(k), \\ \mathcal{E}_c \text{ is semistable} & \text{for all } c \in C(k), \text{ and} \\ \deg(\mathcal{E}_x) = 2g_C - 2. \end{array}$$

7.2. Positivity – global sections of $\mathcal{O}(\Theta)$

The Θ-line bundle on C is now the inverse of the determinant of cohomology. Since vector bundles $\mathcal{E}_c$ and $\mathcal{E}_{c'}$ of the same rank and isomorphic determinant coincide in the Grothendieck group $K(X)$, we see that $\det(R^*p_*(\mathcal{E} \otimes q^*L))^{-1}$ does not depend on the choice of $L \in \mathrm{Pic}^0(X)$. This line bundle is the generalized Θ-line bundle for the family $\mathcal{E}$ of vector bundles on X parameterized by C and therefore denoted by $\mathcal{O}_C(\Theta_C)$ (see Remark 4.6).

Proposition 7.4. *The Θ-line bundle $\mathcal{O}_C(\Theta_C)$ is base point free. In particular, we have $\deg(\mathcal{O}_C(\Theta_C)) \geq 0$.*

Proof. We have seen in Theorem 5.1 that there exists for each $c \in C(k)$ a line bundle $L \in \mathrm{Pic}^0(X)$ such that $H^*(\mathcal{E}_c \otimes L) = 0$. This means that the section of $\mathcal{O}_C(\Theta_C)$ corresponding to L does not vanish at c by Proposition 4.1. $\qquad \square$

7.3. The case of $\deg(\mathcal{O}_C(\Theta_C)) = 0$

Here we prove a special case of Theorem I.4 from Faltings' article [8]. The proof mainly follows the idea given there.

Theorem 7.5. *We have an equivalence of $\deg(\mathcal{O}_C(\Theta_C)) = 0$, and $\mathcal{E}_c \sim_S \mathcal{E}_{c'}$ for all points $c, c' \in C(k)$. In this case the sheaves $\mathcal{E}_x$ and $\mathcal{E}_{x'}$ are isomorphic for all $x, x' \in X(k)$.*

The proof of Theorem 7.5 follows directly from the subsequent results 7.6–7.9.

Lemma 7.6. *If all the vector bundles parameterized by C are S-equivalent, then the degree of $\mathcal{O}_C(\Theta_C)$ is zero.*

Proof. Pick $c \in C(k)$ and consider the semistable vector bundle $E := \mathrm{gr}(\mathcal{E}_c)$. Then there exists a line bundle $L \in \mathrm{Pic}^0(X)$ such that $H^*(X, E \otimes L) = 0$. Since $\mathcal{E}_c \otimes L$ can be filtered with quotients the direct summands of $E \otimes L$, we deduce that $H^*(\mathcal{E}_c \otimes L) = 0$. This holds for all $c \in C(k)$. Thus, by Proposition 4.1 the corresponding global section of $\mathcal{O}_C(\Theta_C)$ is nowhere vanishing. $\qquad \square$

Lemma 7.7. *If the degree of $\mathcal{O}_C(\Theta_C)$ is zero, then there exists a vector bundle F on C such that $\mathcal{E}_x \cong F$ for all points $x \in X(k)$.*

Proof. We take a line bundle L of degree zero on X such that $H^*(\mathcal{E}_c \otimes L) = 0$ for a fixed $c \in C(k)$. Since the degree of $\mathcal{O}_C(\Theta_C)$ is zero we deduce from Proposition 4.1 that $H^*(\mathcal{E}_{c'} \otimes L) = 0$ for points $c' \in C(k)$, in other words $p_*(\mathcal{E} \otimes q^*L) = 0 = R^1 p_*(\mathcal{E} \otimes q^*L)$.

We choose a point $x \in X(k)$. When applying $R^* p_*(\mathcal{E} \otimes q^*\_)$ to the short exact sequence

$$0 \to L \to L(x) \to \left(L(x) \otimes k(x) \cong k(x) \right) \to 0,$$

we deduce from $R^* p_*(\mathcal{E} \otimes q^*L) = 0$ that $\mathcal{E}_x \cong R^* p_*(\mathcal{E} \otimes q^*L(x))$. The function $X(k) \to \mathbb{N}$ which assigns $x' \mapsto h^1(\mathcal{E}_c \otimes L(x - x'))$ is upper semi-continuous. Since it is zero for $x' = x$, we have that for almost all $x' \in X(k)$ the cohomology $H^*(\mathcal{E}_c \otimes L(x - x'))$ is zero. As before, this implies $p_*(\mathcal{E} \otimes q^*L(x - x')) = 0 = R^1 p_*(\mathcal{E} \otimes q^*L(x-x'))$ for almost all x'. Now from this and the short exact sequence

$$0 \to L(x - x') \to L(x) \to k(x') \to 0$$

we deduce that $\mathcal{E}_{x'} \cong R^* p_*(\mathcal{E} \otimes q^*L(x))$. Thus, $\mathcal{E}_x \cong \mathcal{E}_{x'}$ for almost all $x' \in X(k)$. Since $X(k)$ is infinite, this eventually yields the asserted statement. $\qquad \square$

Lemma 7.8. *If the C-vector bundle F of Lemma 7.7 is semistable, then there exists a vector bundle E on X such that $\mathcal{E}_c \cong E$ for all points $c \in C(k)$.*

Proof. We have by Theorem 5.1 a line bundle M on C such that $F \otimes L$ has no cohomology. Therefore, interchanging the role of X and C in Lemma 7.7, we obtain the result.   □

Lemma 7.9. *If the C-vector bundle F of Lemma 7.7 is unstable, then there exist two line bundles L_1 and L_2 on X of degree $g_X - 1$ such that $\mathrm{gr}(\mathcal{E}_c) \cong L_1 \oplus L_2$.*

Proof. Let

$$0 \to M_1 \to F \to M_2 \to 0$$

be the short exact sequence with M_1 the maximal destabilizing line bundle, i.e., $\deg(M_1) > \deg(M_2)$. It follows that $q_* \mathcal{H}om(p^* M_1, \mathcal{E})$ is a line bundle. Thus we have by Proposition 7.1 a short exact sequence

$$0 \to M_1 \boxtimes L_1 \to \mathcal{E} \to M_2 \boxtimes L_2 \to 0$$

inducing the above sequence on all fibers of q.

Since $\mathcal{E}_c$ contains L_1 we conclude from semistability that $\deg(L_1) = g_X - 1 - a$, and $\deg(L_2) = g_X - 1 + a$ for some integer $a \geq 0$. Choosing a line bundle $L \in \mathrm{Pic}^0(X)$ such that $R^* p_*(\mathcal{E} \otimes q^* L) = 0$, we obtain from the above short exact sequence that $R^1 p_*(M_2 \boxtimes (L_2 \otimes L)) = 0$ which implies $h^1(X, L_2 \otimes L) = 0$. Hence, $h^0(X, L_2 \otimes L) = a$. Analogously we obtain that $h^1(X, L_1 \otimes L) = a$. Applying $p_*(q^* L \otimes \_)$ to the above short exact sequence yields

$$p_*(\mathcal{E} \otimes q^* L) = 0 \to M_2^{\oplus a} \xrightarrow{\alpha} M_1^{\oplus a} \to 0 = R^1 p_*(\mathcal{E} \otimes q^* L).$$

Thus, α must be an isomorphism, which is only possible for $a = 0$.   □

8. The construction

8.1. Constructing the moduli space of vector bundles

Now we have everything we need to perform the construction outlined at the beginning.

(1) We start with our nice overparameterizing family $\mathbb{P}(V)$ from Section 3 which parameterizes all semistable vector bundles E on X of rank two with $\det(E) \cong \omega_X$.

(2) On $\mathbb{P}(V)$ we have the Θ-line bundle $\mathcal{O}_{\mathbb{P}(V)}(\Theta)$ and for any $L \in \mathrm{Pic}^0(X)$ a global section $s_L \in H^0(\mathcal{O}_{\mathbb{P}(V)}(\Theta))$ with vanishing divisor

$$\Theta_L = \left\{ [E] \in \mathbb{P}(V) \mid H^*(E \otimes L) \neq 0 \right\}.$$

(3) The intersection $B := \bigcap_{L \in \mathrm{Pic}^0(X)} \Theta_L$ is the base locus of the linear system spanned by the divisors Θ_L. By Raynaud's Theorem 5.1 the base locus is given by

$$B = \left\{ [E] \in \mathbb{P}(V) \mid [E] \text{ is not semistable} \right\}.$$

We can write B as a finite intersection $B = \bigcap_{i=0,\dots,N} \Theta_{L_i}$. We denote the complement of B by $Q := \mathbb{P}(V) \setminus B$. This is the semistable locus.

(4) The global sections $\{s_{L_i}\}_{i=0,\ldots,N}$ define a morphism $\psi\colon Q \to \mathbb{P}^N$. The image of this morphism is a proper subset by Langton's Theorem 6.4. Thus, we do not change the image when passing to the blow up $\tilde{\psi}\colon \tilde{Q} = \mathrm{Bl}_B\mathbb{P}(V) \to \mathbb{P}^N$.

(5) A connected closed curve C is contracted by $\tilde{\psi}$ if and only if all vector bundles parameterized by C are S-equivalent. Thus, the moduli space M_X of all S-equivalence classes of semistable vector bundles of rank two and determinant ω_X is the Stein factorization of $\tilde{\psi}$

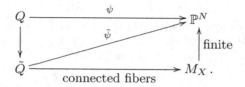

Let us check that M_X is a coarse moduli space.

Points of M_X

First of all each point of $M_X(k)$ corresponds by construction to the image of a bundle parameterized by Q, a semistable bundle. By Theorem 7.5 the S-equivalence class of this bundle is uniquely determined.

Functoriality

Let $\mathcal{E}_S$ be a family of semistable vector bundles on $S \times X$. By Proposition 3.4, we have a covering $S = \cup_{i=1,\ldots,n}S_i$ and morphisms $S_i \to Q$ which induce $\mathcal{E}_{S_i}$. Since the locus $\mathbb{P}(V)_E \subset Q$ is connected (see Proposition 3.3), the invariant functions $S_i \to M_X$ glue along the intersections. Thus, we get a morphism $S \to M_X$.

8.2. Consequences from the construction

Let us illustrate some direct consequences for the moduli space $M_X = \mathrm{SU}_X(2,\omega_X)$.

M_X is unirational. Indeed, we have constructed M_X as the image of some open subset $Q \subset \mathbb{P}(V)$.

The dimension of M_X is $3g-3$. Again we use the morphism $Q \to M_X$. We have $\dim(Q) = \dim\mathbb{P}(V) = 5g - 2$. The fibers are the subschemes $\mathbb{P}(V)_E$ which are open subsets in $\mathbb{P}(\mathrm{Hom}(L^{-1},E)^\vee)$ and therefore of dimension $2g + 1$.

The Θ-line bundle is ample and globally generated. This is a conclusion from Theorems 5.1 and 7.5. This result is the basis for the presented construction and provides us with a finite morphism $M_X \to \mathbb{P}^N$ given by the global section of $\mathcal{O}_{M_X}(\Theta)$.

8.3. Generalization to the case of arbitrary rank and degree

It is the hope of the author to show that the Θ-line bundle $\mathcal{O}_{M_X}(\Theta)$ together with the alternative criterion of Theorem 5.1 for semistability are the key ingredients for the construction. Indeed, even some well-known results, like Langton's result (Theorem 6.4), allow new proofs in this context.

We will next point out how the construction can be generalized to arbitrary rank and degree. Note that we can either fix the determinant or only the degree. The numbers in the following table refer to the table from Part 8.1.

(1) If we fix the determinant, then we can choose again a projective space as our starting point. For fixed degree we have to consider a projective bundle over the Picard torus $\text{Pic}^d(X)$. Grothendieck's Quot scheme (see [9]) also works and was the classical origin. However, it has the disadvantage that it also parameterizes sheaves which are not vector bundles.

(2) Our construction of the Θ-line bundle is just the general construction of Drezet and Narasimhan (see [6]) specialized to our case.

(3) The generalization of Raynaud's Theorem 5.1 to vector bundles of higher rank and arbitrary determinant began in Faltings' article [8, Theorem 1.3]. Popa gave in [25] the best-known explicit bounds. They depend only on the rank of the bundle in question. Thus, passing to a fixed multiple of the generalized Θ-divisor everything works fine.

(4) As in step (2), our proof of Langton's Theorem 6.4 was just a special version of Langton's original result from [18] adapted to the rank two case.

(5) This can be copied verbatim.

Beauville's articles [3] and [4] survey the theory of moduli spaces of vector bundles on algebraic curves with a focus on the generalized Θ-divisor.

9. Prospect to higher dimension

The generalization of Faltings' construction to moduli spaces of coherent sheaves on higher-dimensional varieties started in 1977 when Barth showed in [2] that we can use a construction similar to the one in 8.1 to construct the moduli space of rank two bundles on $\mathbb{P}^2$ by considering jumping lines (see also Le Potier's presentation in [19]).

Let us first introduce the notation to describe Barth's approach. Let E be a rank two vector bundle on $\mathbb{P}^2$ with trivial determinant and second Chern number $n = \int_{\mathbb{P}^2} c_2(E) \in \mathbb{Z}$. The vector bundle E is stable if $H^0(E) = 0$, and semistable when $H^0(E(-1)) = 0$. This easy description of (semi)stability is due to the fact that $\text{Pic}(\mathbb{P}^2) \cong \mathbb{Z}$ and E is of rank two.

The equivalent of Raynaud's theorem (Theorem 5.1) is the following result.

Theorem 9.1 (Grauert–Mülich Theorem). *For a vector bundle E of rank two with $c_1(E) = 0$ on $\mathbb{P}^2$ we have an equivalence*

$$E \text{ is semistable} \iff E|_H \cong \mathcal{O}_H^{\oplus 2} \text{ for a general linear hyperplane } H.$$

Proof. See [24, II, Theorem, 2.1.4]. $\qquad\square$

Note that $E|_H \cong \mathcal{O}_H^{\oplus 2}$ holds if and only if $H^*(E(-1)|_H) = 0$. The linear hyperplanes H undertake the task of the line bundles L in the Picard torus $\text{Pic}^0(X)$. Like the Picard torus they form a nice family – $\mathbb{P}_2$ – the dual projective space of lines in $\mathbb{P}^2$. If we suppose that there exists an overparameterizing scheme Q as before, we can define for any line $l \in \mathbb{P}_2$ a divisor Θ_l given by

$$\Theta_l = \left\{ [E] \in Q \mid H^*(E(-1)|_l) \neq 0 \right\}.$$

Since all lines are rationally equivalent, the line bundle $\mathcal{O}_Q(\Theta) := \mathcal{O}_Q(\Theta_l)$ does not depend on the choice of $l \in \mathbb{P}_2$, and any line defines a global section $s_l \in H^0(\mathcal{O}_Q(\Theta))$ with vanishing divisor Θ_l. For a certain equivalence relation we obtain as before a finite morphism $M \to \mathbb{P}^N$ given by the global sections s_l. This morphism assigns a vector bundle E the divisor

$$\Theta_E := \left\{ l \in \mathbb{P}_2 \mid H^*(E(-1)|_l) \neq 0 \right\}.$$

This is called the *divisor of jumping lines of the vector bundle E*, because along those lines E does not have the expected behavior from Theorem 9.1. The degree of Θ_E is the second Chern number of E.

Replacing lines by conics Hulek obtained an analogous description of vector bundles on $\mathbb{P}^2$ with odd first Chern class in [16].

The author used in [11] restriction theorems and Raynaud's result 5.1 (as well as its generalizations to higher ranks) to obtain a similar construction for moduli spaces of vector bundles on algebraic surfaces. Here (in the rank two case) the lines were replaced by curves in the surfaces with a given line bundle on the curve. This ends in a finite morphism $M \to \mathbb{P}^N$ which was called the *Barth morphism*. In [21] J. Li gave a similar construction which should coincide with ours.

We call a pair (F, E) of coherent sheaves perpendicular when $\mathrm{Ext}^i(F, E) = 0$ for all integers i. If this is the case, then we write $F \perp E$ for such a pair. We can rephrase Proposition 2.7 as the implication: if for a sheaf E there exists a nonzero F such that $F \perp E$, then E is semistable. Theorem 5.1 shows that for a semistable vector bundle E of rank two and degree $2g - 2$ there exists a perpendicular sheaf $F = L^\vee$ which is a line bundle of degree zero. The fact that semistable objects have perpendicular partners allows the definition of semistability in the derived category. For μ-semistability this is derived from restriction theorems and the existence of orthogonal objects on curves. For Gieseker semistability this is shown in [1, Theorem 7.2]. This approach was pursued in [14] and [15].

Acknowledgment

The author wants to thank Luis Álvarez-Cónsul, Oscar García-Prada and Alexander Schmitt, the organizers of the *German/Spanish Workshop on Moduli Spaces of Vector Bundles*, for giving him the opportunity to present this work. Furthermore Alexander Schmitt softly forced the author to finish this article in time.

A first version of these notes was improved thanks to comments of D. Ploog. The author is also very grateful for the comments and suggestions of the referee.

This work has been supported by the SFB/TR 45 "Periods, moduli spaces and arithmetic of algebraic varieties".

References

[1] L. Álvarez-Cónsul, A. King, *A functorial construction of moduli of sheaves*, Invent. math. **168** (2007), 613–666.

[2] W. Barth, *Moduli of vector bundles on the projective plane*, Invent. math. **42** (1977), 63–91.

[3] A. Beauville, *Vector bundles on curves and generalized theta functions: recent results and open problems*, in H. Clemens (ed.) et al., *Current topics in complex algebraic geometry*, Cambridge Univ. Press, New York, 1995.

[4] A. Beauville, *Vector bundles on curves and theta functions*, in *Moduli spaces and arithmetic geometry*, Adv. Stud. Pure Math. **45**, Math. Soc. Japan, Tokyo, 2006, 145–156.

[5] H. Clemens, J. Kollár, S. Mori, *Higher dimensional complex geometry*, Astérisque **166** (1988).

[6] J.-M. Drezet, M.S. Narasimhan, *Groupe de Picard des variétés de modules de fibrés semi-stables sur les courbes algébriques*, Invent. math. **97** (1989), 53–94.

[7] D. Eisenbud, *Commutative Algebra, with a view toward Algebraic Geometry*, Springer-Verlag, New York, 1995.

[8] G. Faltings, *Stable G-bundles and projective connections*, J. Alg. Geom. **2** (1993), 507–568.

[9] A. Grothendieck, *Techniques de construction et théorèmes d'existence en géométrie algébrique, IV. Les schémas des Hilbert*, Séminaire Bourbaki **221** (1960/61).

[10] R. Hartshorne, *Algebraic geometry*, GTM **52**, New York, 1977.

[11] G. Hein, *Duality construction of moduli spaces*, Geometriae Dedicata **75** (1999), 101–113.

[12] G. Hein, *Raynaud's vector bundles and base points of the generalized Theta divisor*, Math. Zeitschrift **257** (2007), 597–611.

[13] G. Hein, *Raynaud vector bundles*, Monatsh. Math. **157** (2009), 233–245.

[14] G. Hein, D. Ploog, *Postnikov-Stability for Complexes*, math.AG/0704.2512.

[15] G. Hein, D. Ploog, *Postnikov-Stability versus Semistability of Sheaves*, math.AG/0901.1554.

[16] K. Hulek, *Stable rank-2 vector bundles on $\mathbb{P}_2$ with c_1 odd*, Math. Ann. **242** (1979), 241–266.

[17] D. Huybrechts, M. Lehn, *The geometry of moduli spaces of shaves*, Aspects of Math. **E 31**, Vieweg, Braunschweig, 1997.

[18] S. Langton, *Valuative criteria for families of vector bundles on algebraic varieties*, Ann. Math. **101** (1975), 88–110.

[19] J. Le Potier, *Fibré déterminant et courbes de saut sur les surfaces algébriques*, London Math. Soc. Lect. Note Ser. **179** (1992), 213–240.

[20] J. Le Potier, *Module des fibrés semistables et fonctions thêta*, Lect. Notes in Pure and Appl. Math. **179** (1996), 83–101.

[21] J. Li, *Algebraic geometric interpretation of Donaldson's polynomial invariants*, J. Diff. Geom. **37** (1993), 417–466.

[22] D. Mumford, *Geometric invariant theory*, Springer-Verlag, Berlin, 1965.

[23] D. Mumford, *Abelian Varieties*, Oxford Univ. Press, London, 1970.

[24] C. Okonek, M. Schneider, H. Spindler, *Vector bundles on complex projective space*, Birkhäuser, Boston, 1980.

[25] M. Popa, *Dimension estimates for Hilbert schemes and effective base point freeness on moduli spaces of vector bundles on curves*, Duke Math. J. **107** (2001), 469–495.

[26] M. Raynaud, *Section des fibrés vectoriels sur une courbe*, Bull. Soc. Math. France **110** (1982), 103–125.

[27] S.S. Shatz, *The decomposition and specialisation of algebraic families of vector bundles*, Comp. Math. **35** (1977), 163–187.

Georg Hein
Universität Duisburg–Essen
Fakultät Mathematik
D-45117 Essen, Germany
e-mail: georg.hein@uni-due.de

Affine Flag Manifolds and Principal Bundles
Trends in Mathematics, 123–153
© 2010 Springer Basel AG

Lectures on the Moduli Stack of Vector Bundles on a Curve

Jochen Heinloth

Abstract. These are lecture notes of a short course on the moduli stack of vector bundles on an algebraic curve. The aim of the course was to use this example to introduce the notion of algebraic stacks and to illustrate how one can work with these objects. Applications given are the (non-)existence of universal families on coarse moduli spaces and the computation of the cohomology of the moduli stack.

Mathematics Subject Classification (2000). 14D23, 14H60.

Keywords. Algebraic stack, moduli space, vector bundle, cohomology.

Introduction

This text consists of my notes for a course on algebraic stacks given at the German–Spanish Workshop on vector bundles on algebraic curves in Essen and Madrid 2007, organized by L. Álvarez-Cónsul, O. García-Prada and A. Schmitt. The aim of the course was to use the example of vector bundles on curves to introduce the basic notions of algebraic stacks and to illustrate how one can work with these objects.

The course consisted of five one-hour lectures and was meant to be introductory. We start with the definition of stacks. In order to get to some interesting applications we chose to give only those parts of the basic theory that are needed in our applications. Also we often chose to give ad hoc definitions, whenever these are easier to digest than the abstract ones. The time constraints had the side effect that I deliberately skipped some of the technical fine print in the first lectures. On the one hand I hope that this makes the subject more accessible, because the ideas are most of the time not so difficult to understand. On the other hand for written notes I feel that some of the fine print should at least be indicated. Since there are excellent references for these points available I will try to include some comments indicating where one can find more information. In order not to distract the reader who wants to get a first idea what the subject is about, I will put these comments in fine print.

Finally, I gave a set of exercises for the course. These are included in the text.

The structure of the lectures is as follows: The first lecture gives the definition of an algebraic stack. The second lecture explains why geometric notions make sense for such stacks and introduces sheaves on stacks. As an example on how to work with these we begin Lecture 3 by proving the innocuous technical result of Laumon–Moret-Bailly that on a noetherian stack any quasi-coherent sheaf is the limit of its coherent subsheaves. We then consider the relation with coarse moduli spaces. We introduce the notion of a gerbe in order to give a simple proof of the classical result on the non existence of a universal family of vector bundles on coarse moduli spaces of bundles when rank and degree are not coprime. This uses the theorem proved in the beginning of the lecture. In the fourth lecture we introduce cohomology of constructible sheaves and, as an example, we indicate how one can compute the cohomology ring of the moduli stack of vector bundles on a curve. In the last lecture we give some ideas how one can deduce results on the cohomology of the coarse moduli spaces.

Of course this material is not original. The basic results on algebraic stacks are explained in the book of Laumon and Moret-Bailly [24], the standard reference on the subject. The results on $\mathbb{G}_m$-gerbes can be found in Lieblich's thesis [26], the application to the non-existence of universal families has been explained in great generality by Biswas and Hoffmann [10]. The calculation of the cohomology ring of the moduli stack owes much to the classical work of Atiyah–Bott [2]. The reformulation in terms of stacks was the subject G. Harder suggested as subject of my Diploma thesis. However we use Beauville's trick here in order to avoid the usage of the Lefschetz trace formula. A more general result is proved in [23].

Also, by now there are several very good introductory notes on the basic definitions on stacks available (for example [14], [18]), each giving an introduction form a different point of view.

Lecture 1: Algebraic stacks

1.1. Motivation and definition

The main motivation to introduce algebraic stacks is that we want to study moduli spaces. One of the simplest problems would be to look for a classifying space for vector bundles, say of rank n. So we would like to have a space $B\mathsf{GL}_n$ such that for any test scheme T

$$\mathrm{Mor}(T, B\mathsf{GL}_n) = \langle \text{ vector bundles of rank } n \text{ on } T \rangle/\text{isomorphism}. \qquad (1.1)$$

This would be nice, because to construct functorial invariants like Chern classes it would then be sufficient to construct cohomology classes of $B\mathsf{GL}_n$.

However, such a space cannot exist, because every vector bundle $\mathcal{E}$ on T is locally trivial, so the map $T \to BGL_n$ corresponding to $\mathcal{E}$ would have to be locally constant so $\mathcal{E}$ would have to be trivial globally.

This is a bit disappointing, since topologists do know a classifying space BGL_n, such that for any space T the homotopy classes of maps $f \colon T \to BGL_n$ correspond to isomorphism classes of vector bundles on T.

Since we do not have a good algebraic replacement for homotopy classes of maps, we have (at least) two other ways to circumvent the problem:

1. Restrict the functor and the expectations on the representing space, i.e., consider coarse moduli spaces. (See Georg Hein's lectures [21].)

2. Don't pass to isomorphism classes in (1.1)!

The second option is used for the definition of stacks. Before giving the definition, recall that the Yoneda lemma (valid in any category) tells us that any scheme X is determined by its functor of points, i.e., X is determined by the functor

$$\mathrm{Mor}(\_, X) \colon \text{Schemes} \to \text{Sets}$$

sending a scheme T to the set $\mathrm{Mor}(T, X)$. Furthermore this functor $\mathrm{Mor}(\_, X)$ is a sheaf, in the sense that a morphism $T \to X$ can be obtained from gluing morphisms on a covering of T.

The definition of a stack follows this idea: We first define a stack to be given by its functor of points. So in the case of BGL_n we just define $BGL_n(T)$ to be the category of vector bundles of rank n on T. Vector bundles can also be obtained by gluing bundles on an open covering, so the general definition of a stack will be that it is a sheaf of categories (more precisely a sheaf of groupoids). In writing down the axioms that should be satisfied by such an assignment we keep the example BGL_n in mind.

In a second step we will then try to see how one can do geometry using such objects.

Definition 1.1. A *stack* is a sheaf of groupoids:

$$\mathcal{M} \colon \text{Sch} \to \text{Groupoids} \subset \text{Categories},$$

i.e., an assignment

1. for any scheme T a category $\mathcal{M}(T)$ in which all morphisms are isomorphisms,
2. for any morphism $f \colon X \to Y$ a functor $f^* \colon \mathcal{M}(Y) \to \mathcal{M}(X)$,
3. for any pair of composable morphisms $X \xrightarrow{f} Y \xrightarrow{g} Z$ a natural transformation $\varphi_{f,g} \colon f^* \circ g^* \Rightarrow (g \circ f)^*$. These transformations have to be associative for composition, in particular we assume this transformation to be the identity if one of the morphisms is the identity,

satisfying the following gluing conditions:

1. (Objects glue) Given a covering[1] $U_i \twoheadrightarrow T$, objects $\mathcal{E}_i \in \mathcal{M}(U_i)$ and isomorphisms $\varphi_{ij} \colon \mathcal{E}_i|_{U_i \cap U_j} \to \mathcal{E}_j|_{U_i \cap U_j}$ satisfying a cocycle condition on 3-fold

---

[1]see the remark following the definition

intersections, there exists an object $\mathcal{E} \in \mathcal{M}(T)$, unique up to isomorphism, together with isomorphisms $\psi_i \colon \mathcal{E}|_{U_i} \to \mathcal{E}_i$ such that $\varphi_{ij} = \psi_j \circ \psi_i^{-1}$.

2. (Morphisms glue) Given a covering $U_i \twoheadrightarrow T$, objects $\mathcal{E}, \mathcal{F} \in \mathcal{M}(T)$ and morphisms $\varphi_i \colon \mathcal{E}|_{U_i} \to \mathcal{F}|_{U_i}$ such that $\varphi_i|_{U_i \cap U_j} = \varphi_j|_{U_i \cap U_j}$, there is a unique morphism $\varphi \colon \mathcal{E} \to \mathcal{F}$ such that $\varphi|_{U_i} = \varphi_i$.

Remark 1.2. In the definition we used the word *covering* to mean one of the following choices: In complex geometry we use the analytic topology. Otherwise we mean a covering in either the étale topology or the fppf topology (i.e., a surjective map $U_i \to X$ which is étale or flat of finite presentation, respectively). In this case the intersection $U_i \cap U_j$ has to be defined as $U_i \cap U_j := U_i \times_X U_j$.

Finally, the notation $\mathcal{E}|_{U_i}$ means the pull back of $\mathcal{E}$ to U_i given by the map $U_i \to X$.

Actually to make the above definition precise, we would need to spell out the canonical isomorphisms of $\mathcal{E}|_{U_i}|_{U_i \cap U_j} \to \mathcal{E}|_{U_j}|_{U_i \cap U_j}$ given by the last part of the bifunctor. This would make the definition more difficult to read.

Also it is sometimes not quite obvious how to define all pull back functors functorially. In fact one often has to make a choice here. This can be avoided using the language of fibred categories as in [24]. This means that similarly to the procedure of replacing a sheaf by its espace étalé one considers the large category $\coprod_{T \in \mathrm{Sch}} \mathcal{M}(T)$ instead of the different $\mathcal{M}(T)$. This has other advantages (see the last remark in [19], VI) but to me the above definition seems easier to digest at first.

Let us collect some examples:

Example 1.3. Let C be a smooth projective curve (a general scheme which is flat of finite type over the base would do here). Then let $\mathrm{Bun}_{n,C}$ be the stack given by[2]

$$\mathrm{Bun}_n := \mathrm{Bun}_{n,C}(T) := \langle \text{vector bundles of rank } n \text{ on } C \times T \rangle.$$

Here the morphisms in the category are isomorphisms of vector bundles and the functors f^* are given by the pull-back of bundles. The gluing conditions are satisfied, by descent for vector bundles ([19], Exposé VIII, Théorème 1.1 and Proposition 1.10).

Similarly Coh_C is the stack of coherent sheaves on C:

$$\mathrm{Coh}_C(T) := \langle \text{coherent sheaves on } C \times T \text{ flat over T} \rangle.$$

Example 1.4. Let G be an affine algebraic group. Then, we denote by

$$BG(T) := \langle \text{principal } G\text{-bundles on } T \rangle$$

the classifying stack of G.

Example 1.5. Let X be a scheme. Then $\underline{X}(T) := \mathrm{Mor}(T, X)$ is a stack. Here we consider the set $\mathrm{Mor}(T, X)$ as a category in which the only morphisms are identities, the pull-back functors f^* for $f \colon S \to T$ being given by composition with f. Such a stack is called a *representable stack*.

---

[2]We denote categories by $\langle \rangle$ to distinguish them from sets.

Example 1.6 (Quotient stacks). Let X be a scheme (say over some field k in order to avoid a flatness condition in what follows) and G be an algebraic group acting on X. Then we define the quotient stack $[X/G]$ by

$$[X/G](T) := \left\langle \begin{matrix} P \xrightarrow{g} X \\ \downarrow p \\ T \end{matrix} \ \middle| \ \begin{matrix} P \to T \text{ is a } G\text{-bundle} \\ P \to X \text{ is a } G\text{-equivariant map} \end{matrix} \right\rangle.$$

Morphisms in this category are isomorphisms of G-bundles commuting with the map to X.

To check that this definition makes sense let us consider the case that there exists a quotient X/G of X by G such that the map $X \to X/G$ is a G-bundle. In this case any diagram

$$\begin{matrix} P & \xrightarrow{g} & X \\ \downarrow p & & \downarrow \\ T & & X/G \end{matrix}$$

defines a map $\bar{g} \colon T \to X/G$ and in this way P becomes canonically isomorphic to the pull-back of the G-bundle $\bar{g}^* X = X \times_{X/G} T$ over T. So in this case the category $[X/G](T)$ is canonically equivalent to the set $X/G(T)$, which we consider as a category in which the only morphisms are the identities of elements.

Remark 1.7. Stacks form a 2-category. Morphisms of stacks are given by functors between the corresponding categories: A morphism $F \colon \mathcal{M} \to \mathcal{N}$ is given by a collection of functors $F_T \colon \mathcal{M}(T) \to \mathcal{N}(T)$ for all T together with for every $f \colon X \to Y$ a natural transformation $F_f \colon F_X \circ f^* \to f^* \circ F_Y$ satisfying an associativity constraint.

A 2-morphism is a morphism between such functors F, G. It is given by a natural transformation $\varphi \colon F_T \to G_T$ for all T, compatible with the pull-back functors f^*.

Example 1.8. 1. The tensor product defines a morphism $\mathrm{Bun}_{n,C} \times \mathrm{Bun}_{m,C} \to \mathrm{Bun}_{nm,C}$ sending a pair of vector bundles $\mathcal{E}, \mathcal{F}$ to $\mathcal{E} \otimes \mathcal{F}$.

2. An example of a 2-morphism is as follows. Consider the identity functor $\mathrm{id} \colon \mathrm{Bun}_{n,C} \to \mathrm{Bun}_{n,C}$. Fix an invertible scalar α. A 2-morphism from $\mathrm{id} \to \mathrm{id}$ is then given by multiplication by α on all objects.

The first important observation is that the examples of stacks given above are indeed moduli spaces:

Lemma 1.9 (Yoneda lemma for stacks). *Let $\mathcal{M}$ be a stack. Then for any scheme T there is a natural equivalence of categories:*

$$\mathrm{Mor}_{\mathrm{Stacks}}(\underline{T}, \mathcal{M}) \cong \mathcal{M}(T).$$

Proof. First note that we can define a functor

$$\mathrm{Mor}_{\mathrm{Stacks}}(\underline{T}, \mathcal{M}) \to \mathcal{M}(T)$$

by sending $F \colon \underline{T} \to \mathcal{M}$ to $F(\mathrm{id}_T) \in \mathcal{M}(T)$.

Conversely, given an object $\mathcal{E} \in \mathcal{M}(T)$ we can define a morphism $F_{\mathcal{E}} \colon \underline{T} \to \mathcal{M}$ by sending $f \in T(S) = \mathrm{Mor}(S, T)$ to $f^*(\mathcal{E}) \in \mathcal{M}(S)$.

Note that the composition $\mathcal{E} \mapsto F_{\mathcal{E}} \mapsto F_{\mathcal{E}}(\mathrm{id}_T) = \mathrm{id}^*\mathcal{E} = \mathcal{E}$ is the identity.

Conversely, let us compute the composition $F \mapsto F(\mathrm{id}_T) \mapsto F_{F(\mathrm{id}_T)}$. We have $F_{F(\mathrm{id}_T)}(f \colon S \to T) = f^*(F(\mathrm{id}_T))$. But $F_f \colon F(f \colon S \to T) \to f^*(F(\mathrm{id}_T))$ then gives a natural isomorphism. $\qquad\qquad\qquad\qquad\qquad\qquad\qquad\qquad\qquad\qquad\qquad\qquad \square$

Because of this lemma we will often simply write T instead of $\underline{T}$.

1.2. How to make this geometric?

In order to make sense of geometric notions for stacks, we look for a notion of charts for an algebraic stack. To see why this could make sense let us begin by computing a fibre product in a simple example:

Take G a smooth group and consider our stack BG, classifying G-bundles. Let $\mathrm{pt} = \mathrm{Spec}(k)$ be a point and $\mathcal{E}$ be a G-bundle on some other scheme X. By the Yoneda lemma, $\mathcal{E}$ defines a morphism $F_{\mathcal{E}} \colon \underline{X} \to BG$ and the trivial bundle defines a morphism $\mathrm{triv} \colon \mathrm{pt} \to BG$:

$$
\begin{array}{c}
\mathrm{pt} \\
\downarrow \text{\scriptsize triv} \\
X \xrightarrow{\ F_{\mathcal{E}}\ } BG.
\end{array}
$$

We want to compute the fibre product of this diagram. For any scheme T this is given by:

$$
X \times_{BG} \mathrm{pt}(T) = \left\langle
\begin{array}{c}
T \\
\end{array}
\ ; \varphi \colon \mathrm{triv} \circ p \xrightarrow{\ \cong\ } F_{\mathcal{E}} \circ f
\right\rangle
$$

$$
= \left\langle (f, p, \varphi) \mid \varphi \colon f^*\mathcal{E} \cong p^*(\mathrm{triv}) = T \times G \right\rangle
$$

$$
= \left\{ (f, s) \mid s \colon T \to f^*\mathcal{E} \ \text{a section} \right\}
$$

$$
= \mathcal{E}(T).
$$

Thus the T-valued points of the fibre product are a set and not only a category and the resulting stack is equivalent to the G-bundle $\mathcal{E}$. This means:

1. For every $F_{\mathcal{E}} \colon X \to BG$ the pull back of the morphism $\mathrm{pt} \to BG$ is the G-bundle $\mathcal{E}$, so $\mathrm{pt} \to BG$ is the universal G-bundle on BG!
2. The map $\mathrm{pt} \to BG$ becomes a smooth surjection after every base-change.

The second point means that we should regard the map $\mathrm{pt} \to BG$ as a smooth covering of BG so we could consider it as an atlas for BG. The existence of such a map will be the main part of the definition of algebraic stacks.

More generally, let $\mathcal{M}$ be any stack and $x\colon X \to \mathcal{M}, y\colon Y \to \mathcal{M}$ be two morphisms. Then for any scheme T

$$X \times_{\mathcal{M}} Y(T) = \left\langle \begin{array}{c} \overset{f}{\nearrow} \overset{X}{} \\ T \\ \underset{g}{\searrow} \underset{Y}{} \end{array} , \varphi\colon f^*x \cong g^*y \right\rangle =: \mathrm{Isom}(x,y)$$

is a sheaf and in all the examples we have seen so far it is even represented by a scheme.

Definition 1.10. A stack $\mathcal{M}$ is called *algebraic* if

1. For all $X \to \mathcal{M}, Y \to \mathcal{M}$ the fibre product $X \times_{\mathcal{M}} Y$ is representable.
2. There exists a scheme $u\colon U \to \mathcal{M}$ such that for all schemes $X \to \mathcal{M}$ the projection $X \times_{\mathcal{M}} U \to X$ is a smooth surjection.
3. The forgetful map $\mathrm{Isom}(u,u) = U \times_{\mathcal{M}} U \to U \times U$ is quasi-compact and separated.

Remark 1.11. The last condition is a technical condition. It implies that $\mathrm{Isom}(x,y)$ is always separated. In particular we are not allowed to consider non-separated group schemes and it also rules out quotients like $[\mathbb{A}^1_{\mathbb{C}}/\mathbb{Q}^{\mathrm{discrete}}]$.

Remark 1.12. There is a second technical problem. In order to get a definition in which algebraicity of a stack can be checked by deformation-theoretic conditions it is more natural to replace the requirement that $X \times_{\mathcal{M}} Y$ is a scheme by the weaker condition that it is an algebraic space. Once one is used to algebraic stacks this will not be a difficult concept, because the definition is exactly the same as the above, if one adds the condition that the stack $\mathcal{M}$ is actually a sheaf, i.e., that all $\mathcal{M}(T)$ are sets. In this context the last technical condition is then needed to make the condition on fibred products to be schemes to be reasonable.

Example 1.13. We have just seen that the stack BG is algebraic. Analogously quotient stacks $[X/G]$ are algebraic, the canonical map $X \to [X/G]$ given by the trivial G-bundle $G \times X \to X$ is an atlas.

The most important example in this course will be the following:

Example 1.14. Let C be a smooth projective curve, and denote by Bun_n the stack of vector bundles of rank n on C. This is an algebraic stack. We know that for two bundles $\mathcal{E}, \mathcal{F}$ on $C \times X$ and $C \times Y$ the sheaf $\mathrm{Isom}(\mathcal{E}, \mathcal{F}) \subset \mathrm{Hom}(\mathcal{E}, \mathcal{F})$ is an open subscheme, and $\mathrm{Hom}(\mathcal{E}, \mathcal{F}) \to X \times Y$ is affine.

Choose an ample bundle $\mathcal{O}(1)$ on C. An atlas of Bun_n is given as follows:

$$U := \coprod_{N \in \mathbb{N}} \left\langle (\mathcal{E}, s_i) \,\middle|\, \begin{array}{l} \mathcal{E} \text{ a bundle on } C \text{ such that } \mathcal{E} \otimes \mathcal{O}_C(N) \text{ is} \\ \text{globally generated, } H^1(C, \mathcal{E} \otimes \mathcal{O}_C(N)) = 0, \\ s_i \text{ a basis of } H^0(C, \mathcal{E} \otimes \mathcal{O}_C(N)) \end{array} \right\rangle.$$

This is a representable functor by the theory of Hilbert- (or Quot-)schemes.

The condition that fibre products are representable gets a name:

Definition 1.15. A morphism $F\colon \mathcal{M} \to \mathcal{N}$ of stacks is called *representable* if for all $X \to \mathcal{N}$ the fibre product $X \times_{\mathcal{N}} \mathcal{M}$ is representable.

As before, in this definition one should again use algebraic spaces to define representability. This will not make a difference in our examples.

Example 1.16. The standard example for a non-representable morphism is the projection $BG \to$ pt. More generally it is not difficult to check that representable morphisms induce injections on the automorphism groups of objects. This condition is actually a sufficient condition for morphisms between algebraic stacks, if one takes the above fine print on the notion of representability into account, i.e., if one uses the larger category of algebraic spaces instead of schemes as representable stacks.

Exercise 1.17. Show that the property of a morphism to be representable is stable under pull-backs.

Exercise 1.18. Show that the fibre product $X \times_{\mathcal{M}} Y$ is representable for all X, Y if and only if the diagonal morphism $\Delta\colon \mathcal{M} \to \mathcal{M} \times \mathcal{M}$ is representable.

The main idea – which will be explained in the next lecture – is that this concept of algebraic stacks allows to translate every notion for schemes that can be checked on a smooth covering into a notion for stacks. For example smoothness, closed and open substacks. But also sheaves and cohomology: We will simply define sheaves to be sheaves on one atlas together with a descent datum to $\mathcal{M}$.

Remark 1.19. One might wonder why one does not replace "smooth" by "flat" in the definition of algebraic stacks. The reason for this is a theorem of Artin ([1], Theorem 6.1) which says that this would not give a more general notion!

Exercise 1.20 (2-Fibred products). For any stack $\mathcal{M}$ one defines its inertia stack $I(\mathcal{M})$ as

$$I(\mathcal{M})(T) \cong \big\langle (t, \varphi) \,|\, t \in \mathcal{M}(T), \varphi \in \mathrm{Aut}(t) \big\rangle.$$

Show that $I(\mathcal{M}) \cong \mathcal{M} \times_{\mathcal{M} \times \mathcal{M}} \mathcal{M}$ where the map $\mathcal{M} \to \mathcal{M} \times \mathcal{M}$ is the diagonal.

Exercise 1.21. Given G an algebraic group and $H \subset G$ a closed subgroup we consider the canonical map $BH \to BG$ mapping any H-bundle to the induced G-bundle. Show that there is a (2-)cartesian diagram:

$$
\begin{array}{ccc}
G/H & \longrightarrow & BH \\
\downarrow & & \downarrow \\
\mathrm{pt} & \longrightarrow & BG.
\end{array}
$$

If $H \subset G$ is normal, then there is a (2-)cartesian diagram

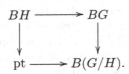

(Of course these diagrams are well known in topology as homotopy fibrations.)

Exercise 1.22 (Points of a stack). Let $\mathcal{M}$ be an algebraic stack. We define its set of points $|\mathcal{M}|$ as the union

$$\left(\amalg_{k \subset K \text{ a field}} \text{Objects}(\mathcal{M}(K)) \right) / \sim$$

where $\sim$ declares $x \in \mathcal{M}(K)$ to be equivalent to $y \in \mathcal{M}(K')$ if there is a field extension K'' containing K and K' such that $x|_{K''} \cong y|_{K''}$.

Let $X \to \mathcal{M}$ be an atlas of $\mathcal{M}$. Show that $|\mathcal{M}| = |X|/\sim_{\mathcal{M}}$ where $\sim_{\mathcal{M}}$ is the equivalence relation defined by $|X \times_{\mathcal{M}} X| \to |X| \times |X|$.

(You might also want to define a Zariski-topology on $|\mathcal{M}|$.)

Lecture 2: Geometric properties of algebraic stacks

2.1. Properties of stacks and morphisms

Recall that the essential point in the definition of algebraicity of a stack $\mathcal{M}$ is the existence of a smooth surjection $u \colon U \twoheadrightarrow \mathcal{M}$ from a scheme U to $\mathcal{M}$. Let us use this to define some first geometric properties of algebraic stacks:

Definition 2.1. An algebraic stack $\mathcal{M}$ is called *smooth* (*resp. normal/reduced/locally of finite presentation/locally noetherian/regular*) if there exists an atlas $u \colon U \twoheadrightarrow \mathcal{M}$ with U being smooth (resp. normal/reduced/locally of finite presentation/locally noetherian/regular).

Note that for schemes this definition gives nothing new, because all the above properties can be checked locally on a smooth covering of a scheme.

Similarly properties of morphisms which can be checked after a smooth base change extend to properties of representable morphisms of algebraic stacks:

Definition 2.2. Let P be a property of morphisms of schemes $f \colon X \to Y$ such that f has P if and only if for some smooth surjective $Y' \to Y$ the induced morphism $f' \colon X \times_Y Y' \to Y'$ has P (e.g., closed immersion, open immersion, affine, finite, proper).

We say that a *representable morphism* $F \colon \mathcal{M} \to \mathcal{N}$ of algebraic stacks *has property P* if for some (equivalently any) atlas $u \colon U \twoheadrightarrow \mathcal{N}$ the morphism $\mathcal{M} \times_{\mathcal{N}} U \to U$ has P.

Remark 2.3. In particular the above definition gives us a notion of closed and open substacks of an algebraic stack.

Let us give some examples:

1. If $\mathcal{M} = [X/G]$ is a quotient stack, then open/closed substacks are of the form $[Y/G]$ where $Y \subset X$ is an open/closed subscheme.
2. In our example Bun_n, the stack of vector bundles on a projective curve, the substack $\mathrm{Bun}_n^{\mathrm{ss}}$ of semistable vector bundles[3] is an open substack, because for any family of bundles, instability is a closed condition.

Finally we can also define properties of arbitrary morphisms of stacks as long as we can check these properties locally in the source and the image of a morphism:

Definition 2.4. Let P be a property of morphisms of schemes $f\colon X \to Y$ such that f has P if and only if there exists some commutative diagram

$$
\begin{array}{ccc}
X' & \xrightarrow{\tilde{f}} & Y' \\
{\scriptstyle\text{smooth}}\downarrow & & \downarrow{\scriptstyle\text{smooth}} \\
X & \xrightarrow{f} & Y
\end{array}
$$

such that $\tilde{f}$ has P. For example being smooth, flat, locally of finite presentation.

Then a *morphism* of algebraic stacks $F\colon \mathcal{M} \to \mathcal{N}$ *has* P if for some atlases $v\colon V \to \mathcal{M}$, $u\colon U \to \mathcal{N}$ there exists a commutative diagram:

$$
\begin{array}{ccc}
V & \xrightarrow{\tilde{f}} & U \\
{\scriptstyle\text{smooth}}\downarrow & & \downarrow{\scriptstyle\text{smooth}} \\
\mathcal{M} & \xrightarrow{F} & \mathcal{N}
\end{array}
$$

such that $\tilde{f}$ has P.

Example 2.5. In geometric invariant theory one often encounters the situation that one has an action of GL_n on a scheme X, such that the center of GL_n acts trivially, i.e., the action actually factors through an action of the group PGL_n on X. In this case the canonical morphism $[X/\mathsf{GL}_n] \to [X/\mathsf{PGL}_n]$ is smooth and surjective, but not representable.

Finally, we claim that our main example of an algebraic stack Bun_n is a smooth stack. One way to do this would be to directly apply the definition above to the atlas given in the last lecture and try to check that this atlas is smooth. However there is an intrinsic way to show smoothness, avoiding the choice of an atlas. This is as follows:

Recall the lifting criterion for smoothness: A morphism of schemes $f\colon X \to Y$ is smooth if and only if f is locally of finite presentation and for all (local) Artin

---

[3]Recall that a vector bundle $\mathcal{E}$ on a curve is called semistable if for all subbundles $\mathcal{F} \subset \mathcal{E}$ we have $\frac{\deg(\mathcal{F})}{\mathrm{rk}(\mathcal{F})} \leq \frac{\deg(\mathcal{E})}{\mathrm{rk}(\mathcal{E})}$.

algebras A with an ideal $I \subset A$ with $I^2 = (0)$ one can complete any diagram:

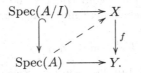

Lemma 2.6. *Let $\mathcal{M}$ be an algebraic stack, locally of finite presentation over $\mathrm{Spec}(k)$ such that the structure morphism $\mathcal{M} \to \mathrm{Spec}(k)$ satisfies the above lifting criterion for smoothness. Then $\mathcal{M}$ is smooth.*

Proof. Let $u\colon U \twoheadrightarrow \mathcal{M}$ be an atlas. We have to show that U is smooth, i.e., that the lifting criterion holds for $U \to \mathrm{Spec}(k)$. So assume that we are given:

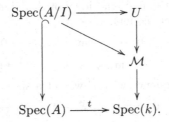

Since $\mathcal{M} \to \mathrm{Spec}(k)$ satisfies the lifting criterion by assumption, we can lift t to $\tilde{t}\colon \mathrm{Spec}(A) \to \mathcal{M}$.

Knowing that u is smooth implies that the projection $U \times_{\mathcal{M}} \mathrm{Spec}(A) \to \mathrm{Spec}(A)$ is a smooth morphism of schemes. So we can also find a lifting in:

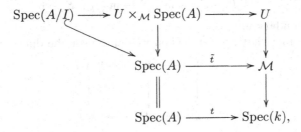

which proves our claim. $\square$

Let us apply this to show that the stack Bun_n is smooth: Giving a morphism $\mathrm{Spec}(A/I) \to \mathrm{Bun}_n$ is the same as giving a family of vector bundles $\overline{\mathcal{E}}$ on $C \times \mathrm{Spec}(A/I)$. We have to check that we can extend any such family to a vector bundle $\mathcal{E}$ on $C \times \mathrm{Spec}(A)$. Denote by $\mathfrak{m}$ the maximal ideal of A and $k = A/\mathfrak{m}$. By induction we can assume that $I = (\nu)$ is generated by one element and that $\nu \cdot \mathfrak{m} = 0$. Denote by $\mathcal{E}_0 := \mathcal{E} \otimes_{A/I} k$.

One way to see that an extension $\mathcal{E}$ exists is to describe $\overline{\mathcal{E}}$ by gluing cocycles and to lift these cocycles. (The lack of the cocycle condition for the lifted elements gives an element in $H^2(C, \mathcal{E}nd(\mathcal{E}_0)) \otimes_k I$. This group is zero, because C is 1-dimensional.) Again one can avoid cocycles here. First, since we assumed that $I = (\nu)$ is generated by one element and that $\nu \cdot \mathfrak{m} = 0$ we have an exact sequence of A/I-modules:

$$0 \to k \xrightarrow{\nu \cdot} \mathfrak{m} \to A/I \to k = A/\mathfrak{m} \to 0. \tag{2.2}$$

Claim 2.7. $\overline{\mathcal{E}}$ extends to a bundle $\mathcal{E}$ on $C \times \mathrm{Spec}(A)$ if and only if the class

$$\mathrm{obs}(\overline{\mathcal{E}}) := (2.2) \otimes_{A/I} \overline{\mathcal{E}} \in \mathrm{Ext}^2(\mathcal{E}_0, \mathcal{E}_0)$$

vanishes.

Of course this condition is automatic here, because $\mathrm{Ext}^2(\mathcal{E}_0, \mathcal{E}_0) = 0$ on a curve C.

Proof. Denote by $\mathfrak{m}_{A/I}$ the maximal ideal of A/I. Let us decompose the sequence (2.2) into two short exact sequences:

$$0 \to k \to \mathfrak{m} \to \mathfrak{m}_{A/I} \to 0$$
$$0 \to \mathfrak{m}_{A/I} \to A/I \to k \to 0.$$

Tensoring the second sequence with $\overline{\mathcal{E}}$ we get a short exact sequence

$$0 \to \mathfrak{m}_{A/I} \otimes \overline{\mathcal{E}} \to \overline{\mathcal{E}} \to \mathcal{E}_0 \to 0.$$

This defines a long exact sequence:

$$\cdots \to \mathrm{Ext}^1(\overline{\mathcal{E}}, \mathcal{E}_0) \to \mathrm{Ext}^1(\mathfrak{m}_{A/I} \otimes \overline{\mathcal{E}}, \mathcal{E}_0) \xrightarrow{\partial} \mathrm{Ext}^2(\mathcal{E}_0, \mathcal{E}_0) \to \cdots.$$

The class $\mathrm{obs}(\overline{\mathcal{E}})$ is then (by definition) given by the image of

$$0 \to \mathcal{E}_0 \to \mathfrak{m} \otimes \overline{\mathcal{E}} \to \mathfrak{m}_{A/I} \otimes \overline{\mathcal{E}} \to 0$$

under the boundary map ∂.

Assume first that $\mathcal{E}$ exists. Then we can tensor the diagram

$$
\begin{array}{ccc}
k & \longrightarrow \mathfrak{m} & \longrightarrow \mathfrak{m}_{A/I} \\
\downarrow & \downarrow & \downarrow \\
k & \longrightarrow A & \longrightarrow A/I
\end{array}
$$

with $\mathcal{E}$:

$$
\begin{array}{ccc}
\mathcal{E}_0 & \longrightarrow \mathfrak{m} \otimes \overline{\mathcal{E}} & \longrightarrow \mathfrak{m}_{A/I} \otimes \overline{\mathcal{E}} \\
\downarrow & \downarrow & \downarrow \\
\mathcal{E}_0 & \longrightarrow \mathcal{E} & \longrightarrow \overline{\mathcal{E}}.
\end{array}
$$

This shows that $\mathrm{obs}(\overline{\mathcal{E}})$ is zero, because of the long exact sequence above.

Conversely given such a diagram one can reconstruct the A-module structure on $\mathcal{E}$ by defining multiplication with an element in $\mathfrak{m}$ by the composition $\mathcal{E} \to \overline{\mathcal{E}} \to \mathfrak{m} \otimes \overline{\mathcal{E}} \to \mathcal{E}$. $\square$

2.2. Sheaves on stacks

Recall that descent for quasi-coherent sheaves on schemes says that given a scheme X and $U \twoheadrightarrow X$ a smooth surjective morphism (one could replace smooth by fppf here) then we have an equivalence of categories

$$\text{Qcoh}(X) \xrightarrow{\cong} \left\langle \begin{array}{l} \mathcal{F} \text{ quasi-coherent sheaf on } U \text{ together with a descent datum:} \\ \varphi\colon \text{pr}_1^*\mathcal{F} \xrightarrow{\cong} \text{pr}_2^*\mathcal{F} \text{ on } U \times_X U + \text{cocycle condition} \end{array} \right\rangle.$$

(See [19], Exposé VIII, Théorème 1.1.)

Definition 2.8. A *quasi-coherent sheaf* $\mathcal{F}$ on an algebraic stack $\mathcal{M}$ is the datum consisting of:

1. For all smooth maps $x\colon X \to \mathcal{M}$, where X is a scheme, a quasi-coherent sheaf $\mathcal{F}_{X,x}$ on X.
2. For all diagrams

$$\begin{array}{ccc} V & \xrightarrow{\quad f \quad} & U \\ & \underset{v}{\searrow} \quad \underset{\mathcal{M}}{} \quad \underset{u}{\swarrow} & \end{array}$$

together with an isomorphism $\varphi\colon u \circ f \to v$ an isomorphism $\vartheta_{f,\varphi}\colon f^*\mathcal{F}_{U,u} \to \mathcal{F}_{V,v}$ compatible under composition.

Remark 2.9.

1. The category of quasi-coherent sheaves on $\mathcal{M}$ can also be described as the category of sheaves on some atlas together with a descent datum.
2. Since the functors f^*, f_* commute with flat base change, we immediately get such functors for representable $F\colon \mathcal{M} \to \mathcal{N}$.
3. We can always define F_* as a limit, i.e.,

$$\Gamma(\mathcal{M}, \mathcal{F}) := \left\{ s_{U,u} \in H^0(U, \mathcal{F}_{U,u}) \mid \vartheta_{f,\varphi}(s_{U,u}) = s_{V,v} \right\}.$$

And again this can also be computed on a single atlas.

Example 2.10.

1. The structure sheaf $\mathcal{O}_\mathcal{M}$ of an algebraic stack is given by $\mathcal{O}_{\mathcal{M},U,u} = \mathcal{O}_U$. Similarly we can define ideal sheaves of closed substacks, by defining it to be given by the ideal sheaf of the preimage of the substack in any atlas.
2. Continuing the first example, given a smooth, closed substack $\mathcal{N} \subset \mathcal{M}$ of a smooth algebraic stack the normal bundle of $\mathcal{N}$ is a vector bundle, given by the normal bundle computed on any presentation.
3. There is a universal vector bundle $\mathcal{E}_{\text{univ}}$ on $C \times \text{Bun}_n$, simply because any morphism $T \to \text{Bun}_n$ defines a bundle on $C \times T$.
4. To give a line bundle on Bun_n is the same as a functorial assignment of a line bundle to any family of vector bundles. An example is given by the determinant of cohomology $\det(H^*(C, \mathcal{E}))$. (See G. Hein's lectures [21].)

Remark 2.11. Using coherent sheaves we can also perform local constructions on stacks, i.e., blowing up substacks, taking the projective bundle of a vector bundle, taking normalizations or the reduced substack: We can do this on any smooth atlas and then use descent to define the corresponding object over any $T \to \mathcal{M}$.

For quotient stacks $[X/G]$ this just means to do the corresponding construction on X and observe that the G-action extends to the scheme obtained.

To put this definition in the general framework one should of course spell out an explicit Grothendieck topology in order to obtain all the standard functorialities. This requires some careful work. Also some natural sheaves (like the cotangent bundle) do not satisfy the condition that the $\vartheta_{f,\varphi}$ are isomorphisms so it is natural to drop this condition. The first written results in this direction appeared in the language of simplicial schemes [15]. A reference for the results on stacks is [24] together with corrections by Olsson [27].

Exercise 2.12 (A normalization). The group $\mathbb{Z}/2\mathbb{Z}$ acts on $\mathbb{A}^2$ by interchanging the coordinates. Let $X \subset \mathbb{A}^2$ denote the union of the two coordinate axis and consider the stack $\mathcal{M} := [X/(\mathbb{Z}/2\mathbb{Z})]$. Note that the inclusion of $\{0\} \subset X$ defines a closed embedding $B\mathbb{Z}/2\mathbb{Z} \subset [X/(\mathbb{Z}/2\mathbb{Z})]$. Show that the normalization $\mathcal{M}^{\mathrm{norm}} \cong \mathbb{A}^1$ is a scheme.

Exercise 2.13 (A blow-up). Let the multiplicative group $\mathbb{G}_m$ act on $\mathbb{A}^2$ via $t.(x,y) = (tx, t^{-1}y)$ and consider the quotient stack $[\mathbb{A}^2/\mathbb{G}_m]$. Again the inclusion $\{0\} \subset \mathbb{A}^2$ defines a closed substack $B\mathbb{G}_m \subset \mathbb{A}^2$.

Calculate the blow-up $\mathrm{Bl}_{\mathcal{M}}(B\mathbb{G}_m)$ of $B\mathbb{G}_m$ in $\mathcal{M}$. Show that an open subset of the exceptional fibre is isomorphic to $B\mu_2$. (Here $\mu_2 \subset \mathbb{G}_m$ is the subgroup of elements of square 1, so this is just ± 1 if we are not in characteristic 2.)

Exercise 2.14. Let ζ be a primitive 6th root of unity. Let $\mathbb{Z}/6$ act on $\mathbb{A}^2$ by $n.(x,y) := (\zeta^{2n}x, \zeta^{3n}y)$. Calculate the inertia stack $I([\mathbb{A}^2/(\mathbb{Z}/6)])$ and describe its irreducible components. (These are sometimes called *sectors* in physics-related literature.)

Lecture 3: Relation with coarse moduli spaces

Before describing some applications of stacks to classical questions I want to give a sample theorem on sheaves on algebraic stacks which appeared in [24]. This theorem might look completely innocuous, or even boring at first sight. However it turns out to have surprising applications:

Theorem 3.1 ([24], Proposition 15.4.). *Let $\mathcal{M}$ be a noetherian algebraic stack. Then any quasi-coherent sheaf on $\mathcal{M}$ is the filtered limit of its coherent subsheaves.*

Corollary 3.2. *Any representation of a smooth noetherian algebraic group is the union of its finite-dimensional subrepresentations.*

Proof of the corollary. Take $\mathcal{M} = BG$. Then by definition a quasi-coherent sheaf on BG is the same as a sheaf on $\mathrm{Spec}(k)$, i.e., a vector space together with an action of G. $\square$

Proof. Let $\mathcal{F}$ be a quasi-coherent sheaf on $\mathcal{M}$. Choose an atlas $u \colon U \twoheadrightarrow \mathcal{M}$ of $\mathcal{M}$. In particular $\mathcal{F}_{u,U} = u^*\mathcal{F}$ is quasi-coherent on U. In particular this sheaf is the union of its quasi-coherent subsheaves $u^*\mathcal{F} = \varinjlim \mathcal{G}_i$, where $\mathcal{G}_i$ are coherent on U. In particular we have $\mathcal{F} \hookrightarrow u_*u^*\mathcal{F} = \varinjlim u_*\mathcal{G}_i$.

Define $\mathcal{F}_i := \mathcal{F} \cap u_*\mathcal{G}_i$ so that $\mathcal{F} = \varinjlim \mathcal{F}_i$. To see that the $\mathcal{F}_i$ are coherent consider the diagram:

$$
\begin{array}{ccc}
\mathcal{F}_i := \mathcal{F} \cap u_*\mathcal{G}_i & \hookrightarrow & u_*\mathcal{G}_i \\
\big\uparrow & & \big\uparrow \\
\mathcal{F} & \hookrightarrow & u_*u^*\mathcal{F}.
\end{array}
$$

By adjunction we get:

$$
\begin{array}{ccc}
u^*\mathcal{F}_i & \longrightarrow & \mathcal{G}_i \\
\big\uparrow & & \big\downarrow \\
u^*\mathcal{F} & \xrightarrow{\ \mathrm{id}\ } & u^*\mathcal{F}.
\end{array}
$$

In particular $u^*\mathcal{F}_i$ is a subsheaf of $\mathcal{G}_i$. Thus $\mathcal{F}_i$ is coherent. $\qquad\square$

Exercise 3.3. (If you know the proof that any representation of an algebraic group G over a field is the union of its finite-dimensional subrepresentations.) Rewrite the proof that any quasi-coherent sheaf on a noetherian stack is the filtered inductive limit of coherent subsheaves in the case of BG explicitly on the standard presentation pt $\to BG$ in order to see that the argument is a generalization of the argument you know.

Corollary 3.4. *Let $\mathcal{M}$ be a smooth, noetherian algebraic stack and $\mathcal{U} \subset \mathcal{M}$ an open substack. Let $\mathcal{L}_\mathcal{U}$ be a line bundle on $\mathcal{U}$. Then there exists a line bundle $\mathcal{L}$ on $\mathcal{M}$ such that $\mathcal{L}|_\mathcal{U} \cong \mathcal{L}_\mathcal{U}$.*

I learnt this corollary from Lieblich's thesis [26]. Note that we cannot argue with divisors here, because the example of BG already shows that a line bundle on a stack does not necessarily have meromorphic sections.

Proof. The sheaf $j_*\mathcal{L}_\mathcal{U}$ is quasi-coherent. Thus we can write $j_*\mathcal{L}_\mathcal{U} = \varinjlim \mathcal{F}_i$ for some coherent sheaves $\mathcal{F}_i$. This implies that we can even find $\mathcal{F}_i$ such that $\mathcal{F}_i|_U = \mathcal{L}_\mathcal{U}$. Then the double dual $(\mathcal{F}_i^\vee)^\vee$ is a reflexive sheaf of rank 1 on a smooth stack. So it has to be a line bundle. (Again, this result holds for stacks, because we can check it on a smooth atlas.) $\qquad\square$

3.1. Coarse moduli spaces

Definition 3.5. Let $\mathcal{M}$ be an algebraic stack. An algebraic space M together with a map $p \colon \mathcal{M} \to M$ is called *coarse moduli space* for $\mathcal{M}$ if

1. For all schemes X and morphisms $q\colon \mathcal{M} \to X$ there exists a unique morphism $M \to X$ making

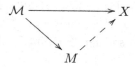

   commutative.
2. For all algebraically closed fields $\overline{K}$ we have $\mathcal{M}(\overline{K})/\text{isomorphism} = M(\overline{K})$.

If M only satisfies the first condition, it is called *categorial coarse moduli space*.

Remark 3.6. Categorial coarse moduli spaces can be very small. If $\mathcal{M} = [\mathbb{A}^n / \mathbb{G}_m]$, the quotient of the affine space by the multiplication by non-zero scalars, then $\mathcal{M} \to \text{pt}$ is a categorial quotient.

Example 3.7. Let $\mathrm{Bun}_n^{\text{stable}}$ be the moduli stack of stable bundles on a curve. Then the coarse moduli space of stable bundles M^{stable} constructed by geometric invariant theory (see G. Hein's lectures [21]) is a coarse moduli space for $\mathrm{Bun}_n^{\text{stable}}$. The GIT construction shows that $\mathrm{Bun}_n^{\text{stable}} = [X/\mathsf{GL}_N]$ (for some scheme X) and constructs $M^{\text{stable}} = X/\mathsf{PGL}_N$. In particular we get a map $\mathrm{Bun}_n^{\text{stable}} \to M^{\text{stable}}$ which satisfies the stronger property that for any $T \to M^{\text{stable}}$ there exists an étale covering $T' \twoheadrightarrow T$ such that the map $T' \to M^{\text{stable}}$ lifts to $T' \to X$ and therefore it lifts to $\mathrm{Bun}_n^{\text{stable}}$.

Note that in the above example all geometric fibres of the map $\mathrm{Bun}_n^{\text{stable}} \to M^{\text{stable}}$ are isomorphic to $B\mathbb{G}_m$. This corresponds to the fact that the automorphism group of a stable bundle consists only of scalars. Such a morphism is called a gerbe. Let us give the definition:

Definition 3.8. A morphism $F\colon \mathcal{M} \to \mathcal{N}$ of algebraic stacks is called a *gerbe* over $\mathcal{N}$ if

1. F is locally surjective, i.e., for any $T \to \mathcal{N}$ there exists a covering $T' \twoheadrightarrow T$ such that the morphism $T' \to \mathcal{N}$ can be lifted to $\mathcal{M}$. (In other words, any object of $\mathcal{N}$ locally comes from an object of $\mathcal{M}$.)
2. All objects in a fibre are locally isomorphic: If $t_1, t_2 \colon T \to \mathcal{M}$ are two objects with $F(t_1) \cong F(t_2)$ then there exists a covering $T' \twoheadrightarrow T$ such that $t_1|_{T'} \cong t_2|_{T'}$.

$F\colon \mathcal{M} \to \mathcal{N}$ is called a $\mathbb{G}_m$-*gerbe* if for all $t\colon T \to \mathcal{M}$ the relative automorphism group $\mathrm{Aut}(t/\mathcal{N})$ is canonically isomorphic to $\mathbb{G}_m(T)$ (equivalently: $I(\mathcal{M}) \times_{I(\mathcal{N})} \mathcal{N} \cong \mathbb{G}_m \times \mathcal{N}$).

Example 3.9. We have just seen that $\mathrm{Bun}_n^{\text{stable}} \to M^{\text{stable}}$ is a $\mathbb{G}_m$-gerbe. More generally if $\mathcal{N} = [X/\mathsf{PGL}_n]$ then $\mathcal{M} := [X/\mathsf{GL}_n] \to [X/\mathsf{PGL}_n]$ is a $\mathbb{G}_m$-gerbe.

So a $\mathbb{G}_m$-gerbe on a scheme X can be thought of as a $B\mathbb{G}_m$-bundle over X.

A notion of triviality of such a bundle is useful:

Lemma 3.10. *For a $\mathbb{G}_m$-gerbe $F\colon \mathcal{M} \to \mathcal{N}$ the following are equivalent:*

1. *The morphism $F\colon \mathcal{M} \to \mathcal{N}$ has a section.*
2. *$\mathcal{M} \cong B\mathbb{G}_m \times \mathcal{N}$.*
3. *There is a line bundle of weight[4] 1 on $\mathcal{M}$.*

A gerbe satisfying the above conditions is called neutral.

Remark 3.11 (Weight). Since for any $u\colon U \to \mathcal{M}$ we have $\mathbb{G}_m \subset \operatorname{Aut}(U \to \mathcal{M})$, the $\vartheta_{\mathrm{id},\alpha}$ for $\alpha \in \mathbb{G}_m(U)$ define a $\mathbb{G}_m$-action on $\mathcal{F}_{U,u}$, i.e., a direct sum decomposition $\mathcal{F}_{U,u} = \oplus_{n\in\mathbb{Z}}\mathcal{F}_{U,u}^n$ such that $\mathcal{F}^i$ is the subsheaf on which $\mathbb{G}_m$ acts via multiplication with the ith power.

Since this decomposition is canonical it defines $\mathcal{F} = \oplus \mathcal{F}^n$. A sheaf is called of *weight i* if $\mathcal{F} = \mathcal{F}^i$.

Proof. By the remark, 2. implies 3., because the universal bundle on $B\mathbb{G}_m$ is of weight 1. The implication 2. $\Rightarrow$ 1. is also clear.

Let us show that 1. $\Rightarrow$ 3.. Let $s\colon \mathcal{N} \to \mathcal{M}$ be a section. Then $\mathcal{N} \times_\mathcal{M} \mathcal{N} = \operatorname{Aut}(s) = \mathbb{G}_m \times \mathcal{N}$, i.e., the section $s\colon \mathcal{N} \to \mathcal{M}$ makes $\mathcal{N}$ into a $\mathbb{G}_m$-bundle of weight one on $\mathcal{M}$.

Finally 3. $\Rightarrow$ 2.: Let $\mathcal{L}$ be a line bundle of weight 1 on $\mathcal{M}$ and denote by $\mathcal{L}^\circ$ the corresponding $\mathbb{G}_m$-bundle. Since $B\mathbb{G}_m$ is the classifying stack of line bundles this defines a morphism $\mathcal{M} \to B\mathbb{G}_m$ such that $\mathcal{L}^\circ = \mathrm{pt} \times_{B\mathbb{G}_m} \mathcal{M}$. So we get a cartesian diagram:

$$
\begin{array}{ccc}
\mathcal{L}^\circ & \longrightarrow & \mathcal{N} = \mathcal{N} \times \mathrm{pt} \\
\downarrow & & \downarrow \\
\mathcal{M} & \longrightarrow & \mathcal{N} \times B\mathbb{G}_m.
\end{array}
$$

This implies that $\mathcal{M} \to \mathcal{N} \times B\mathbb{G}_m$ is locally surjective, all objects in a fibre are locally isomorphic, because this already holds for $\mathcal{M} \to \mathcal{N}$ so this is a gerbe. However the map is also an isomorphism on automorphism groups, so it must be an isomorphism. $\qquad\square$

Let us apply these notions in order to study when a Poincaré family exists on the coarse moduli space $M_n^{d,\mathrm{stable}}$ of stable vector bundles of rank n and degree d. Recall that a Poincaré family is a vector bundle on $C \times M_n^{d,\mathrm{stable}}$ such that the fibre over every point of $M_n^{d,\mathrm{stable}}$ lies in the isomorphism class of bundles defined by this point. So such a bundle is the same thing as a section of the map $\mathrm{Bun}_n^{d,\mathrm{stable}} \to M_n^{d,\mathrm{stable}}$.

With these preparations we can now show the following result of Ramanan ([28]), reproved by Drezet and Narasimhan ([13], Théorème G).

---

[4]see below

Corollary 3.12. *Assume that the genus g of our curve C is bigger than 1.*

1. *If $(n, d) = 1$ then there exists a Poincaré family on the coarse moduli space $M_n^{d,\text{stable}}$ of stable vector bundles on C.*
2. *If $(n, d) \neq 1$ then there is no open subset $U \subset M_n^{d,\text{stable}}$ (with $U \neq \varnothing$) such that there exists a Poincaré family on $C \times U$.*

Proof. The first part is well known: For any point $c \in C$ the restriction of the universal bundle $\mathcal{E}$ on $C \times \text{Bun}_n^d$ to $c \times \text{Bun}_n^d$ has weight 1, so $\det(\mathcal{E}|_{c \times \text{Bun}_n^d})$ has weight n.

Next, note that by transport of structure for any bundle $\mathcal{E}$ the scalar automorphisms act as scalar automorphisms on the cohomology groups $H^i(C, \mathcal{E})$. Thus, by the Riemann–Roch theorem the bundle $\det(H^*(C, \mathcal{E}))$ has weight $d + (g - 1)n$ so that there exist a product of these two bundles having weight 1. By Lemma 3.10, this means that the map $\text{Bun}_n^{d,\text{stable}} \to M_n^d$ has a section, so that we can pull back the universal bundle by this section to obtain a Poincaré bundle on M_n^d.

For the second part let $(n, d) = (k) \neq 1$. Assume there was a Poincaré family on some non-empty open subset $U \subset M_n^d$. This would mean that there is a section $U \to \text{Bun}_n^d$. By Lemma 3.10, there would then exist a line bundle of weight 1 on $\mathcal{U} := U \times_{M_n^d} \text{Bun}_n^d \subset \text{Bun}_n^d$. Since Bun_n^d is smooth, we can apply the corollary from the beginning of the lecture to extend this line bundle to any noetherian substack of Bun_n^d. And this would still have weight 1.

Now pick a stable bundle $\mathcal{E}$ of rank n/k and degree d/k (the assumption on the genus of C assures that stable bundles exist, see the remark following this proof). Then the bundle $\mathcal{E}^{\oplus k}$ defines a point of Bun_n^d. We have $\text{Aut}(\mathcal{E}^{\oplus k}) = \text{GL}_k$ and thus we find $B\text{GL}_k \subset \text{Bun}_n^d$. However there is no line bundle of weight 1 on $B\text{GL}_k \to B\text{PGL}_k$, because this would imply that $B\text{GL}_k = B\text{PGL}_k \times B\mathbb{G}_m$ but $\text{GL}_k \not\cong \text{PGL}_k \times \mathbb{G}_m$. So we found a contradiction. $\square$

A much more general statement of this type can be found in an article of Biswas and Hoffmann [10].

Remark 3.13. In the above proof we used the existence of (semi-)stable bundles on curves of genus > 1. This is a classical result (see for example [29]), but again one can also rephrase this in terms of the moduli stack. Namely in the next lecture we will see that the substack of instable bundles has positive codimension in the stack of all bundles if the genus of C is > 0, so in particular semistable bundles have to exist. A similar argument works for the strictly semistable locus if the genus of C is at least two.

Exercise 3.14. Let M be a scheme and $\mathcal{M} \to M$ be a $\mathbb{G}_m$-gerbe. Let $E \to \mathcal{M}$ be a vector bundle of weight 1. Denote the complement of the zero-section of E by $E^\circ \subset E$. Show that $E^\circ \to M$ is a bundle of projective spaces, i.e., there is a smooth covering $U \to M$ such that $E^\circ|_U \cong U \times \mathbb{P}^{n-1}$.

Lecture 4: Cohomology of Bun_n^d

Next we want to calculate the cohomology of some étale sheaves on stacks. (If you prefer to work in an analytic category these would just correspond to constructible sheaves in the analytic topology.) The aim of this lecture is on the one hand to give an impression of some techniques which help to do such computations and on the other hand to show that the results are often much nicer than the corresponding results for coarse moduli spaces.

To avoid introducing more theory, we use the same working definition as before:

Definition 4.1. A *sheaf* on an algebraic stack $\mathcal{M}$ is a collection of sheaves $\mathcal{F}_{U,u}$ for all $u\colon U \to \mathcal{M}$ together with compatible morphisms $\vartheta_{f,\varphi}\colon f^{-1}\mathcal{F}_{U,u} \to \mathcal{F}_{V,v}$ for all $v\colon V \to \mathcal{M}$, $f\colon V \to U$ and $\varphi\colon u \circ f \to v$.

A sheaf is called *cartesian* if all $\vartheta_{f,\varphi}$ are isomorphisms.

Remark 4.2. As before it turns out that the cohomology groups $H^*(\mathcal{M},\mathcal{F})$ can be calculated from an atlas. Given an atlas $u\colon U \twoheadrightarrow \mathcal{M}$ we denote by $U_p := U \times_\mathcal{M} \cdots \times_\mathcal{M} U$ the $(p+1)$-fold product of U over $\mathcal{M}$. Then there exists a spectral sequence (see, e.g., [27]) with

$$E_1^{p,q} := H^q(U_p,\mathcal{F}_{U_p}) \Rightarrow H^{p+q}(\mathcal{M},\mathcal{F}).$$

From this one can immediately conclude that the Künneth formula and Gysin sequences [11] also exist for stacks, simply by applying the formulas for the U_p.

Example 4.3. Let us compute $H^*(B\mathbb{G}_m,\mathbb{Q}_\ell)$. The quotient map $\mathbb{A}^n - 0 \to \mathbb{P}^{n-1}$ is a $\mathbb{G}_m$-bundle. Thus we get a cartesian diagram:

$$
\begin{array}{ccc}
\mathbb{A}^n - 0 & \longrightarrow & \mathrm{pt} \\
\downarrow & & \downarrow \\
\mathbb{P}^{n-1} & \xrightarrow{\ p\ } & B\mathbb{G}_m.
\end{array}
$$

From this we see that the fibres of p are $\mathbb{A}^n - 0$. Thus we get that

$$\mathbf{R}^i p_* \mathbb{Q}_\ell = \begin{cases} \mathbb{Q}_\ell & i = 0 \\ 0 & i < 2n - 1. \end{cases}$$

The Leray spectral sequence for p therefore implies $H^i(B\mathbb{G}_m) \cong H^i(\mathbb{P}^{n-1})$ for $i < 2n - 1$. Thus we find that $H^*(B\mathbb{G}_m,\mathbb{Q}_\ell) = \mathbb{Q}_\ell[c_1]$ is a polynomial ring in one generator of degree 2.

Example 4.4. Replacing $\mathbb{A}^n - 0$ in the previous example by the space $\mathrm{Mat}(n,N)_{\mathrm{rk}=n}$ of $n \times N$ matrices of rank n one can show that $H^*(B\mathrm{GL}_n,\mathbb{Q}_\ell) = \mathbb{Q}_\ell[c_1,\ldots,c_n]$ is a polynomial ring, with generators c_i (the universal Chern classes) of degree $2i$.

As an application of this computation one can define the Chern classes of a vector bundle as the pull back of the c_i under the morphism to $B\mathrm{GL}_n$ which is defined by the vector bundle.

Although we will not use it, I would like to mention Behrend's generalization of the Lefschetz trace formula for algebraic stacks over finite fields. This gives another way to interpret the above examples. Recall that for a smooth variety X over a finite field $k = \mathbb{F}_q$ there is a natural action of the Frobenius on the cohomology groups $H^i(X_{\overline{k}}, \mathbb{Q}_\ell)$. And the Lefschetz trace formula (see [11], p. 88), says[5] that

$$\#X(\mathbb{F}_q) = q^{\dim(X)} \sum_i (-1)^i \operatorname{Trace}(\operatorname{Frob}, H^i(X_{\overline{k}}, \mathbb{Q}_\ell)).$$

Furthermore, the Weil conjectures imply that for a smooth, proper variety the eigenvalues of Frob on $H^i(X_{\overline{k}}, \mathbb{Q}_\ell)$ have absolute value $q^{-i/2}$ and that this is enough to recover the dimension of the cohomology groups from the knowledge of $\#X(\mathbb{F}_{q^n})$ for sufficiently many n.

Behrend showed [5], [6] that a similar trace formula also holds for a large class of smooth algebraic stacks $\mathcal{M}$ over $\mathbb{F}_q$, including quotient stacks and Bun_n, namely that

$$\sum_{x \in \mathcal{M}(\mathbb{F}_q)/\mathrm{iso}} \frac{1}{\# \operatorname{Aut}(x)(\mathbb{F}_q)} = q^{\dim(\mathcal{M})} \sum_i (-1)^i \operatorname{Trace}(\operatorname{Frob}, H^i(\mathcal{M}_{\overline{k}}, \mathbb{Q}_\ell)).$$

For example for $B\mathbb{G}_m$ the left-hand side is $1/(q-1) = q^{-1}(1 + q^{-1} + q^{-2} + \cdots)$. The fact that the coefficients of the powers of q in this expansion are all equal to 1 already suggests that the cohomology should have a single generator in each even degree.

Exercise 4.5. Check Behrend's trace formula for the stacks $B(\mathbb{Z}/2)$ and $B(\mathbb{Z}/n)$ over the finite field $\mathbb{F}_p$.

Remark 4.6. In contrast to the preceding exercise the proof of the trace formula uses a reduction to quotients by GL_n. This trick is quite useful in other contexts as well, because GL_n-bundles are locally trivial for the Zariski topology.

Our aim of this lecture is to compute the cohomology $H^*(\operatorname{Bun}_n^d, \mathbb{Q}_\ell)$. First note that the Gysin sequence implies that the cohomology in low degrees does not change if one removes a substack of high codimension. Therefore

$$H^*(\operatorname{Bun}_n^d, \mathbb{Q}_\ell) = \lim_{\mathcal{U} \subset \operatorname{Bun}_n^d \text{ finite type}} H^*(\mathcal{U}, \mathbb{Q}_\ell).$$

And for each fixed degree $*$ the limit on the right-hand side becomes stationary for sufficiently large $\mathcal{U}$, i.e., $H^i(\operatorname{Bun}_n^d, \mathbb{Q}_\ell) = H^i(\mathcal{U}, \mathbb{Q}_\ell)$ if the codimension of the complement of $\mathcal{U}$ is larger than $i/2$.

Let us first recall the construction of the so-called *Atiyah–Bott classes* in the cohomology of Bun_n^d:

---

[5]This is usually formulated for cohomology with compact supports. In that case the leading factor $q^{\dim(X)}$ disappears.

We have already seen that by definition of sheaves on stacks, there is a universal family of vector bundles $\mathcal{E}_{\text{univ}}$ on $C \times \text{Bun}_n^d$. In particular this bundle defines a morphism $\text{univ} \colon C \times \text{Bun}_n^d \to B\text{GL}_n$ and we set

$$c_i(\mathcal{E}_{\text{univ})} := \text{univ}^* c_i \in H^*(C \times \text{Bun}_n^d, \mathbb{Q}_\ell) = H^*(C, \mathbb{Q}_\ell) \otimes H^*(\text{Bun}_n^d, \mathbb{Q}_\ell).$$

We choose a basis $1 \in H^0(C, \mathbb{Q}_\ell), (\gamma_j)_{j=1,\dots,2g} \in H^1(C, \mathbb{Q}_\ell), [\text{pt}] \in H^2(C, \mathbb{Q}_\ell)$ in order to decompose these Chern classes:

$$c_i(\mathcal{E}_{\text{univ}}) = 1 \otimes a_i + \sum_{j=1}^{2g} \gamma_j \otimes b_i^j + [\text{pt}] \otimes f_i$$

for some $a_i, b_i^j, f_i \in H^*(\text{Bun}_n^d, \mathbb{Q}_\ell)$.

Using these classes we can give the main theorem of this lecture. This was first proved by Atiyah–Bott in an analytic setup using equivariant cohomology. An algebraic proof was first given by Bifet–Ghione–Laetizia [9]. The proof we will explain here is a variant of [22]. To simplify notations we will write $H^*(X, \overline{\mathbb{Q}}_\ell)$ for the cohomology of X computed over the algebraic closure of the ground field.

Theorem 4.7. *The cohomology of Bun_n^d is freely generated by the Atiyah–Bott classes:*

$$H^*(\text{Bun}_n^d, \overline{\mathbb{Q}}_\ell) = \overline{\mathbb{Q}}_\ell[a_1, \dots, a_n] \otimes \bigwedge [b_i^j]_{\substack{i=1,\dots n \\ j=1,\dots 2g}} \otimes \overline{\mathbb{Q}}_\ell[f_2, \dots, f_n].$$

A similar result holds for moduli spaces of principal bundles (see [2] for an analytic proof and [23] for an algebraic version).

We want to indicate a proof of the above result.

4.1. First step: Independence of the generators

We first want show that the Atiyah–Bott classes generate a free subalgebra of the cohomology. Let us consider the simplest case $n = 1$. Denote by Pic_C^d the Picard scheme of C, which is a coarse moduli space for Bun_1^d. Over an algebraically closed field there exists a Poincaré bundle on $C \times \text{Pic}_C^d$ and thus 3.10 implies that:

$$\text{Bun}_1^d \cong \text{Pic}_C^d \times B\mathbb{G}_m.$$

Furthermore we know that Pic_C^d is isomorphic to the Jacobian of C, which is an abelian variety and its cohomology is the exterior algebra on $H^1(C, \overline{\mathbb{Q}}_\ell)$. Thus

$$H^*(\text{Bun}_1^d, \overline{\mathbb{Q}}_\ell) \cong H^*(\text{Pic}_C^d, \overline{\mathbb{Q}}_\ell) \otimes H^*(B\mathbb{G}_m, \overline{\mathbb{Q}}_\ell) \cong \bigwedge [b_1^j] \otimes \overline{\mathbb{Q}}_\ell[a_1].$$

For $n > 1$ for any partition $d = \sum_{i=1^n} d_i$ with $d_i \in \mathbb{Z}$ consider the map

$$\bigoplus_{\underline{d}} \colon \prod_{i=1}^n \text{Bun}_1^{d_i} \to \text{Bun}_n^d$$

$$(\mathcal{L}_i) \mapsto \mathcal{L}_1 \oplus \cdots \oplus \mathcal{L}_n$$

We know that the Chern classes of a direct sum of line bundles are given by the elementary symmetric polynomials σ_i in the Chern classes of the line bundles. Write $c_1(\mathcal{L}_{\mathrm{univ}}^{d_i}) := 1 \otimes A_i + \sum \gamma_j \otimes B_i^j + [\mathrm{pt}] \otimes d_i$. Then we have:

$$\oplus_{\underline{d}}^*\big(c_i(\mathcal{E}_{\mathrm{univ}})\big) = \sigma_i\big(c_1(\mathcal{L}_{\mathrm{univ}}^{d_1}), \dots, c_1(\mathcal{L}_{\mathrm{univ}}^{d_n})\big)$$

$$= \sigma_i(A_i, \dots, A_n) + \sum_{j,k} \gamma_j \otimes \partial_k \sigma_i(A_1, \dots, A_n) B_k^j$$

$$+ \sum_{j+m=2g+1, k, l} [\mathrm{pt}] \otimes \partial_k \partial_l \sigma(A_1, \dots, A_n) B_k^j B_l^m$$

$$+ \sum_{j,k} [\mathrm{pt}] \otimes \partial_k \sigma_i(A_1, \dots, A_n) d_k.$$

Thus taking the union over all $\underline{d} = (d_1, \dots, d_n)$, we get a commutative diagram:

$$
\begin{array}{ccc}
H^*(\mathrm{Bun}_n^d, \overline{\mathbb{Q}}_\ell) & \longrightarrow & \prod_{\underline{d}, \sum d_i = d} \overline{\mathbb{Q}}_\ell[A_1, \dots, A_n] \otimes \bigwedge[B_i^j]_{i,j} \\
\uparrow & & \uparrow {\scriptstyle D_i \mapsto (d_i)_{\underline{d}}} \\
\overline{\mathbb{Q}}_\ell[a_i, f_i] \otimes \bigwedge[b_i^j] & \hookrightarrow & \overline{\mathbb{Q}}_\ell[A_i] \otimes \bigwedge[B_i^j] \otimes \overline{\mathbb{Q}}_\ell[D_1, D_2, \dots, D_n] / \sum D_i = d.
\end{array}
$$

Here the lower horizontal map is given by

$$a_i \mapsto \sigma_i(A_1, \dots, A_n), \quad b_i^j \to \sum_k \partial_k \sigma_i(A_1, \dots, A_n) B_k^j,$$

and f_i is mapped to the last two summands of our computation above, replacing the constants d_k by variables D_k. This map is injective, because the $\partial_k \sigma_i$ are linearly independent (this is equivalent to the fact that the map $\mathbb{A}^n \to \mathbb{A}^n / S_n$ is generically étale).

The right vertical arrow is the evaluation of D_i. This is injective, because we evaluate at all integers simultaneously.

This shows that the left vertical arrow must be injective as well.

4.2. Second step: Why is it the whole ring?

One way to see this is to use Beauville's trick[6] to show that the Atiyah–Bott classes generate the cohomology of some coarse moduli spaces: If X is a smooth projective scheme then the Künneth components of the diagonal $[\Delta] \subset H^*(X \times X) \cong H^*(X) \otimes H^*(X)$ generate $H^*(X)$. (This is not difficult. Note however that this does not seem to make sense for stacks, because the diagonal morphism is not an embedding – look at the example of $B\mathbb{G}_m$.)

Let again $\mathcal{E}_{\mathrm{univ}}$ denote the universal bundle on $C \times \mathrm{Bun}_n^d$ and consider the sheaf $\mathcal{H}om(p_{12}^* \mathcal{E}_{\mathrm{univ}}, p_{13}^* \mathcal{E}_{\mathrm{univ}})$ on $C \times \mathrm{Bun}_n^{d,\mathrm{stable}} \times \mathrm{Bun}_n^{d,\mathrm{stable}}$.

---

[6] I think Beauville quotes Ellingsrud and Strømme.

The complex $\mathbf{R}p_{23,*}\mathcal{H}om(p_{12}^*\mathcal{E}_{\text{univ}}, p_{13}^*\mathcal{E}_{\text{univ}})$ can be represented by a complex $[K_0 \xrightarrow{d^1} K_1]$ of vector bundles on $\text{Bun}_n^{d,\text{stable}} \times \text{Bun}_n^{d,\text{stable}}$. Since there are no homomorphisms between non-isomorphic stable vector bundles of the same rank and degree we know that the map d^1 has maximal rank outside the diagonal $\Delta \subset \text{Bun}_n^{d,\text{stable}} \times \text{Bun}_n^{d,\text{stable}}$.

Thus we can apply the Porteous formula ([17], Chapter 14.4) (if we know that $\text{codim}\,\Delta = \chi(K_0 \to K_1)+1$) to see that the top Chern class $c_{\text{top}}(K_0 \to K_1) = [\Delta]$. On the other hand we can use the Riemann–Roch theorem to compute

$$\text{ch}\big(\mathbf{R}p_{23,*}\mathcal{H}om(p_{12}^*\mathcal{E}_{\text{univ}}, p_{13}^*\mathcal{E}_{\text{univ}})\big) = p_{23,*}\big(\text{pr}_C^*(\text{Todd}(C)) \cdot \text{ch}(\mathcal{E}_{\text{univ}}) \cdot \text{ch}(\mathcal{E}_{\text{univ}}^\vee)\big).$$

The right-hand side of this formula is given in terms of the Atiyah–Bott classes. Together with the Porteous formula, this gives an expression of $[\Delta]$ in terms of the Atiyah–Bott classes. However this does only work for stable bundles and to use the trick one also needs to know that $\text{Bun}_n^{d,\text{stable}}$ is a $\mathbb{G}_m$-gerbe over a smooth projective variety.

To get into such a situation one can use parabolic bundles: Pick a finite set of points $S = \{p_1, \ldots, p_N\} \in C$. Then one defines the stack of parabolic bundles:

$$\text{Bun}_{n,S}^d(T) \quad := \quad \big\langle \mathcal{E} \in \text{Bun}_n^d(T),$$

$$(\mathcal{E}_{1,p} \subsetneq \cdots \subsetneq \mathcal{E}_{n,p} = \mathcal{E}|_{p\times T})_{p\in S} \text{ a full flag of subspaces}\big\rangle.$$

Forgetting the flags defines a morphism forget: $\text{Bun}_{n,S}^d \to \text{Bun}_n^d$, the fibres of which are products of flag manifolds $\prod_{p\in S}\text{Flag}_n$. The theorem of Leray–Hirsch says that for such a fibration we have:

$$H^*(\text{Bun}_{n,S}^d) = H^*(\text{Bun}_n^d) \otimes \bigotimes_{p\in S} H^*(\text{Flag}_n).$$

In particular $H^*(\text{Bun}_n^d)$ is generated by the Atiyah–Bott classes if and only if $H^*(\text{Bun}_{n,S}^d)$ is generated by the Atiyah–Bott classes and the Chern classes defined by the flags $\mathcal{E}_{i,p}$. We call this collection of classes the *canonical classes*.

Now one can argue as follows (all these steps require some care):

1. There exist open substacks $\text{Bun}_{n,S}^{d,\alpha\text{-stable}} \subset \text{Bun}_{n,S}^d$ of α-stable bundles, depending on some parameter α. If α is chosen well, this substack has a projective coarse moduli space $M_{n,S}^d$ and the map $\text{Bun}_{n,S}^{d,\alpha\text{-stable}} \to M_{n,S}^d$ is a $\mathbb{G}_m$-gerbe.

2. We can do Beauville's trick for parabolic bundles using homomorphisms respecting the flags instead of arbitrary homomorphisms of vector bundles. Thus $H^*(\text{Bun}_{n,S}^{d,\text{stable}})$ is generated by the canonical classes.

3. The codimension of the instable bundles $\text{Bun}_{n,S}^{d,\text{inst}} \subset \text{Bun}_{n,S}^d$ goes to ∞ for $N \to \infty$ (and well-chosen stability parameters).

Putting these results together we get a proof of the theorem.

Lecture 5: The cohomology of the coarse moduli space (coprime case)

In this lecture we want to continue our study of geometric properties of the stack Bun_n^d in order to give some more phenomena that can occur when studying algebraic stacks. As aim of the lecture we also want to explain why the results of the previous lecture are useful, even if one is only interested in coarse moduli spaces. Namely, we want to deduce a description of the cohomology of the coarse moduli space from the results of the previous lecture. To do this we will study the part of the moduli stack that parameterizes instable bundles. This stack has a natural stratification. We will see what the strata look like and we will analyze the Gysin sequence for this stratification.

We begin with the Harder–Narasimhan "stratification"[7] of Bun_n^d:

Proposition 5.1 (Harder–Narasimhan filtration). *Let $\mathcal{E}$ be a vector bundle on C, defined over an algebraically closed field. Then there exists a canonical filtration*

$$0 \subsetneq \mathcal{E}_1 \subsetneq \cdots \subsetneq \mathcal{E}_s = \mathcal{E}$$

such that for all i we have:

1. $\mu(\mathcal{E}_i) := \frac{\deg(\mathcal{E}_i)}{\text{rk}(\mathcal{E}_i)} > \mu(\mathcal{E}_{i+1})$. *($\mu(\mathcal{E})$ is called the* slope *of $\mathcal{E}$.)*
2. $\mathcal{E}_{i+1}/\mathcal{E}_i$ *is a semistable vector bundle.*

We denote by $t(\mathcal{E}) := (n_i, d_i)_i$ the type of instability of $\mathcal{E}$.

The proof of the proposition proceeds by induction, observing that a subsheaf of maximal slope has to be semistable and that for two such subsheaves, their sum also satisfies this condition.

Remark 5.2.   1. The type of instability $t(\mathcal{E})$ defines a convex polygon, with vertices (n_i, d_i). Here convexity is guaranteed by the first condition above. This is called the *Harder–Narasimhan polygon (HN-polygon)* of $\mathcal{E}$.
2. For any $\mathcal{F} \subset \mathcal{E}$ the point $(\text{rk}(\mathcal{F}), \deg(\mathcal{F}))$ lies below the polygon of $\mathcal{E}$.
3. If one has a family of vector bundles on C, then the HN-polygon can only get bigger under specialization, because the closure of a subsheaf in the generic fibre defines a subsheaf in the special fibre.

In particular for any $T \rightarrow \text{Bun}_n^d$ given by a family $\mathcal{E}$ we get a canonical decomposition $T = \cup_{t \text{ polygon}} T^t$ into locally closed subschemes such that T^t consists of those points such that the Harder–Narasimhan polygon of the corresponding bundle is of type t. Since this is canonical it defines a decomposition of $\text{Bun}_n^d = \cup_t \text{Bun}_n^{d,t}$ and, by the third point of the above remark, the substack $\text{Bun}_n^{d,\leq t} = \cup_{t' \leq t} \text{Bun}_n^{d,t} \subset \text{Bun}_n^d$ is open.

---

[7] We put quotation marks here, in order to warn the reader that the closure of a stratum need not be a union of strata if $n > 2$ (see Example 5.11).

We can describe this more precisely:

1. Given a type $t = (n_i, d_i)_i$ we define the stack of filtered bundles
$$\text{Filt}^{\underline{d}}_{\underline{n}}(T) := \langle \mathcal{E}_1 \subset \mathcal{E}_2 \subset \cdots \subset \mathcal{E}_s | \deg(\mathcal{E}_i) = d_i, \text{rk}(\mathcal{E}_i) = n_i \rangle.$$
This is an algebraic stack and the forgetful morphism
$$\text{Filt}^{\underline{d}}_{\underline{n}} \to \text{Bun}^d_n$$
is representable, by the theory of Quot-schemes.

2. There is a morphism $\text{Filt}^{\underline{d}}_{\underline{n}} \to \prod_i \text{Bun}^{d_i - d_{i-1}}_{n_i - n_{i-1}}$ that maps the filtered bundle $\mathcal{E}_\bullet$ to its subquotients $\mathcal{E}_i / \mathcal{E}_{i-1}$.

3. There is an open substack $\text{Filt}^{\text{ss}}_{\underline{n}, \underline{d}} \subset \text{Filt}_{\underline{n}, \underline{d}}$ defined by the condition that the subquotients $\mathcal{E}_i / \mathcal{E}_{i-1}$ are semistable.

Given a filtered bundle $\mathcal{E}_\bullet$ we define $\mathcal{E}nd(\mathcal{E}_\bullet) \subset \mathcal{E}nd(\mathcal{E}_s)$ to be the subgroup of those endomorphisms respecting the filtration, i.e., those φ such that $\varphi(\mathcal{E}_i) \subset \mathcal{E}_i$ for all i.

Proposition 5.3. *If a type $t = (\underline{n}, \underline{d})$ is a convex polygon, then the forgetful map $\text{Filt}^{\underline{d}, \text{ss}}_{\underline{n}} \to \text{Bun}^d_n$ is an immersion.*

The normal bundle $\mathcal{N}_{\text{forget}}$ to forget is $\mathbf{R}^1 p_* (\mathcal{E}nd(\mathcal{E}_{\text{univ}}) / \mathcal{E}nd(\mathcal{E}_{\bullet, \text{univ}}))$.

The first point has a nice corollary:

Corollary 5.4. *The Harder–Narasimhan filtration of a vector bundle $\mathcal{E}$ on $C \times \text{Spec}(K)$ is defined over K and not only after passing to the algebraic closure.*

Remark 5.5. The above proposition implies that all Harder–Narasimhan strata are smooth stacks. However, for any $k \in \mathbb{Z}$ there are only finitely many strata of dimension $\geq k$. Moreover, any bundle $\mathcal{E}$ of rank $n > 1$ admits subbundles of rank 1. However given an extension $\mathcal{L} \to \mathcal{E} \to \mathcal{E}/\mathcal{L}$ we can find a family over $\mathbb{A}^1 \times C$ such that the restriction to $\mathbb{G}_m$ is the constant family $\mathcal{E} \times \mathbb{G}_m$, but the fibre over 0 is $\mathcal{L} \oplus \mathcal{E}/\mathcal{L}$. In particular no point of Bun_n is closed.

To prove the proposition we need to introduce the tangent stack of an algebraic stack:

Recall that for a scheme X the tangent space TX can be defined as the scheme representing the functor given on affine schemes by $TX(\text{Spec}(A)) := X(\text{Spec}(A[\varepsilon]/\varepsilon^2))$. We can do the same for stacks:

Definition 5.6. The *tangent stack* $T\mathcal{M}$ to an algebraic stack $\mathcal{M}$ is the stack given on affine schemes by
$$T\mathcal{M}(\text{Spec}(A)) := \mathcal{M}(\text{Spec}(A[\varepsilon]/\varepsilon^2)).$$

Remark 5.7. $T\mathcal{M}$ is an algebraic stack, given an atlas $u: U \twoheadrightarrow \mathcal{M}$ the canonical map $TU \to T\mathcal{M}$ is an atlas for $T\mathcal{M}$.

Example 5.8. $TB G = [\text{pt}/TG]$. Note that the tangent space to a group is again a group. This is immediate from the above definition of TG.

Example 5.9. The fibre of the tangent stack $T\operatorname{Bun}_n^d$ at a bundle $\mathcal{E}$ on C is by definition the groupoid of extensions $\tilde{\mathcal{E}}$ of $\mathcal{E}$ to $C \times \operatorname{Spec}(k[\varepsilon]/\varepsilon^2)$. As in our proof of smoothness of Bun_n^d this can be described as follows: We have an exact sequence of $k[\varepsilon]/\varepsilon^2$-modules $k \to k[\varepsilon]/\varepsilon^2 \to k$. Thus an extension $\tilde{\mathcal{E}}$ of $\mathcal{E}$ gives an extension

$$0 \to \mathcal{E} \to \tilde{\mathcal{E}} \to \mathcal{E} \to 0.$$

And conversely such an extension of vector bundles defines a $k[\varepsilon]/\varepsilon^2$-module structure on $\tilde{\mathcal{E}}$, multiplication by ε being given by the composition $\tilde{\mathcal{E}} \to \mathcal{E} \to \tilde{\mathcal{E}}$. The automorphisms of such an extension are given by $\mathcal{H}om(\mathcal{E}, \mathcal{E}) = H^0(C, \mathcal{E}nd(\mathcal{E}))$. Thus we see that:

$$T_{\mathcal{E}} \operatorname{Bun}_n^d = \left[H^1(C, \mathcal{E}nd(\mathcal{E})) / H^0(C, \mathcal{E}nd(\mathcal{E})) \right]$$

where the quotient is taken by letting the additive group H^0 act trivially on H^1.

The same computation holds for $\operatorname{Filt}_{\underline{n}}^d$ if one replaces $\mathcal{E}nd(\mathcal{E})$ by $\mathcal{E}nd(\mathcal{E}_\bullet)$. Now note:

1. The uniqueness of the Harder–Narasimhan flag of a bundle $\mathcal{E}$ is equivalent to the statement that the fibre of forget over $\mathcal{E}$ consists of a single point.
2. The map $T\operatorname{Filt}_{\underline{n}}^d \to T\operatorname{Bun}_n^d$ at $\mathcal{E}$ can be computed from the cohomology sequence:

$$H^0(C, \mathcal{E}nd(\mathcal{E})) \hookrightarrow H^0(C, \mathcal{E}nd(\mathcal{E}_\bullet)) \to 0 = H^0(C, \mathcal{E}nd(\mathcal{E})/\mathcal{E}nd(\mathcal{E}_\bullet))$$
$$H^1(C, \mathcal{E}nd(\mathcal{E})) \hookrightarrow H^1(C, \mathcal{E}nd(\mathcal{E}_\bullet)) \to H^1(C, \mathcal{E}nd(\mathcal{E})/\mathcal{E}nd(\mathcal{E}_\bullet)).$$

Here we used that $H^0(C, \mathcal{E}nd(\mathcal{E})/\mathcal{E}nd(\mathcal{E}_\bullet)) = 0$ because there are no homomorphisms from a semistable bundle to a semistable bundle of smaller slope.

This implies the proposition (pointwise), because an unramified map (i.e., inducing an injection on the tangent spaces) whose fibres are points is an immersion.

Remark 5.10. In the preceding computation we could replace the bundle $\mathcal{E}$ by any family parametrized by an affine scheme $\operatorname{Spec}(A)$. For any family $T \to \operatorname{Bun}_n^d$ (of finite type) write $\mathbf{R}p_* \mathcal{E}nd(\mathcal{E}) = [K_0 \to K_1]$ as a complex of vector bundles on T. Using the computation for affine schemes it is easy to see that the pull back of $T\operatorname{Bun}_n^d$ to T is given by the quotient stack $[K_0/K_1]$. This then proves the last part of the proposition.

Example 5.11. We briefly consider the case $n = 3, d = 1$, in order to indicate why the closure of a HN-stratum need not be a union of strata. We consider strata of bundles such that the HN-filtration consists of a single subsheaf, namely the strata of type $t_1 = (n_1 = 1, d_1 = 1)$ and $t_2 = (n_1 = 2, d_1 = 2)$. Since the HN-polygon of t_1 lies below t_2, the closure of $\operatorname{Bun}_{3,1}^{t_1}$ can contain elements of $\operatorname{Bun}_{3,1}^{t_2}$, but by Remark 5.2, Point 3, any such specialization $\mathcal{E}$ will contain a subsheaf $\mathcal{L}$ of rank 1 and degree 1. Since the destabilizing subbundle $\mathcal{E}_1$ of $\mathcal{E}$ is a vector bundle of rank 2 and degree 2, $\mathcal{L}$ will be contained in $\mathcal{E}_1$, so that $\mathcal{E}_1$ is semi-stable but not stable.

Thus, in case there exist stable bundles of rank 2 and degree 2 on C, i.e., if $g \geq 2$, the closure of $\mathrm{Bun}_{3,1}^{t_1}$ cannot contain the whole of $\mathrm{Bun}_{3,1}^{t_2}$.

Let us give a concrete example to show that the closure indeed intersects $\mathrm{Bun}_{3,1}^{t_2}$: The stratum $\mathrm{Bun}_{3,1}^{t_2}$ is non-empty because it contains direct sums of line bundles $\mathcal{E} = \mathcal{L}_1 \oplus \mathcal{L}_2 \oplus \mathcal{L}_3$ with $\deg(\mathcal{L}_1) = \deg(\mathcal{L}_2) = 1$ and $\deg(\mathcal{L}_3) = -1$.

Moreover, $\mathrm{Ext}^1(\mathcal{L}_2, \mathcal{L}_3) = H^1(C, \mathcal{L}_3 \otimes \mathcal{L}_2^{-1})$. By the theorem of Riemann–Roch, this is a vector space of dimension $2 - 1 + g > 0$. Thus there exist non-trivial extensions $\mathcal{L}_3 \to \mathcal{E}_2 \to \mathcal{L}_2$ and such an extension $\mathcal{E}_2$ cannot contain subbundles of positive degree, since such a subbundle would have to split the extension. So we find a bundle $\mathcal{L}_1 \oplus \mathcal{E}_2$ in $\mathrm{Bun}_{3,0}^{t_1}$ that can be degenerated into $\mathcal{E} = \mathcal{L}_1 \oplus \mathcal{L}_2 \oplus \mathcal{L}_3$.

The reader may consult the article [16] for a complete discussion of the HN-stratification over curves of genus one and a further counterexample on curves of higher genus.

Corollary 5.12. *There is a Gysin sequence:*

$$\cdots \to H^{*-\mathrm{codim}}(\mathrm{Filt}^{t,\mathrm{ss}}) \to H^*(\mathrm{Bun}_n^{d,\leq t}) \to H^*(\mathrm{Bun}_n^{d,<t}) \to \cdots.$$

To prove that this sequence splits we need a lemma:

Lemma 5.13. *Let* $p\colon \widetilde{\mathcal{M}} \to \mathcal{M}$ *be a* $\mathbb{G}_m$*-gerbe and* $\mathcal{E}$ *be a vector bundle of weight* $w \neq 0$ *on* $\widetilde{\mathcal{M}}$*. Then*

$$H^*(\widetilde{\mathcal{M}}) = H^*(\mathcal{M})[c_1(\mathcal{E})]$$

and the top Chern class $c_{\mathrm{top}}(\mathcal{E})$ *is not a zero divisor in* $H^*(\widetilde{\mathcal{M}})$*.*

Proof. First note that the map $(p, \det(\mathcal{E}))\colon \widetilde{\mathcal{M}} \to \mathcal{M} \times B\mathbb{G}_m$ is a $\mu_{\mathrm{rk}(\mathcal{E}) \cdot w}$-gerbe. For finite groups G the cohomology $H^*(BG, \overline{\mathbb{Q}}_\ell)$ vanishes. Therefore the Leray spectral sequence for $(p, \det(\mathcal{E}))$ shows that $H^*(\widetilde{\mathcal{M}}, \overline{\mathbb{Q}}_\ell) \cong H^*(\mathcal{M})[c_1]$. This is the first claim.

Now let $x\colon \mathrm{Spec}(k) \to \mathcal{M}$ be any geometric point and $\tilde{x}\colon \mathrm{Spec}(k) \to \widetilde{\mathcal{M}}$ a lift of x. Then $\mathrm{Spec}(k) \times_{\mathcal{M}} \widetilde{\mathcal{M}} = B\mathbb{G}_m$ canonically. Using this we get a map:

$$m\colon B\mathbb{G}_m \xrightarrow{i} \widetilde{\mathcal{M}} \xrightarrow{\det(\mathcal{E})} B\mathbb{G}_m.$$

And this composition is given by raising to the $(\mathrm{rk}(\mathcal{E}) \cdot w)$th power.

Thus $i^*(c_{\mathrm{top}}(\mathcal{E})) = (w \cdot c_1)^n$. Writing $c_{\mathrm{top}}(\mathcal{E}) = \sum_{i=0}^{\mathrm{rk}(\mathcal{E})} \beta_i c_1^i$ we see that $\beta_{\mathrm{rk}(\mathcal{E})}$ is a non-zero constant and this proves the second claim. $\qquad\square$

Corollary 5.14. *The Gysin sequence:*

$$\cdots \to H^{*-\mathrm{codim}}(\mathrm{Filt}^{t,\mathrm{ss}}) \to H^*(\mathrm{Bun}_n^{d,\leq t}) \to H^*(\mathrm{Bun}_n^{d,<t}) \to \cdots$$

splits.

In particular $H^*(\mathrm{Bun}_n^{d,\mathrm{ss}})$ *is a quotient of* $H^*(\mathrm{Bun}_n^d)$ *and thus generated by the Atiyah–Bott classes.*

Proof. The composition $H^{*-\mathrm{codim}}(\mathrm{Filt}^{t,\mathrm{ss}}) \to H^*(\mathrm{Bun}_n^{d,\leq t}) \to H^*(\mathrm{Filt}^{t,\mathrm{ss}})$ is given by the cup product with $c_{\mathrm{top}}(\mathcal{N}_{\mathrm{forget}})$ which is injective. $\qquad\square$

Remark 5.15. The cohomology of Filt^t can also be computed:

$$H^*(\text{Filt}^d_{\underline{n}}) = \bigotimes_i H^*(\text{Bun}^{d_i}_{n_i}).$$

The same holds for the semistable part.

This remark implies that the cohomology of the instable part which occurs in the Gysin sequence can be described in terms of the cohomology of moduli stacks of bundles of smaller rank. This gives an inductive procedure to compute $H^*(\text{Bun}^{d,\text{ss}}_n)$ for all n, d. Furthermore, in case that $(n, d) = 1$ this space is a $\mathbb{G}_m$-gerbe over the coarse moduli space, so we get a recursive formula for the cohomology of the coarse moduli space as well, by Lemma 5.13.

However, since the recursive formula contains a sum over all possible types of instability, the result will not look very pleasant and we will not write it down. To resolve the recursive formula is a quite difficult combinatorical problem. This was first solved by Zagier [30], and in the more general situation of G-bundles this was solved by Laumon and Rapoport [25].

In the special case of vector bundles of rank 2 these difficulties disappear, so let us give the result in this simple case.

In order to cope with the formulae, let us introduce the Poincaré series of an algebraic stack $\mathcal{X}$ with finite-dimensional cohomology groups:

$$P(\mathcal{X}, t) := \sum_{i=0}^{\infty} \dim\big(H^i(\mathcal{X})\big) t^i.$$

More generally, one can also use this formula to define $P(H^*, t)$ for any graded algebra H^* such that the graded pieces H^i are finite-dimensional. For example $P(\overline{\mathbb{Q}}_\ell[z], t) = \frac{1}{1 - t^{\deg(z)}}$ and $P(H_1^* \otimes H_2^*, t) = P(H_1^*, t) P(H_2^*, t)$. This implies that

$$P(\text{Bun}_{n,d}, t) = \frac{\prod_{i=1}^n (1 + t^{2i-1})^{2g}}{\prod_{i=1}^n (1 - t^{2i}) \prod_{i=1}^{n-1} (1 - t^{2i})}.$$

Theorem 5.16. *The Poincaré series of the moduli stacks of semi-stable bundles of rank 2 are:*

$$P(\text{Bun}^{\text{ss}}_{2,d}, t) = \begin{cases} \frac{(1+t)^{2g}}{(1-t^2)^2(1-t^4)}\big((1+t^3)^{2g} - t^{2g}(1+t)^{2g}\big), & \text{if } d \text{ is odd} \\ \frac{(1+t)^{2g}}{(1-t^2)^2(1-t^4)}\big((1+t^3)^{2g} - t^{2g+2}(1+t)^{2g}\big), & \text{if } d \text{ is even} \end{cases}.$$

For odd d we have:

$$P(M_2^{d,\text{stable}}, t) = \frac{(1+t)^{2g}}{(1-t^2)(1-t^4)}\big((1+t^3)^{2g} - t^{2g}(1+t)^{2g}\big).$$

Remark 5.17. It is a nice exercise to check these formulae for $g = 0$ using an explicit description of $\text{Bun}^{\text{ss}}_{2,d}$.

Proof. Remark 5.15 implies that

$$P(\text{Filt}^{(i,d-i)}_{(1,2)}, t) = P(\text{Bun}^i_1, t) P(\text{Bun}^{d-i}_1, t) = \frac{((1+t)^{2g})^2}{(1-t^2)^2},$$

which is independent of i and d.

To compute the codimension of a HN-stratum, we recall that the fibre of the normal bundle to the stratum $\mathrm{Bun}_{2,d}^{(1,i)}$ at $\mathcal{E} \in \mathrm{Bun}_{2,d}^{(1,i)}$ is $H^1(C, \mathcal{E}nd(\mathcal{E})/\mathcal{E}nd(\mathcal{E}_\bullet))$. By the Riemann–Roch theorem, the dimension of this vector space is $-((d-2i)+1-g) = g-1+2i-d$. Now we apply Corollary 5.14:

$$P(\mathrm{Bun}_{n,d}^{\mathrm{ss}}, t) = P(\mathrm{Bun}_{n,d}, t) - \sum_{i > \frac{d}{2}} t^{2(g-1+2i-d)} P(\mathrm{Filt}_{(1,2)}^{(i,d-i)}, t)$$

$$= P(\mathrm{Bun}_{n,d}, t) - P(\mathrm{Bun}_1^0, t)^2 t^{2g-2} \sum_{i > \frac{d}{2}} t^{2(2i-d)}.$$

Thus for odd d we find:

$$P(\mathrm{Bun}_{n,d}^{\mathrm{ss}}, t) = \frac{(1+t)^{2g}(1+t^3)^{2g}}{(1-t^2)^2(1-t^4)} - \frac{((1+t)^{2g})^2}{(1-t^2)^2} \frac{t^{2g}}{1-t^4},$$

and for even d we have:

$$P(\mathrm{Bun}_{n,d}^{\mathrm{ss}}, t) = \frac{(1+t)^{2g}(1+t^3)^{2g}}{(1-t^2)^2(1-t^4)} - \frac{((1+t)^{2g})^2}{(1-t^2)^2} \frac{t^{2g+2}}{1-t^4}.$$

To deduce the statement for the coarse moduli space we note that for $(2,d) = 1$ we have seen that $\mathrm{Bun}_{2,d}^{\mathrm{ss}} = M_2^{d,\mathrm{stable}} \times B\mathbb{G}_m$ (Corollary 3.12 and Lemma 3.10). We also know $P(B\mathbb{G}_m, t) = P(\overline{\mathbb{Q}}_\ell[c_1], t) = \frac{1}{1-t^2}$. This proves the theorem. □

Remark 5.18. Instead of the Poincaré polynomial, one can also consider much more refined invariants of the moduli space and the moduli stack. This was suggested by Behrend and Dhillon [7], building on earlier work [4], [8] and [12].

Exercise 5.19. In the lecture we used the following construction: Let $E_0 \to E_1$ be a map of vector bundles on a stack $\mathcal{M}$, then the quotient stack $[E_1/E_0]$ is an algebraic stack. Prove this by giving a presentation.

Now if $E_0' \to E_1'$ is another map of vector bundles and $(E_0 \to E_1) \to (E_0' \to E_1')$ a morphism of complexes which is a quasi-isomorphism, then this map induces an isomorphism $[E_1/E_0] \to [E_1'/E_0']$.

Acknowledgement

First of all I would like to thank the organizers of the workshops for the opportunity to give the lectures and A. Schmitt in particular for his encouragement leading to this write-up. I thank N. Hoffmann for many discussions around stacks. I thank H. Esnault and E. Viehweg for their generous support and the unique atmosphere of their group.

References

[1] M. Artin, *Versal deformations and algebraic stacks*, Invent. Math. **27** (1974), 165–189.

[2] M.F. Atiyah, R. Bott, *The Yang-Mills equations over Riemann surfaces*, Phil. Trans. R. Soc. Lond. A **308** (1983), 523–615.

[3] A. Beauville, *Sur la cohomologie de certains espaces de modules de fibrés vectoriels* in *Geometry and Analysis* (Bombay, 1992), 37–40, Tata Inst. Fund. Res., Bombay, 1995.

[4] K.A. Behrend, *The Lefschetz Trace Formula for the Moduli Space of Principal Bundles*, PhD thesis, Berkeley, 1990, 96 pp., available at http://www.math.ubc.ca/~behrend/thesis.html.

[5] K.A. Behrend, *The Lefschetz trace formula for algebraic stacks*, Invent. Math. **112** (1993), 127–49.

[6] K.A. Behrend, *Derived ℓ-adic categories for algebraic stacks*, Mem. Amer. Math. Soc. **163** (2003), no. 774, viii+93 pp.

[7] K.A. Behrend, A. Dhillon, *On the motive of the stack of bundles*, Adv. Math. **212** (2007), 617–644.

[8] K.A. Behrend, A. Dhillon, *Connected components of moduli stacks of torsors via Tamagawa numbers*, Can. J. Math. **61** (2009), 3–28.

[9] E. Bifet, F. Ghione, M. Letizia, *On the Abel–Jacobi map for divisors of higher rank on a curve*, Math. Ann. **299** (1994), 641–672.

[10] I. Biswas, N. Hoffmann, *The line bundles on moduli stacks of principal bundles on a curve*, preprint, arXiv:0805.2915.

[11] P. Deligne, *Cohomologie Étale (SGA 4½)*, Lecture Notes in Mathematics **569**, Springer Verlag 1977.

[12] A. Dhillon, *On the cohomology of moduli of vector bundles and the Tamagawa number of* SL_n, Can. J. Math. **58** (2006), 1000–1025.

[13] J.-M. Drezet, M.S. Narasimhan, *Groupe de Picard des variétés de modules de fibrés semi-stables sur les courbes algébriques*, Invent. Math. **97** (1989), 53–94.

[14] B. Fantechi, *Stacks for everybody*, European Congress of Mathematics, Vol. I (Barcelona, 2000), 349–359, Progr. Math., 201, Birkhäuser, Basel, 2001.

[15] E.M. Friedlander, *Étale homotopy of simplicial schemes*, Annals of Mathematics Studies, 104, Princeton University Press, Princeton, N.J., University of Tokyo Press, Tokyo, 1982, vii+190 pp.

[16] R. Friedman, J.W. Morgan, *On the converse to a theorem of Atiyah and Bott*, J. Algebr. Geom. **11** (2002), 257–292.

[17] W. Fulton, *Intersection theory*, second edition, Springer Verlag, 1998.

[18] T. Gómez, *Algebraic stacks*, Proc. Indian Acad. Sci. Math. Sci. **111** (2001), no. 1, 1–31.

[19] A. Grothendieck et. al., *Revêtements étales et groupe fondamental (SGA 1)*, Documents Mathématiques (Paris), 3, Société Mathématique de France, Paris, 2003, xviii+327 pp.

[20] G. Harder, M.S. Narasimhan, *On the cohomology groups of moduli spaces of vector bundles on curves*, Math. Ann. **212** (1975), 215–48.

[21] G. Hein, *Faltings' construction of the moduli space*, this volume.

[22] J. Heinloth, *Über den Modulstack der Vektorbündel auf Kurven*, Diploma thesis, Bonn, 1998, 64 pp, available at http://www.uni-essen.de/~hm0002/.

[23] J. Heinloth, A.H.W. Schmitt, *The Cohomology Ring of Moduli Stacks of Principal Bundles over Curves*, preprint, available at http://www.uni-essen.de/~hm0002/.

[24] G. Laumon, L. Moret-Bailly, *Champs algébriques*, Ergebnisse der Mathematik und ihrer Grenzgebiete, 3. Folge, 39, Springer-Verlag, Berlin, 2000, xii+208 pp.

[25] G. Laumon, M. Rapoport, *The Langlands lemma and the Betti numbers of stacks of G-bundles on a curve*, Intern. J. Math. **7** (1996), 29–45.

[26] M. Lieblich, *Moduli of twisted sheaves*, Duke Math. J. **138** (2007), no. 1, 23–118.

[27] M. Olsson, *Sheaves on Artin stacks*, J. reine angew. Math. **603** (2007), 55–112.

[28] S. Ramanan, *The moduli space of vector bundles on an algebraic curve*, Math. Ann. **200** (1973), 69–84.

[29] C.S. Seshadri, *Fibrés vectoriels sur les courbes algébriques*, Astérisque **96** (1982).

[30] D. Zagier, *Elementary aspects of the Verlinde formula and of the Harder–Narasimhan–Atiyah–Bott formula*, Proceedings of the Hirzebruch 65 Conference on Algebraic Geometry (Ramat Gan, 1993), Israel Math. Conf. Proc., 9, Bar-Ilan Univ., Ramat Gan, 1996, 445–462.

Jochen Heinloth
University of Amsterdam
Korteweg-de Vries Institute for Mathematics
Plantage Muidergracht 24
NL-1018 TV Amsterdam, The Netherlands
e-mail: J.Heinloth@uva.nl

Affine Flag Manifolds and Principal Bundles
Trends in Mathematics, 155–163
© 2010 Springer Basel AG

On Moduli Stacks of G-bundles over a Curve

Norbert Hoffmann

Abstract. Let C be a smooth projective curve over an algebraically closed field k of arbitrary characteristic. Given a linear algebraic group G over k, let $\mathcal{M}_G$ be the moduli stack of principal G-bundles on C. We determine the set of connected components $\pi_0(\mathcal{M}_G)$ for smooth connected groups G.

Mathematics Subject Classification (2000). 14D20, 14F05.

Keywords. Principal bundle, algebraic curve, moduli stack.

1. Introduction

Let C be a smooth projective algebraic curve over an algebraically closed field k. This text explains some basic properties of the moduli stack $\mathcal{M}_G$ of algebraic principal G-bundles on C, for a linear algebraic group G over k. The arguments given are purely algebraic, and valid in any characteristic.

The stack $\mathcal{M}_G$ is algebraic in the sense of Artin, and locally of finite type over k. Moreover, $\mathcal{M}_G$ is smooth if G is smooth. The main purpose of this paper is to determine the set of connected components $\pi_0(\mathcal{M}_G)$ if G is smooth and connected. It turns out that the unipotent radical of G doesn't matter for this. In the case where G is reductive, Theorem 5.8 gives a canonical bijection between $\pi_0(\mathcal{M}_G)$ and the fundamental group $\pi_1(G)$, the latter being defined in terms of the root system; cf. Definition 5.4.

This statement is well established folklore, and thus not a new result. But the published literature seems to contain no proof of it in full generality, covering also the case of positive characteristic $\mathrm{char}(k) = p > 0$. For simply connected G, the result is proved in [6]; the general case is treated, from a different point of view, in the apparently unpublished preprint [11].

The proof given here is based on the maps $\mathcal{M}_G \to \mathcal{M}_H$ induced by group homomorphisms $G \to H$. In particular, it uses criteria for lifting H-bundles to G-bundles if H is a quotient of G. Corollary 3.4 states that this is always possible

The author was supported by the SFB 647: Raum – Zeit – Materie.

if G, H, and the kernel are smooth and connected; this little observation might be of independent interest.

After recalling the algebraicity of $\mathcal{M}_G$ in Section 2, these lifting problems are studied in Section 3. Based on them, the standard deformation theory argument for smoothness of $\mathcal{M}_G$ is recalled in Section 4. Finally, Section 5 contains the results mentioned above about connected components of $\mathcal{M}_G$.

2. Algebraicity

Throughout this text, we fix an algebraically closed base field k and an irreducible smooth projective curve C/k. We denote by $\mathcal{M}_G$ the moduli stack of principal G-bundles E on C, where $G \subseteq \mathrm{GL}_n$ is a linear algebraic group.

Remark 2.1. More precisely, $\mathcal{M}_G$ is given as a prestack over k by the groupoid $\mathcal{M}_G(S)$ of principal G-bundles on $C \times_k S$ for each k-scheme S. This prestack is indeed a stack: the required descent for G-bundles is a special case of the standard descent for affine morphisms since G is affine.

Remark 2.2. More generally, one can consider the moduli stack $\mathcal{M}_{\mathcal{G}}$ of principal bundles under a relatively affine group scheme $\mathcal{G}$ over C. We will use only the special case where $\mathcal{G} = V$ is (the underlying additive group scheme of) a vector bundle on C. Here principal V-bundles correspond to vector bundle extensions

$$0 \longrightarrow V \longrightarrow E \longrightarrow \mathcal{O}_C \longrightarrow 0,$$

so their moduli stack $\mathcal{M}_V$ is the stack quotient of the affine space $\mathrm{H}^1_{\mathrm{Zar}}(C, V)$ modulo the trivial action of the additive group $\mathrm{H}^0_{\mathrm{Zar}}(C, V)$. In particular, we see that $\mathcal{M}_V$ is a smooth connected Artin stack in this case.

Given a morphism of linear algebraic groups $\phi \colon G \to H$, extending the structure group of principal G-bundles to H defines a 1-morphism

$$\phi_* \colon \mathcal{M}_G \longrightarrow \mathcal{M}_H.$$

Fact 2.3. *If $\iota \colon H \hookrightarrow G$ is a closed embedding, then the 1-morphism of stacks $\iota_* \colon \mathcal{M}_H \to \mathcal{M}_G$ is representable and locally of finite type.*

Proof. (cf. [15, 3.6.7]) The homogeneous space G/H exists by Chevalley's theorem [5, III, §3, Théorème 5.4]; more precisely, G is a principal H-bundle over some quasiprojective variety $X = G/H$. Given a principal G-bundle $\pi \colon E \to C \times_k S$, reductions of its structure group to H correspond bijectively to sections of the associated bundle $\pi_X \colon E \times^G X \to C \times_k S$ with fiber X.

This means that the fiber product of S and $\mathcal{M}_H$ over $\mathcal{M}_G$ is the functor from S-schemes to sets that sends $f \colon T \to S$ to the sections of $f^* \pi_X$. This functor is representable by some locally closed subscheme of an appropriate relative Hilbert scheme, which is locally of finite type over S.   □

By an *algebraic stack* over k, we always mean an Artin stack that is locally of finite type over k (but not necessarily quasi-compact). For example, the moduli stack $\mathcal{M}_V$ for a vector bundle V on C is algebraic, according to Remark 2.2.

Fact 2.4. *If G is a linear algebraic group, then $\mathcal{M}_G$ is an algebraic stack.*

Proof. (cf. [15, 3.6.6.]) In the case $G = \mathrm{GL}_n$, this is well known, cf. [12, 4.14.2.1]. The general case $G \hookrightarrow \mathrm{GL}_n$ then follows from the previous fact. $\square$

3. Lifting principal bundles

We say that a short sequence of linear algebraic groups

$$1 \longrightarrow K \longrightarrow G \overset{\pi}{\longrightarrow} H \longrightarrow 1 \tag{3.1}$$

is *exact* if π is faithfully flat and K is the kernel of π. Then H acts on K by conjugation in G. Given a principal H-bundle F on C, we denote by

$$K^F := K \times^H F := (K \times F)/H$$

the corresponding twisted group scheme over C with fiber K.

Proposition 3.1. *Suppose that (3.1) is a short exact sequence of linear algebraic groups, with K commutative. Let F be a principal H-bundle on C.*

   i) *There is a canonical obstruction class $\mathrm{ob}_F \in \mathrm{H}^2_{\mathrm{fppf}}(C, K^F)$, which vanishes if and only if $F \cong \pi_* E$ for some principal G-bundle E on C.*
   ii) *If ob_F vanishes, then the fiber of $\pi_* \colon \mathcal{M}_G \to \mathcal{M}_H$ over the point F is 1-isomorphic to the moduli stack $\mathcal{M}_{K^F}$ of principal K^F-bundles.*

$$\begin{array}{ccc} \mathcal{M}_{K^F} & \longrightarrow & \mathcal{M}_G \\ \downarrow & & \downarrow{\scriptstyle\pi_*} \\ \mathrm{Spec}(k) & \overset{F}{\longrightarrow} & \mathcal{M}_H. \end{array}$$

Proof. The lifts of F to G-bundles E form a stack $\mathcal{K}_F$ over C, which is more precisely given by the following groupoid $\mathcal{K}_F(X)$ for each C-scheme $f \colon X \to C$:

  • Its objects are principal G-bundles $\mathcal{E}$ on X together with isomorphisms $\pi_*(\mathcal{E}) \cong f^*(F)$ of principal H-bundles on X.
  • Its morphisms are isomorphisms of principal G-bundles on X which are compatible with the identity on $f^*(F)$.

If F is trivial, then a lift of F to a principal G-bundle is nothing but a principal K-bundle, so $\mathcal{K}_F$ is just the classifying stack $BK \times C$ in this case. In any case, F is fppf-locally trivial, so $\mathcal{K}_F$ is an fppf-gerbe over C, whose band is the common automorphism group scheme K^F of all (local) lifts of F. The class of this gerbe in $\mathrm{H}^2_{\mathrm{fppf}}(C, K^F)$ is the required obstruction ob_F; cf. [7, IV, Théorème 3.4.2].

If ob_F vanishes, then the gerbe $\mathcal{K}_F \to C$ admits a section, so $\mathcal{K}_F$ is the classifying stack $B(K^F)$ over C by [12, Lemme 3.21]. Thus sections $C \to \mathcal{K}_F$ are nothing but principal K^F-bundles on C; this implies ii). $\square$

Remark 3.2. In the above situation, suppose that K is central in G. Given a principal G-bundle E with $\pi_* E \cong F$, we can explicitly describe a 1-isomorphism between $\mathcal{M}_{K^F} = \mathcal{M}_K$ and the fiber of π_* over $[F]$ as follows:

The multiplication $\mu \colon K \times G \to G$ is a group homomorphism, so it induces a 1-morphism $\mu_* \colon \mathcal{M}_K \times \mathcal{M}_G \to \mathcal{M}_G$. Its restriction $\mu_*(\_, [E]) \colon \mathcal{M}_K \to \mathcal{M}_G$ is then a 1-isomorphism onto the fiber of π_* over $[F]$.

Remark 3.3. Up to now, we have not used the assumption $\dim(C) = 1$. Using it, one can show that the obstruction ob_F vanishes in the following two cases:

i) Assume $K \cong \mathbb{G}_a^r$, and that the action $H \to \mathrm{Aut}(K)$ factors through GL_r. (The latter is automatic for $K \cong \mathbb{G}_a$, since $\mathrm{Aut}(K) \cong \mathbb{G}_m$ in this situation. But for $r > 1$ and $\mathrm{char}(k) = p > 0$, this is actually a condition.) Then K^F is a vector bundle on C, and

$$\mathrm{H}^2_{\mathrm{fppf}}(C, K^F) = \mathrm{H}^2_{\mathrm{\acute{e}t}}(C, K^F) = \mathrm{H}^2_{\mathrm{Zar}}(C, K^F) = 0$$

due to [8, Théorème 11.7], [10, Exposé VII, Proposition 4.3], and the assumption $\dim(C) = 1$.

ii) Assume $K \cong \mathbb{G}_m^r$, and that H is connected. Then $\mathrm{Aut}(K) \cong \mathrm{GL}_r(\mathbb{Z})$ is discrete, so the action of H on K is trivial. Thus K^F is just the split torus $\mathbb{G}_m^r$ over C, and $\mathrm{H}^2_{\mathrm{fppf}}(C, K^F) = \mathrm{H}^2_{\mathrm{\acute{e}t}}(C, K^F) = 0$ by Tsen's theorem.

Corollary 3.4. *If* $1 \to K \longrightarrow G \overset{\pi}{\longrightarrow} H \to 1$ *is a short exact sequence of smooth connected linear algebraic groups, then* $\pi_* \colon \mathcal{M}_G \to \mathcal{M}_H$ *is surjective.*

Proof. Choose a Borel subgroup B_G in G. Then $B_H := \pi(B_G)$ is a Borel subgroup in H due to [3, Proposition (11.14)]. Every principal H-bundle F on C admits a reduction of its structure group to B_H by [6, Theorem 1 and Remark 2.e].

The identity component $B_K^0 \subseteq B_K$ of the intersection $B_K := K \cap B_G$ is a Borel subgroup in K due to [3, Proposition (11.14)] again. As B_K^0 is normal in B_K, it follows that B_K is contained in the normalizer of B_K^0 in K, which is just B_K^0 itself by [3, Theorem (11.15)]. Thus $B_K^0 = B_K$, and the sequence $1 \to B_K \to B_G \to B_H \to 1$ is again exact. Replacing the given exact sequence by this one, we may assume without loss of generality that the three groups G, H and K are all solvable.

Using induction on $\dim(K)$, we may then assume $\dim(K) = 1$, which means $K \cong \mathbb{G}_a$ or $K \cong \mathbb{G}_m$. In this situation, the obstruction against lifting principal H-bundles on C to principal G-bundles vanishes by Remark 3.3. This shows that the induced 1-morphism $\pi_* \colon \mathcal{M}_G \to \mathcal{M}_H$ is indeed surjective. $\square$

4. Smoothness

From now on, we will concentrate on *smooth* linear algebraic groups G over k. Then every principal G-bundle is étale-locally trivial.

Proposition 4.1. *If the group* G *is smooth, then the stack* $\mathcal{M}_G$ *is also smooth.*

Proof. (See [1, 4.5.1 and 8.1.9] for a different presentation of similar arguments.) We verify that $\mathcal{M}_G$ satisfies the infinitesimal criterion for smoothness.

Let a pair $(A, \mathfrak{m})$ and $(\tilde{A}, \tilde{\mathfrak{m}})$ of local artinian k-algebras with residue field k be given, such that $A = \tilde{A}/(\nu)$ for some $\nu \in \tilde{A}$ with $\tilde{\mathfrak{m}} \cdot \nu = 0$. We have to show that every principal G-bundle $\mathcal{E}$ on $C \otimes_k A$ can be extended to $C \otimes_k \tilde{A}$.

We define a functor G_A from k-schemes to groups by $G_A(S) := G(S \otimes_k A)$. Then G_A is a smooth linear algebraic group, and the infinitesimal theory of group schemes [5, II, §4, Théorème 3.5] yields an exact sequence

$$1 \longrightarrow \mathfrak{g} \longrightarrow G_{\tilde{A}} \longrightarrow G_A \longrightarrow 1$$

where $\mathfrak{g}$ is (the underlying additive group of) the Lie algebra of G.

As C and $C \otimes_k A$ are homeomorphic for the étale topology, the étale-locally trivial principal G-bundle $\mathcal{E}$ on $C \otimes_k A$ corresponds to a principal G_A-bundle $\mathcal{E}$ on C. Using Proposition 3.1 and Remark 3.3.i, we can lift this G_A-bundle to a principal $G_{\tilde{A}}$-bundle on C. This yields the required G-bundle on $C \otimes_k \tilde{A}$. $\square$

Corollary 4.2. *If* $1 \to K \longrightarrow G \xrightarrow{\pi} H \to 1$ *is a short exact sequence of smooth linear algebraic groups, then* $\pi_* \colon \mathcal{M}_G \to \mathcal{M}_H$ *is also smooth.*

Proof. We know already that $\mathcal{M}_G$ and $\mathcal{M}_H$ are smooth over k, so it suffices to show that the 1-morphism $\pi_* \colon \mathcal{M}_G \to \mathcal{M}_H$ is submersive.

Let E be a principal G-bundle on C, with induced H-bundle $F := \pi_*(E)$. Given an extension of F to a principal H-bundle $\mathcal{F}$ on $C \otimes_k k[\varepsilon]$ with $\varepsilon^2 = 0$, we have to extend E to a principal G-bundle $\mathcal{E}$ on $C \otimes_k k[\varepsilon]$ such that the identity $\pi_*(E) = F$ can be extended to an isomorphism $\pi_*(\mathcal{E}) \cong \mathcal{F}$.

The given datum $(E, F, \mathcal{F})$ corresponds to a principal $(G \times_H H_{k[\varepsilon]})$-bundle on C. Using the exact sequence of groups

$$1 \longrightarrow \mathfrak{k} := \mathrm{Lie}(K) \longrightarrow G_{k[\varepsilon]} \longrightarrow G \times_H H_{k[\varepsilon]} \longrightarrow 1,$$

we can lift it to a principal $G_{k[\varepsilon]}$-bundle on C, according to Proposition 3.1 and Remark 3.3.i. This extends E to a G-bundle $\mathcal{E}$ on $C \otimes_k k[\varepsilon]$, as required. $\square$

5. Connected components

In this section, we suppose that the linear algebraic group G is smooth and connected. The aim is to describe the set of connected components $\pi_0(\mathcal{M}_G)$.

Proposition 5.1. *If* $1 \to U \to G \to H \to 1$ *is a short exact sequence of smooth connected linear algebraic groups with U unipotent, then* $\pi_0(\mathcal{M}_G) = \pi_0(\mathcal{M}_H)$.

Proof. The induced 1-morphism $\mathcal{M}_G \to \mathcal{M}_H$ is smooth by Corollary 4.2, and surjective by Corollary 3.4. We have to show that its fibers are connected.

Let $B_H \subseteq H$ be a Borel subgroup. Every principal H-bundle on C admits a reduction of its structure group to B_H by [6, Theorem 1 and Remark 2.e]. Replacing H by B_H and G by the inverse image B_G of B_H if necessary, we may thus assume that G and H are solvable.

Using induction on $\dim(U)$, we may then moreover assume $U \cong \mathbb{G}_a$. In this situation, the fibers in question have the form $\mathcal{M}_L$ for line bundles L on C, according to Proposition 3.1.ii); see also Remark 3.3.i). Hence these fibers are connected due to Remark 2.2.   □

In particular, $\pi_0(\mathcal{M}_G) = \pi_0(\mathcal{M}_{G/G_u})$, where $G_u \subseteq G$ denotes the unipotent radical. Thus it suffices to determine the set $\pi_0(\mathcal{M}_G)$ for reductive groups G.

Given any torus $T \cong \mathbb{G}_m^r$ over k, we denote its cocharacter lattice by

$$X_*(T) := \mathrm{Hom}(\mathbb{G}_m, T) \cong \mathbb{Z}^r.$$

Sending line bundles to their degree defines a bijection $\pi_0(\mathcal{M}_{\mathbb{G}_m}) \xrightarrow{\sim} \mathbb{Z}$, since the Jacobian $\mathrm{Pic}^0(C)$ is connected. Thus we obtain an induced canonical bijection

$$\pi_0(\mathcal{M}_T) \xrightarrow{\sim} X_*(T).$$

If T appears in a central extension of smooth connected linear algebraic groups

$$1 \longrightarrow T \longrightarrow G \xrightarrow{\pi} H \longrightarrow 1,$$

then the multiplication $\mu \colon T \times G \to G$ is a group homomorphism, and

$$\mu_* \colon \pi_0(\mathcal{M}_T) \times \pi_0(\mathcal{M}_G) \longrightarrow \pi_0(\mathcal{M}_G)$$

is an action of the group $\pi_0(\mathcal{M}_T)$ on the set $\pi_0(\mathcal{M}_G)$.

Remark 5.2. Actually the group stack $\mathcal{M}_T$ acts on $\mathcal{M}_G$, and $\pi_* \colon \mathcal{M}_G \to \mathcal{M}_H$ is a torsor under this action; see [2, Section 5.1]. But we won't use these stack notions here, since all we need can readily be said in more elementary language.

Proposition 5.3. *In the above situation, $\pi_0(\mathcal{M}_H) = \pi_0(\mathcal{M}_G)/\pi_0(\mathcal{M}_T)$.*

Proof. The induced 1-morphism $\pi_* \colon \mathcal{M}_G \to \mathcal{M}_H$ is surjective by Corollary 3.4, and smooth by Corollary 4.2. In particular, π_* is open; its fibers are all isomorphic to $\mathcal{M}_T$ by Proposition 3.1.ii). These properties imply the proposition as follows:

Since π_* is surjective, it induces a surjective map $\pi_0(\mathcal{M}_G) \to \pi_0(\mathcal{M}_H)$. As it is invariant under the action of $\pi_0(\mathcal{M}_T)$, it descends to a surjective map

$$\pi_0(\mathcal{M}_G)/\pi_0(\mathcal{M}_T) \longrightarrow \pi_0(\mathcal{M}_H).$$

To check that this map is also injective, let $\pi_0(\mathcal{M}_G) = \coprod_i X_i$ be the decomposition into $\pi_0(\mathcal{M}_T)$-orbits. It correspond to a decomposition $\mathcal{M}_G = \coprod_i \mathcal{U}_i$ into open substacks. Due to Remark 3.2, each fiber of π_* is contained in a single $\mathcal{U}_i$, so the images $\pi_*(\mathcal{U}_i) \subseteq \mathcal{M}_H$ are still disjoint. As π_* is open, $\pi_*(\mathcal{U}_i)$ is open in $\mathcal{M}_H$. They form a decomposition of $\mathcal{M}_H$, since π_* is surjective. Hence different $\pi_0(\mathcal{M}_T)$-orbits in $\pi_0(\mathcal{M}_G)$ map to different components of $\mathcal{M}_H$.   □

Now suppose that the smooth and connected linear algebraic group G over k is reductive. Choosing a maximal torus $T_G \subseteq G$, let

$$X_{\mathrm{coroots}} \subseteq X_*(T_G)$$

denote the subgroup generated by the coroots of G.

Definition 5.4. The fundamental group of G is $\pi_1(G) := X_*(T_G)/X_{\text{coroots}}$.

Note that the Weyl group of (G, T_G) acts trivially on $\pi_1(G)$. Hence this fundamental group does not depend on the choice of the maximal torus T_G, up to a *canonical* isomorphism. G is called *simply connected* if $\pi_1(G)$ is trivial.

Remark 5.5. If $k = \mathbb{C}$, then $\pi_1(G)$ coincides with the usual topological fundamental group $\pi_1^{\text{top}}(G)$ of G as a complex Lie group. If more generally $\text{char}(k) = 0$, then $\pi_1(G)$ coincides with $\pi_1^{\text{top}}(G \otimes_k \mathbb{C})$ for every embedding $k \hookrightarrow \mathbb{C}$.

Remark 5.6. i) Due to [4], each finite quotient $\pi_1(G) \twoheadrightarrow \mathbb{Z}/n_1 \times \cdots \times \mathbb{Z}/n_r$ corresponds to a central isogeny $\widetilde{G} \twoheadrightarrow G$. Its kernel is isomorphic to $\mu_{n_1} \times \cdots \times \mu_{n_r}$.

ii) In particular, étale isogenies $\widetilde{G} \twoheadrightarrow G$ correspond to finite quotients of $\pi_1(G)$ whose order is not divisible by the characteristic of k.

iii) If G is semisimple, then $\pi_1(G)$ itself is finite. The corresponding central isogeny $\widetilde{G} \twoheadrightarrow G$ is called the *universal covering* of G.

Remark 5.7. i) Denote by $\pi_1^{\text{ét}}(G)$ the étale fundamental group of G, and by $\hat{\pi}_1(G)$ the profinite completion of $\pi_1(G)$. Let $\pi_1^{\text{ét}}(G) \twoheadrightarrow \pi_1^{\text{ét}}(G)'$ and $\hat{\pi}_1(G) \twoheadrightarrow \hat{\pi}_1(G)'$ be identities if $\text{char}(k) = 0$, and the largest prime-to-p quotients if $\text{char}(k) = p > 0$. Then Remark 5.6.ii) implies that $\pi_1^{\text{ét}}(G)'$ is canonically isomorphic to $\hat{\pi}_1(G)'$.

To verify this, one has to show, for every connected scheme X together with a finite étale morphism $\pi \colon X \to G$ such that $\deg(\pi)$ is not divisible by $\text{char}(k)$, that there is a group structure on X such that π is an isogeny. This can be checked like the analogous statement in topology, using the Künneth formula

$$\pi_1^{\text{ét}}(G \times G)' = \pi_1^{\text{ét}}(G)' \times \pi_1^{\text{ét}}(G)'$$

proved in [9, Exposé XIII, Proposition 4.6] and [13, Proposition 4.7].

ii) Suppose $\text{char}(k) = p > 0$. Then each finite quotient of $\pi_1(G)$ which is a p-group corresponds to a purely inseparable central isogeny $\widetilde{G} \twoheadrightarrow G$. On the other hand, the p-part of $\pi_1^{\text{ét}}(G)$ is huge and in particular non-abelian; cf. for example [14]. Thus the p-parts of $\hat{\pi}_1(G)$ and of $\pi_1^{\text{ét}}(G)$ don't seem to be related.

Theorem 5.8. *If the linear algebraic group G over k is smooth, connected, and reductive, then one has a canonical bijection $\pi_0(\mathcal{M}_G) \cong \pi_1(G)$.*

Proof. We partly follow [6, Proposition 5], where the connectedness of $\mathcal{M}_G$ for simply connected G is proved. Another reference is [11, Proposition 3.15].

Let $B_G \subseteq G$ be a Borel subgroup containing the maximal torus T_G. Then $\pi_0(\mathcal{M}_{T_G}) = \pi_0(\mathcal{M}_{B_G})$ by Proposition 5.1. The inclusion $B_G \hookrightarrow G$ induces a map

$$X_*(T_G) = \pi_0(\mathcal{M}_{T_G}) = \pi_0(\mathcal{M}_{B_G}) \longrightarrow \pi_0(\mathcal{M}_G). \tag{5.1}$$

This map is surjective, because every principal G-bundle on C admits a reduction of its structure group to B_G by [6, Theorem 1 and Remark 2.e].

We claim that this map (5.1) is constant on cosets modulo X_{coroots}. Given a coroot $\alpha \in X_*(T_G)$ of G and a cocharacter $\delta \in X_*(T_G)$, it suffices to show that δ and $\delta + \alpha$ have the same image in $\pi_0(\mathcal{M}_G)$. As the inclusion $T_G \hookrightarrow G$ factors through the subgroup of semisimple rank one $G_\alpha \subseteq G$ given by α, we may assume without loss of generality that G has semisimple rank one. Splitting off any direct factor $\mathbb{G}_m$ of G reduces us to the cases $G \cong \mathrm{SL}_2$, $G \cong \mathrm{GL}_2$, or $G \cong \mathrm{PGL}_2$.

To deal with these three cases, we choose a closed point $P \in C(k)$. Let L and L' be invertible sheaves on C; in the case $G \cong \mathrm{SL}_2$, we assume $L \otimes L' \cong \mathcal{O}_C(P)$. For every line ℓ in the two-dimensional vector space $L_P \oplus L'_P$, its inverse image subsheaf $E_\ell \subseteq L \oplus L'$ defines a G-bundle on C; thus we obtain a $\mathbb{P}^1$-family of G-bundles on C. This family connects the two G-bundles defined by $L(-P) \oplus L'$ and by $L \oplus L'(-P)$, which come from the maximal torus $T_G \subseteq G$. Thus we see that the elements δ and $\delta + \alpha$ of $X_*(T_G) = \pi_0(\mathcal{M}_{T_G})$ indeed have the same image in $\pi_0(\mathcal{M}_G)$. Hence the above map (5.1) descends to a surjective map

$$\pi_1(G) = X_*(T_G)/X_{\mathrm{coroots}} \longrightarrow \pi_0(\mathcal{M}_G). \tag{5.2}$$

Note that this map does not depend on the choice of the maximal torus $T_G \subseteq G$. Thus it is functorial in G, in the sense that the diagram

$$
\begin{array}{ccc}
\pi_1(G) & \longrightarrow & \pi_0(\mathcal{M}_G) \\
\varphi_* \downarrow & & \downarrow \varphi_* \\
\pi_1(H) & \longrightarrow & \pi_0(\mathcal{M}_H)
\end{array}
$$

commutes for every homomorphism $\varphi \colon G \to H$ of smooth, connected, reductive algebraic groups.

Finally, we have to show that this canonical map (5.2) is injective. We first consider the case where the commutator subgroup $[G, G] \subseteq G$ is simply connected. Then $\pi_1(G) = \pi_1(G/[G,G])$, so the required injectivity for G follows by functoriality from the already verified injectivity for the torus $G/[G,G]$.

Next we consider the case where G is semisimple, so $\pi_1(G)$ is finite. Let μ be the kernel of the universal covering $\widetilde{G} \twoheadrightarrow G$. We choose an embedding $\mu \hookrightarrow T$ into a torus T, and denote by $\widehat{G}$ the pushout of linear algebraic groups

$$
\begin{array}{ccc}
\mu & \longrightarrow & \widetilde{G} \\
\downarrow & & \downarrow \\
T & \longrightarrow & \widehat{G}.
\end{array}
$$

By construction, $\widehat{G}$ is smooth, connected, reductive, and $[\widehat{G}, \widehat{G}] = \widetilde{G}$ is simply connected. Moreover, we have an exact sequence

$$1 \longrightarrow T \longrightarrow \widehat{G} \longrightarrow G \longrightarrow 1.$$

Using Proposition 5.3, the injectivity for G follows from the injectivity for $\widehat{G}$, which has already been proved in the previous case.

Finally, we consider the case where G is reductive. If $\pi \colon G \twoheadrightarrow H$ is a central isogeny, then the induced map $\pi_1(G) \to \pi_1(H)$ is injective; hence we may replace G by H without loss of generality. We take $H := G/[G,G] \times G/Z_G$, where $Z_G \subseteq G$

is the center. Splitting off the torus $G/[G, G]$ reduces us to the case where G is of adjoint type. This is covered by the previous case. □

References

[1] K. Behrend. *The Lefschetz Trace Formula for the Moduli Stack of Principal Bundles.* PhD thesis, Berkeley, 1991. http://www.math.ubc.ca/~behrend/thesis.html.

[2] I. Biswas and N. Hoffmann. The Line Bundles on Moduli Stacks of Principal Bundles on a Curve. *Documenta Math.* 15:35–72, 2010.

[3] A. Borel. *Linear algebraic groups.* New York – Amsterdam: W.A. Benjamin, 1969.

[4] C. Chevalley. Les isogénies. Séminaire C. Chevalley 1956–1958: Classification des groupes de Lie algébriques, Exposé 18. Paris: Secrétariat mathématique, 1958.

[5] M. Demazure and P. Gabriel. *Groupes algébriques. Tome* I. Amsterdam: North-Holland Publishing Company, 1970.

[6] V.G. Drinfeld and C. Simpson. *B*-structures on *G*-bundles and local triviality. *Math. Res. Lett.*, 2(6):823–829, 1995.

[7] J. Giraud. *Cohomologie non abelienne.* Grundlehren, Band 179. Berlin-Heidelberg-New York: Springer-Verlag, 1971.

[8] A. Grothendieck. Le groupe de Brauer. III: Exemples et complements. Dix exposés sur la cohomologie des schémas, Advanced Studies Pure Math. 3, 88-188, 1968.

[9] A. Grothendieck et al. *SGA 1: Revêtements étales et groupe fondamental.* Lecture Notes in Mathematics, Vol. 224. Springer-Verlag, Berlin, 1971.

[10] A. Grothendieck et al. *SGA 4: Théorie des topos et cohomologie étale des schémas. Tome* 2. Lecture Notes in Mathematics, Vol. 270. Springer-Verlag, Berlin, 1972.

[11] Y.I. Holla. Parabolic reductions of principal bundles. Preprint math.AG/0204219. Available at http://www.arXiv.org.

[12] G. Laumon and L. Moret-Bailly. *Champs algébriques.* Ergebnisse der Mathematik und ihrer Grenzgebiete. 3. Folge, Band 39. Berlin: Springer, 2000.

[13] F. Orgogozo. Altérations et groupe fondamental premier à *p. Bull. Soc. Math. Fr.*, 131(1):123–147, 2003.

[14] M. Raynaud. Revêtements de la droite affine en caractéristique $p > 0$ et conjecture d'Abhyankar. *Invent. Math.*, 116(1-3):425–462, 1994.

[15] C. Sorger. Lectures on moduli of principal G-bundles over algebraic curves. in: L. Göttsche (ed.), Moduli Spaces in Algebraic Geometry (Trieste, ICTP, 1999), 1-57. Available at http://users.ictp.it/~pub_off/lectures/vol1.html.

Norbert Hoffmann
Mathematisches Institut der Freien Universität
Arnimallee 3
D-14195 Berlin, Germany
e-mail: norbert.hoffmann@fu-berlin.de

Affine Flag Manifolds and Principal Bundles
Trends in Mathematics, 165–202
© 2010 Springer Basel AG

Clifford Indices for Vector Bundles on Curves

Herbert Lange and Peter E. Newstead

Abstract. For smooth projective curves of genus $g \geq 4$, the Clifford index is an important invariant which provides a bound for the dimension of the space of sections of a line bundle. This is the first step in distinguishing curves of the same genus. In this paper we generalise this to introduce Clifford indices for semistable vector bundles on curves. We study these invariants, giving some basic properties and carrying out some computations for small ranks and for general and some special curves. For curves whose classical Clifford index is two, we compute all values of our new Clifford indices.

Mathematics Subject Classification (2000). Primary: 14H60; Secondary: 14F05, 32L10.

Keywords. Semistable vector bundle, Clifford index, gonality, Brill–Noether theory.

1. Introduction

Let C be a smooth projective curve of genus $g \geq 4$ defined over an algebraically closed field of characteristic 0. If L is a line bundle on C with space of sections $H^0(L)$ of dimension $h^0(L) \geq 2$, then evaluation of sections defines a morphism $C \to \mathbb{P}^{(h^0(L)-1)}$. These morphisms yield much information about the geometry of C, in particular about the possible projective embeddings of C and the syzygies resulting from such embeddings. It is therefore important to obtain precise upper bounds on $h^0(L)$ in terms of the degree of L. These upper bounds depend on the curve, not just on the value of g, and a first measure of the possible bounds is given by the Clifford index of C, whose definition runs as follows. We consider line bundles L on C and define the *Clifford index* γ_1 of C by

$$\gamma_1 = \min_{L} \left\{ \deg L - 2(h^0(L) - 1) \mid h^0(L) \geq 2, h^1(L) \geq 2 \right\}$$

Both authors are members of the research group VBAC (Vector Bundles on Algebraic Curves). The second author would like to thank the Department Mathematik der Universität Erlangen–Nürnberg for its hospitality.

or equivalently

$$\gamma_1 = \min_L \left\{ \deg L - 2(h^0(L) - 1) \mid h^0(L) \geq 2, \deg L \leq g - 1 \right\}.$$

(The equivalence of the two definitions follows from Serre duality. The reason for requiring $g \geq 4$ is to ensure the existence of line bundles as in the definition.)

It is natural to generalise this to higher rank and in particular to semistable (or stable) vector bundles. The restriction to semistable bundles is natural as it allows for restrictions on the dimension of the space of sections essentially identical to those which exist for line bundles. In particular Clifford's Theorem has been extended to semistable bundles; the simplest proof of this is due to G. Xiao and appears as [8, Theorem 2.1]. Semistable bundles also arise naturally in the study of syzygies in connection with conjectures of Green and Lazarsfeld [23, 24] (in this context, see [44]). Moreover the moduli spaces of semistable bundles are objects of interest in their own right. We may note that the existence of semistable bundles with specified numbers of sections has recently been used in two papers [25, 13] which obtain new information on the base locus of the generalised theta-divisor; the first of these papers extends results of Arcara [1], Popa [36] and Schneider [39], while the second uses also the strange duality theorem, recently proved in [4, 5, 30].

A key rôle in some of these papers is played by the evaluation sequence

$$0 \to E_L^* \to H^0(L) \otimes \mathcal{O} \to L \to 0, \tag{1.1}$$

where L is a line bundle generated by its sections. For our purposes the key issue is the stability of E_L, which was proved for $L = K_C$ in [35] and for $\deg L \geq 2g + 1$ by Ein and Lazarsfeld [20]. Subsequently David Butler [11, 12] considered more generally the sequence

$$0 \to M_{V,E}^* \to V \otimes \mathcal{O} \to E \to 0, \tag{1.2}$$

where E is a vector bundle and V is a linear subspace of $H^0(E)$ which generates E. (The construction of (1.2) has come to be known as the *dual span construction*.) Butler showed [11, Theorem 1.2] that, if $V = H^0(E)$ and E is semistable (stable) of slope $\geq 2g$, then $M_{V,E}$ is semistable (stable) with a minor exception in the case of stability when the slope is equal to $2g$. Results for line bundles of smaller degree have been obtained in [12, 7, 6, 10]. There is an important conjecture of Butler (which we discuss briefly in the final section) that the bundles $M_{V,E}$ are semistable "in general".

The Brill–Noether locus $B(n, d, k)$ in the moduli space of stable bundles of rank n and degree d on C is comprised of those bundles E for which $h^0(E) \geq k$. These loci have been studied for around 20 years and a good deal is known about their non-emptiness. However almost all the results are either true for any C [8, 9, 32, 33] or for general C only [41, 42, 43]. Precise results are known for hyperelliptic curves [9, Section 6] and for bielliptic curves [3] but little has been done on other special curves. The main exceptions to this are papers of R. Re [38] and V. Mercat [34] and a recent paper by L. Brambila-Paz and A. Ortega [10].

These papers use only the classical Clifford index γ_1 which is defined using line bundles, although in many respects [34] is the starting point for our investigations.

In [2], E. Ballico gave five definitions of Clifford indices for vector bundles but did not develop the concept to any significant extent. We give two definitions, both using semistable bundles (whereas Ballico used indecomposable and stable bundles). Our definitions differ from those of Ballico in other respects as well, in that we do not assume that our bundles are generated by their sections (although in fact most of our examples are so generated) and we consider only bundles whose slope is at most $g - 1$ (whereas Ballico requires only that $h^1(E) \neq 0$). In fact we define, for any vector bundle E of rank n and degree d,

$$\gamma(E) := \frac{1}{n}\left(d - 2(h^0(E) - n)\right) = \mu(E) - 2\frac{h^0(E)}{n} + 2,$$

where $\mu(E) = \frac{d}{n}$. (Ballico defines $\mathrm{Cliff}(E)$ in the same way but without the scaling factor $\frac{1}{n}$.) We then define γ_n to be the minimum value of $\gamma(E)$ for semistable bundles E of rank n with $h^0(E) \geq n + 1$ and $\mu(E) \leq g - 1$. Our second index γ'_n is defined similarly, but with $h^0(E) \geq 2n$. For line bundles, the two definitions coincide and reduce to the classical Clifford index. The use of semistable, rather than stable, bundles is likely to give better specialisation properties, although the question of Clifford indices for stable bundles is also of interest and will undoubtedly be investigated further in the future. It may be noted that there are results in the literature giving bounds on $h^0(E)$ for indecomposable bundles [40] and also for bundles of rank ≤ 3 [15, 28] in terms of degrees of stability, but, for the reasons stated earlier, we feel that semistable bundles form the most natural context for these ideas.

Another natural question to ask is why we use at least $n + 1$ independent sections for the definition of γ_n. The first reason is that the question of non-emptiness of Brill–Noether loci has been completely solved for $h^0(E) \leq n$ (see [8, 9]) and depends only on the genus of C. More fundamentally, the existence of semistable bundles with $h^0(E) \geq n+1$ is closely linked with the non-emptiness of certain Quot schemes and the existence of stable maps from C to a Grassmannian (see [37]), both of which have implications for the geometry of C.

We now summarise the contents of the paper. In Section 2, we give the definitions of γ_n and γ'_n and obtain some elementary properties. In Section 3, we relate our invariants to the conjecture of Mercat, which is a strengthened version of the assertion that $\gamma'_n = \gamma_1$. We then make some deductions from the results of [34], including an almost complete determination of the values of γ_n for $n \geq g - 3$ (Theorem 3.6).

Section 4 is the central section of the paper. In it, we introduce the invariants

$$d_r := \min\left\{ \deg L \mid L \text{ a line bundle on } C \text{ with } h^0(L) \geq r + 1 \right\},$$

which form the *gonality sequence* of C; these invariants play an important rôle in the theory of special curves and are completely known in many cases. We describe the properties of these invariants which we need later in the paper. We then introduce the dual span construction (1.1). We verify Butler's conjecture in the case of line bundles of degree d_n under certain conditions on the gonality sequence (Proposition 4.9). This allows us to prove our first main theorem.

Theorem 4.15. *Let E be a semistable bundle of rank n and degree d_n.*

(a) *If $\frac{d_p}{p} \geq \frac{d_{p+1}}{p+1}$ for all $p < n$ and $d_n \neq nd_1$, then*

$$h^0(E) \leq n + 1$$

and there exist semistable bundles of rank n and degree d_n with $h^0 = n + 1$.

(b) *If $d_n = nd_1$, then*

$$h^0(E) \leq 2n$$

and there exist semistable bundles of rank n and degree d_n with $h^0 = 2n$.

(c) *If $\frac{d_p}{p} \geq \frac{d_n}{n}$ for all $p < n$ and E is stable, then*

$$h^0(E) \leq n + 1.$$

As a corollary (Corollary 4.16) we show that Mercat's conjecture holds for semistable bundles of rank n and degree $\leq d_n$, again under certain conditions on the gonality sequence. We also complete the computation of γ_n for $n \geq g - 3$ (Theorem 4.21). For a curve with $\gamma_1 = 2$, this allows us to compute all values of γ_n (Corollary 4.22). We complete the section by obtaining an upper bound for γ_n and lower bounds for $\gamma(E)$ dependent on the existence of certain subbundles.

In Section 5, we prove the following two theorems for bundles of rank 2.

Theorem 5.1. $\gamma_2 = \min\left\{\gamma_2', \frac{d_2}{2} - 1\right\}$.

Theorem 5.2. $\gamma_2' \geq \min\left\{\gamma_1, \frac{d_4}{2} - 2\right\}$.

These theorems yield the precise formula $\gamma_2 = \min\left\{\gamma_1, \frac{d_2}{2} - 1\right\}$ (Corollary 5.3). In particular, γ_2 is not determined by γ_1.

In Sections 6 and 7, we extend this partially to ranks 3, 4 and 5, obtaining the following results.

Theorem 6.1. *Suppose $\frac{d_2}{2} \geq \frac{d_3}{3}$. Then*

$$\gamma_3 = \min\left\{\gamma_3', \frac{1}{3}(d_3 - 2)\right\}.$$

Theorem 6.2. *If $\frac{d_3}{3} \geq \frac{d_4}{4}$, then*

$$\gamma_4 = \min\left\{\gamma_4', \frac{1}{4}(d_4 - 2), \frac{1}{2}(d_2 - 2)\right\}.$$

Theorem 7.3. *If $\frac{d_p}{p} \geq \frac{d_{p+1}}{p+1}$ for $1 \leq p \leq 4$, then*

$$\gamma_5 \geq \min\left\{ \gamma_5', \frac{1}{2}(d_2 - 2), \frac{1}{5}(d_5 - 2), \frac{1}{5}(d_1 + 2d_2 - 6), \frac{1}{5}(d_1 + d_4 - 4), \right.$$

$$\left. \frac{1}{5}(2d_1 + d_3 - 6), \frac{1}{5}(3d_1 + d_2 - 8), \frac{1}{5}(d_2 + d_3 - 5) \right\}.$$

This last result looks weaker than we would hope, but we show that for a general curve it gives the following much more precise result.

Corollary 7.4 *Let C be a general curve. Then*

$$\gamma_5 = \min\left\{ \gamma_5', \frac{1}{5}\left(g - \left\lceil\frac{g}{6}\right\rceil + 3\right) \right\}.$$

In Section 8 we consider smooth plane curves. In this case we know the gonality sequence precisely by a theorem of Noether. Although such curves do not satisfy all the conditions mentioned earlier, we are able to carry out the same analysis and to obtain good results for $n \leq 5$.

In the final section, we discuss some problems.

Our main arguments depend on a result of Paranjape and Ramanan [35, Lemma 3.9] and on Mercat's paper [34] as well as on the dual span construction. We have also made much use of results on special curves due to Gerriet Martens and his collaborators. We are grateful to him for some useful discussions and for drawing our attention to a number of papers.

Throughout the paper C will be a smooth curve of genus $g \geq 4$ defined over an algebraically closed field of characteristic 0. We recall that, for a vector bundle E of rank n and degree d, the *slope* $\mu(E)$ is defined by $\mu(E) := \frac{d}{n}$.

We are grateful to the referee for pointing out the reference [2].

2. Definition of γ_n and γ_n'

Let C be a smooth projective curve of genus $g \geq 4$. For any vector bundle E of rank n and degree d on C consider

$$\gamma(E) := \frac{1}{n}\left(d - 2(h^0(E) - n)\right) = \mu(E) - 2\frac{h^0(E)}{n} + 2.$$

The proof of the following lemma is a simple computation.

Lemma 2.1. $\gamma(K_C \otimes E^*) = \gamma(E)$. $\qquad\qquad\qquad\qquad\square$

For any positive integer n we define the following invariants of C:

$$\gamma_n := \min_E \left\{ \gamma(E) \,\middle|\, \begin{array}{l} E \text{ semistable of rank } n \\ h^0(E) \geq n+1, \ \mu(E) \leq g-1 \end{array} \right\}$$

and

$$\gamma_n' := \min_E \left\{ \gamma(E) \,\middle|\, \begin{array}{l} E \text{ semistable of rank } n \\ h^0(E) \geq 2n, \ \mu(E) \leq g-1 \end{array} \right\}.$$

Note that $\gamma_1 = \gamma_1'$ is the usual Clifford index of the curve C. We say that E *contributes to* γ_n (respectively γ_n') if E is semistable of rank n with $\mu(E) \leq g-1$ and $h^0(E) \geq n+1$ (respectively $h^0(E) \geq 2n$). If in addition $\gamma(E) = \gamma_n$ (respectively $\gamma(E) = \gamma_n'$), we say that E *computes* γ_n (respectively γ_n').

Lemma 2.2. *If $p|n$, then $\gamma_n \leq \gamma_p$ and $\gamma_n \leq \gamma_n' \leq \gamma_p'$.*

Proof. Let E be a vector bundle computing γ_p. Then $\gamma(\bigoplus_{i=1}^{\frac{n}{p}} E) = \gamma_p$ which gives the first assertion. It is obvious that $\gamma_n \leq \gamma_n'$. The proof of the last inequality is the same as the proof of the first statement. $\square$

Lemma 2.3.

$$0 \leq \gamma_n \leq \frac{1}{n}\left(g - \left[\frac{g}{n+1}\right] + n - 2\right).$$

Proof. If E is a vector bundle computing γ_n, we have by [8, Theorem 2.1] that $h^0(E) \leq \frac{\deg E}{2} + n$, which implies $\gamma(E) \geq 0$. Hence $\gamma_n \geq 0$.

From [12, Theorem 2] or [7, Proposition 4.1 (ii)] we know that on a general curve there exist semistable vector bundles of rank n and degree $d = g - \left[\frac{g}{n+1}\right] + n$ with $h^0(E) \geq n+1$. Since $g \geq 4$, we have $\mu(E) \leq g-1$. By semicontinuity this is valid on any curve C. Hence

$$\gamma_n \leq \frac{1}{n}\left(g - \left[\frac{g}{n+1}\right] + n - 2\right). \qquad \square$$

Corollary 2.4. *Suppose $g \geq 7$. For a general curve C and every $n \geq 3$, we have*

$$\gamma_n < \gamma_1.$$

Proof. On a general curve we know $\gamma_1 = \left[\frac{g-1}{2}\right]$. According to Lemma 2.3 it suffices to show

$$\frac{1}{n}\left(g - \left[\frac{g}{n+1}\right] + n - 2\right) < \left[\frac{g-1}{2}\right]. \qquad (2.1)$$

Suppose $n \geq 3$. Since $\frac{g}{n+1} - 1 < \left[\frac{g}{n+1}\right]$ and $\frac{g}{2} - 1 \leq \left[\frac{g-1}{2}\right]$, this is implied by

$$\frac{1}{n}\left(g - \frac{g}{n+1} + 1 + n - 2\right) \leq \frac{g}{2} - 1,$$

which is equivalent to

$$g \geq \frac{(4n-2)(n+1)}{(n+1)(n-2)+2} = 4 + \frac{6n-2}{n^2-n}.$$

This is valid for $g \geq 7$. $\square$

Remark 2.5. Corollary 2.4 remains valid for $g = 5$ and for $g = 6$, $n \geq 4$ (for $g = 5$, $n \leq 6$ one needs to check directly in (2.1)). The corollary is also valid for $n = 2$, provided $g \geq 7$, $g \neq 8$. In fact, for $g \geq 9$ the same proof works. The case $g = 7$ can be checked from (2.1).

Proposition 2.6.

(a) *If $\gamma_1 = 0$ or 1, then for all n,*

$$\gamma_n = \gamma'_n = \gamma_1.$$

(b) *If $\gamma_1 \geq 2$, then $\gamma_n \geq 1$ for all n.*

Proof. (a) By Lemma 2.3 we have $\gamma_n \geq 0$. So the result for $\gamma_1 = 0$ follows by Lemma 2.2. Suppose $\gamma_1 = 1$. If $\gamma_n < 1$, then there exists a semistable bundle E with $h^0(E) \geq n + 1$ and degree $d \leq n(g - 1)$ such that

$$d - 2(h^0(E) - n) < n.$$

So

$$h^0(E) > \frac{d + n}{2}.$$

If $d \geq n$, this contradicts [38]. If $d < n$, then $h^0(E) < n$ by [8]. So $\gamma_n \geq 1$ and hence $\gamma_n = \gamma'_n = \gamma_1$.

(b) The argument in the proof of (a) for $\gamma_1 = 1$ is valid also for $\gamma_1 \geq 2$. □

Corollary 2.7. *If $\gamma_1 \geq 1$, then $\lim_{n \to \infty} \gamma_n = 1$.*

Proof. This follows from Proposition 2.6 and Lemma 2.3. □

3. Mercat's conjecture

We want to relate the invariants γ'_n with Mercat's conjecture (see [34]), which can be stated as follows:

Conjecture 3.1. *Let E be a semistable vector bundle of rank n and degree d.*

(i) *If $\gamma_1 + 2 \leq \mu(E) \leq 2g - 4 - \gamma_1$, then $h^0(E) \leq \frac{d - \gamma_1 n}{2} + n$.*

(ii) *If $1 \leq \mu(E) \leq \gamma_1 + 2$, then $h^0(E) \leq \frac{1}{\gamma_1 + 1}(d - n) + n$.*

Lemma 3.2. *Conjecture 3.1 (i) is equivalent to*

(i′) *If $\gamma_1 + 2 \leq \mu(E) \leq 2g - 4 - \gamma_1$, then $\gamma(E) \geq \gamma_1$.*

Proof. Suppose (i) holds. Then

$$\gamma(E) = \mu(E) - 2\frac{h^0(E)}{n} + 2 \geq \mu(E) - 2\frac{\frac{d - \gamma_1 n}{2} + n}{n} + 2 = \gamma_1.$$

The converse implication follows by the same computation. □

Proposition 3.3. *Let $n \geq 2$ be an integer.*

(a) *Conjecture 3.1 implies the equality $\gamma'_n = \gamma_1$.*

(b) *The equality $\gamma'_n = \gamma_1$ implies Conjecture 3.1 (i).*

Proof. (a) Assume Conjecture 3.1 holds and suppose E contributes to γ'_n. According to Lemma 2.2 we have to show that $\gamma(E) \geq \gamma_1$.

If $\mu(E) \geq \gamma_1 + 2$, Lemma 3.2 implies the assertion. So suppose $1 \leq \mu(E) < \gamma_1 + 2$. Then by (ii),

$$h^0(E) \leq \frac{1}{\gamma_1 + 1}(d - n) + n < \frac{1}{\gamma_1 + 1}\left(n(\gamma_1 + 2) - n\right) + n = 2n,$$

a contradiction.

(b) Assume that $\gamma'_n = \gamma_1$ and consider a semistable vector bundle E of rank n with $\gamma_1 + 2 \leq \mu(E) \leq 2g - 4 - \gamma_1$. By Lemma 3.2 we have to show that $\gamma(E) \geq \gamma_1$. In view of Lemma 2.1 we can assume that $\mu(E) \leq g - 1$.

If $h^0(E) \geq 2n$, then $\gamma(E) \geq \gamma'_n = \gamma_1$ by assumption. If $h^0(E) < 2n$, then

$$\gamma(E) = \mu(E) - 2\frac{h^0(E)}{n} + 2 > \gamma_1 + 2 - 2\frac{2n}{n} + 2 = \gamma_1. \qquad \square$$

Remark 3.4. If Conjecture 3.1 (ii) holds and $1 \leq \mu(E) < \gamma_1 + 2$, then $h^0(E) < 2n$. So E does not contribute to γ'_n.

Proposition 3.5. *If $\gamma_1 \geq 2$, then $\gamma'_n \geq 2$ for all n.*

Note that $\gamma_1 \geq 2$ implies $g \geq 5$.

Proof. We use [34, Theorem 1]. Let E be a semistable bundle of rank n and degree d. If $d < (2 + \frac{2}{g-4})n$, then by [34, Theorem 1 (ii)] we have

$$
\begin{aligned}
h^0(E) &\leq \frac{1}{g-2}(d - n) + n \\
&< \frac{1}{g-2}\left(n + \frac{2n}{g-4}\right) + n \\
&= n\left(1 + \frac{1}{g-4}\right) \leq 2n.
\end{aligned}
$$

So E does not contribute to γ'_n. Now [34, Theorem 1 (i)] implies $\gamma'_n \geq 2$. $\qquad \square$

We can use Mercat's results of [32], [33] and [34] to obtain the following theorem.

Theorem 3.6. *Let C be a curve with Clifford index $\gamma_1 \geq 2$.*

(a) *If $n > g$, then*

$$\gamma_n = 1 + \frac{g - 2}{n};$$

(b) *If $n = g$, then*

$$\gamma_n \begin{cases} = 2 - \frac{2}{g} & \text{for} \quad g \geq 6, \\ \geq \frac{7}{5} & \text{for} \quad g = 5; \end{cases}$$

(c) *If $n = g - 1$, then*

$$\gamma_n = 2 - \frac{2}{g - 1};$$

(d) *If $g - 3 \leq n \leq g - 2$, then*

$$\gamma_n \geq 2 - \frac{1}{n};$$

(e) *If $n \leq g - 4$, then*

$$\gamma_n \geq 2.$$

Proof. Let E be a semistable bundle of rank n and degree d. So $\mu = \mu(E) = \frac{d}{n}$. We consider 4 cases:

Case 1: $1 < \mu < 2$: By [32], $h^0(E) \leq n + \frac{1}{g}(d - n)$ and so

$$\gamma(E) \geq \frac{1}{n}\left(d - \frac{2}{g}(d - n)\right).$$

Case 2: $\mu = 2$: By [33], $h^0(E) \leq n + [\frac{n}{g-1}]$ and so

$$\gamma(E) \geq \frac{1}{n}\left(2n - 2\left[\frac{n}{g-1}\right]\right).$$

Case 3: $2 < \mu < 2 + \frac{2}{g-4}$: By [34, Theorem 2.1 (ii)], $h^0(E) \leq n + \left[\frac{1}{g-2}(d - n)\right]$ and so

$$\gamma(E) \geq \frac{1}{n}\left(d - 2\left[\frac{d-n}{g-2}\right]\right). \tag{3.1}$$

Case 4: $2 + \frac{2}{g-4} \leq \mu$: By [34, Theorem 2.1 (i)],

$$\gamma(E) \geq 2. \tag{3.2}$$

In Case 1 the right-hand side is an increasing function of d. So we need to look for the smallest d in the given range for which a bundle E exists with $h^0(E) \geq n + 1$. We must have $d = n + g$ and this is in the required range if $n > g$ and then for such E,

$$\gamma(E) = 1 + \frac{g-2}{n} < 2. \tag{3.3}$$

When $n > g$, such E always exists (see [32]).

By [33], Case 2 always occurs provided $n \geq g - 1$ and gives us bundles E with

$$\gamma(E) = 2 - \frac{2}{n}\left[\frac{n}{g-1}\right] < 2. \tag{3.4}$$

It remains to deal with Case 3. The smallest value of the right-hand side of (3.1) is given by one of the following three possibilities:

- $d = 2n + 1$ and $n \geq g - 3$,
- $d = 2n + 2$ and $g - 2$ divides $n + 2$,
- $d = n + g - 2$.

If none of these possibilities occurs within the range $2 < \mu < 2 + \frac{2}{g-4}$, then Case 3 does not arise.

For $d = 2n + 1$ we get

$$\gamma(E) \geq 2 - \frac{1}{n}\left(2\left[\frac{n+1}{g-2}\right] - 1\right) \tag{3.5}$$

and we require $2n + 1 < n(2 + \frac{2}{g-4})$, i.e., $n > \frac{g-4}{2}$ which is true since $n \geq g - 3$.

For $d = 2n + 2$ we get

$$\gamma(E) \geq 2 - \frac{2}{n}\left(\frac{n+2}{g-2} - 1\right) \tag{3.6}$$

and we require that $g - 2$ divides $n + 2$ and $2n + 2 < n(2 + \frac{2}{g-4})$, i.e., $n > g - 4$.

For $d = n + g - 2$ we get

$$\gamma(E) \geq 1 + \frac{g-4}{n} \tag{3.7}$$

and we require $2n < n + g - 2 < 2n + \frac{2n}{g-4}$, i.e., $n = g - 3$. In this case (3.7) gives the same inequality as (3.5) and hence can be ignored.

If $n > g$, the right-hand side of (3.3) is less than or equal to the right-hand sides of (3.4), (3.5) and (3.6). So for $n > g$ we obtain

$$\gamma_n \geq \min\left\{2, 1 + \frac{g-2}{n}\right\} = 1 + \frac{g-2}{n}$$

and this can be attained by a bundle E of degree $n + g$ with $h^0(E) = n + 1$.

For $n = g \geq 6$ we get from (3.2), (3.4), (3.5) and (3.6),

$$\gamma_n \geq \min\left\{2, 2 - \frac{2}{g}, 2 - \frac{1}{g}\right\} = 2 - \frac{2}{g}$$

and this can be attained by a bundle E of degree $2g$ with $h^0(E) = g + 1$. For $n = g = 5$ we get

$$\gamma_5 \geq \min\left\{2, 2 - \frac{2}{5}, 2 - \frac{3}{5}\right\} = \frac{7}{5}.$$

For $n = g - 1$ we get from (3.2), (3.4), (3.5) and (3.6),

$$\gamma_{g-1} \geq \min\left\{2, 2 - \frac{2}{g-1}, 2 - \frac{1}{g-1}\right\} = 2 - \frac{2}{g-1}$$

and the bound is attained by a bundle of degree $2g - 2$ with $h^0(E) = g$. (In fact, the unique such semistable bundle is the dual span of the canonical bundle K_C [33, Theorem 1]).

For $n = g - 2$ or $g - 3$ we get from (3.2) and (3.5),

$$\gamma_n \geq 2 - \frac{1}{n}\left(2\left[\frac{n+1}{g-2}\right] - 1\right) = 2 - \frac{1}{n}.$$

For $n \leq g-4$ none of the inequalities (3.3) to (3.6) applies. So there is no semistable E of rank $n \leq g - 4$ with $\mu(E) < 2 + \frac{2}{g-4}$ and $h^0(E) \geq n+1$. Hence

$$\gamma(E) \geq 2$$

by (3.2). □

Remark 3.7. Note that Theorem 3.6 (a) gives a more precise version of Corollary 2.7.

Proposition 3.8. *If $\gamma_1 \geq 3$, then*

$$\gamma_2 \geq \min\left\{\gamma_1, \frac{\gamma_1}{2}+1\right\} \quad and \quad \gamma_2' \geq \min\left\{\gamma_1, \frac{\gamma_1}{2}+2\right\}.$$

In particular, $\gamma_2' = \gamma_1$ for $\gamma_1 \leq 4$.

Proof. Suppose E is semistable of rank 2 and degree d. If $3\gamma_1 - 1 \leq d \leq 2g - 2$, then by [34, Corollary 4.4],

$$\gamma(E) \geq \gamma_1.$$

If $d \leq 3\gamma_1 - 2$, then by [34, Lemma 4.3], $h^0(E) \leq \frac{d-\gamma_1}{4} + 2$. So

$$\gamma(E) \geq \frac{d + \gamma_1}{4}.$$

In the last case E can contribute to γ_2 only if $d \geq \gamma_1 + 4$ and to γ_2' only if $d \geq \gamma_1 + 8$. This gives the assertion. □

4. The invariants d_r

For any positive integer r we define the invariant d_r of the curve C by

$$d_r := \min\left\{\deg L \mid L \text{ a line bundle on } C \text{ with } h^0(L) \geq r+1\right\}.$$

Note that d_1 is the gonality of C, d_2 is the minimal degree of a non-degenerate rational map of the curve C into the projective plane etc. We refer to the sequence $d_1, d_2, \ldots$ as the *gonality sequence* of C. We say that L *computes* d_r if $\deg L = d_r$ and $h^0(L) \geq r+1$. We say also that d_r *computes* γ_1 if $\gamma_1 = d_r - 2r$.

Remark 4.1. The gonality sequence is usually defined only for those r for which $d_r \leq g - 1$ (see [17, Digression (3.5)]), but for our purposes the definition above is more convenient. If $k = d_1$ computes γ_1 (i.e., $d_1 = \gamma_1 + 2$), the curve C is called *k-gonal.*

Lemma 4.2.

(a) $d_r < d_{r+1}$ for all r;
(b) if L computes d_r, then $h^0(L) = r + 1$ and L is generated;
(c) $d_{r+s} \leq d_r + d_s$ for any $r, s \geq 1$;
(d) $d_r \leq r(g - 1)$.

Proof. (a) and (b) are obvious. (c) Suppose L computes d_r and M computes d_s. The map

$$H^0(L) \otimes H^0(M) \to H^0(L \otimes M)$$

satisfies the hypotheses of the Hopf lemma. Hence

$$h^0(L \otimes M) \geq h^0(L) + h^0(M) - 1 = r + s + 1.$$

(d) follows from (c) and the fact that $d_1 \leq g - 1$, since $g \geq 4$. $\square$

Lemma 4.3. *If $d_r + d_s = d_{r+s}$, then $d_n = nd_1$ for all $n \leq r + s$.*

Proof. Suppose L computes d_r and M computes d_s. Then $h^0(L \otimes M) \geq r + s + 1$ as in the proof of Lemma 4.2 (c). On the other hand, $\deg(L \otimes M) = d_r + d_s = d_{r+s}$. If $h^0(L \otimes M) > r + s + 1$, then $\deg(L \otimes M) \geq d_{r+s+1}$ which contradicts Lemma 4.2 (a). So

$$h^0(L \otimes M) = r + s + 1.$$

It now follows from [21, Corollary 5.2] (see also [18, Lemma 1.8]) that there exists a line bundle N with $h^0(N) \geq 2$ such that

$$L \simeq N^r \quad \text{and} \quad M \simeq N^s.$$

Hence

$$d_r = \deg L = r \deg N \geq rd_1.$$

By Lemma 4.2 (c), we have $d_r = rd_1$ and similarly $d_s = sd_1$. So $d_{r+s} = (r + s)d_1$. Then

$$d_{r+s} = (r + s)d_1 = nd_1 + (r + s - n)d_1 \geq d_n + d_{r+s-n} \geq d_{r+s}.$$

So the inequalities must all be equalities. In particular $d_n = nd_1$. $\square$

Remark 4.4.

(a) Clifford's theorem implies that $d_r \geq 2r$ for $r \leq g-1$; moreover $d_{g-1} = 2g - 2$.

(b) Riemann–Roch implies that $d_r = r + g$ for $r \geq g$.

(c) Brill–Noether theory implies that $d_r \leq g - \left\lceil \frac{g}{r+1} \right\rceil + r$ for all r and for a general curve we have

$$d_r = g - \left\lceil \frac{g}{r+1} \right\rceil + r.$$

So for a general curve we know the gonality sequence. Indeed, for our purposes, we can define a general curve to be one which has this gonality sequence.

Remark 4.5.

(a) If C is hyperelliptic, then

$$d_r = \begin{cases} 2r & \text{for} \quad r \leq g - 1, \\ r + g & \text{for} \quad r \geq g. \end{cases}$$

(b) If C is trigonal, then

$$d_r = \begin{cases} 3r & 1 \le r \le \left[\frac{g-1}{3}\right], \\ r + g - 1 - \left[\frac{g-r-1}{2}\right] & \text{for} \quad \left[\frac{g-1}{3}\right] < r \le g - 1, \\ r + g & r \ge g. \end{cases}$$

This follows from Maroni's theory (see [31, Proposition 1]) and, for $r > \left[\frac{g-1}{3}\right]$, Serre duality and Riemann–Roch.

(c) If C is a general k-gonal curve, $k \ge 4$, then by [26, Theorem 3.1],

$$d_r = kr \quad \text{for} \quad 1 \le r \le \frac{1}{k-2}\left[\frac{g-4}{2}\right].$$

For $k = 4$, this can be marginally improved by [19, Theorem 4.3.2] and then extended using Serre duality and Riemann–Roch to give

$$d_r = \begin{cases} 4r & 1 \le r \le \left[\frac{g-1}{4}\right], \\ r + g - 1 - \left[\frac{g-r-1}{3}\right] & \text{for} \quad \left[\frac{g-1}{4}\right] < r \le g - 1, \\ r + g & r \ge g, \end{cases}$$

except when $g \equiv 0 \mod 4$, in which case $d_{\frac{g}{4}} = g - 1$ (see [18, Proposition 3.3]).

(d) If C is bielliptic of genus $g \ge 5$, then C is tetragonal, but its gonality sequence is quite different from that of (c). In fact,

$$d_r = \begin{cases} 2r + 2 & 1 \le r \le g - 3, \\ 2g - 3 & \text{for} \quad r = g - 2, \\ 2g - 2 & r = g - 1, \\ r + g & r \ge g. \end{cases}$$

Lemma 4.6. $\qquad d_r \ge \min\{\gamma_1 + 2r, g + r - 1\}.$

Proof. Let L be a line bundle computing d_r. Then $h^1(L) = r + g - d_r$. If $d_r < g + r - 1$, then L contributes to γ_1 and $d_r \ge \gamma_1 + 2r$. $\qquad\square$

Lemma 4.7. $\gamma_1 = d_r - 2r$, where $r = 1$, except in the following cases:

(a) *if C is a smooth plane curve, then $r = 2$;*

(b) *if C is exceptional in the sense of [22], then $r \le \frac{g+2}{4}$ and for $r \le 9$ we have $r = \frac{g+2}{4}$.*

Proof. This is a consequence of the results of [22]. $\qquad\square$

The following lemma is a restatement of [35, Lemma 3.9]

Lemma 4.8. *Let E be a vector bundle of rank n with $h^0(E) \ge n + s$, $s \ge 1$. Suppose that E has no proper subbundle N with $h^0(N) > \mathrm{rk} N$. Then*

$$\deg E \ge d_{ns}.$$

Proof. Let E be as in the statement of the lemma. Note that $h^0(E^*) = 0$, for otherwise the kernel N of a non-zero homomorphism $E \to \mathcal{O}_C$ would contradict the hypothesis. [35, Lemma 3.9] now implies that $h^0(\det E) \geq ns + 1$. So $\deg E = \deg \det E \geq d_{ns}$. $\qquad\square$

Suppose the line bundle L computes d_n. Define a line bundle E_L of rank n and degree d_n by the exact sequence

$$0 \to E_L^* \to H^0(L) \otimes \mathcal{O}_C \to L \to 0. \tag{4.1}$$

As mentioned in the introduction, this is a special case of the *dual span construction* [12]. Note that $\mu(E_L) = \frac{d_n}{n} \leq g - 1$ by Lemma 4.2 (d).

Proposition 4.9.

(a) $h^0(E_L^*) = 0$;

(b) E_L is generated;

(c) if $n \geq (>) g$, then E_L is semistable (stable);

(d) if $\frac{d_p}{p} \geq \frac{d_n}{n}$ for all $p < n$, then E_L is semistable;

(e) if $\frac{d_p}{p} > \frac{d_n}{n}$ for all $p < n$, then E_L is stable.

Proof. (a) and (b) are obvious. (c) If $n \geq g$, then $d_n = n + g$ by Remark 4.4 (b). Now apply [11, Theorem 1.2] for the case of a line bundle.

(d) Let M be a quotient bundle of E_L of rank $p < n$. It follows from (a) and (b) that M is generated with $h^0(M^*) = 0$. So $h^0(M) \geq p + 1$. Choose a $(p+1)$-dimensional subspace V of $H^0(M)$ which generates M. Then we have an exact sequence

$$0 \to \det M^* \to V \otimes \mathcal{O}_C \to M \to 0.$$

Dualizing this, we see that $\det M$ is a line bundle with $h^0(\det M) \geq p + 1$. So

$$\deg M = \deg \det M \geq d_p.$$

Under the hypothesis of (d), $\mu(M) \geq \mu(E_L)$. Since this is true for all quotient bundles of E_L, this proves that E_L is semistable.

For (e) the proof proceeds as for (d), but now we have $\mu(M) > \mu(E_L)$ for every proper quotient bundle M of E_L. Hence E_L is stable. $\qquad\square$

Remark 4.10. It follows from Remark 4.4 (a) and (b) that the hypothesis of Proposition 4.9 (d) is satisfied for all $n \geq g$ and similarly the hypothesis of Proposition 4.9 (e) is satisfied for all $n > g$. So these two parts of the proposition are generalisations of part (c). Note also that $\frac{d_{g-1}}{g-1} = \frac{d_g}{g} = 2$ by the same remark.

Proposition 4.11. *Suppose E is a semistable vector bundle of rank n with $\deg E < \frac{nd_p}{p}$ for all $p \leq n$. Then $h^0(E) \leq n$.*

Proof. Suppose $h^0(E) \geq n+1$. If E possesses no proper subbundle N with $h^0(N) \geq \mathrm{rk}N + 1$, then $\deg E \geq d_n$ by Lemma 4.8, contradicting the hypotheses.

So let N be a subbundle of rank p of E minimal with respect to the property $h^0(N) \geq p+1$. Then Lemma 4.8 applies to N and

$$\frac{\deg N}{p} \geq \frac{d_{p(h^0(N)-p)}}{p} \geq \frac{d_p}{p} > \frac{\deg E}{n},$$

contradicting the semistability of E. □

Corollary 4.12. *Suppose $\frac{d_p}{p} \geq \frac{d_n}{n}$ for $p < n$ and E is a semistable vector bundle of rank n with $\deg E < d_n$. Then $h^0(E) \leq n$.* □

Corollary 4.13. *If $d_n = nd_1$, then $\gamma_n = \gamma_n'$.*

Proof. If $d_n = nd_1$, then, as in the proof of Lemma 4.3 we have $d_p = pd_1$ for all $p \leq n$.

Now suppose that E contributes to γ_n, but not to γ_n'. Then $d \geq d_n$ by Corollary 4.12 and $h^0(E) \leq 2n - 1$. So

$$\gamma(E) \geq \frac{1}{n}\left(nd_1 - 2(n-1)\right) = d_1 - 2 + \frac{2}{n} > \gamma_1.$$

Hence

$$\gamma_n \geq \min\{\gamma_n', \gamma_1\} = \gamma_n'.$$

It follows from Lemma 2.2 that $\gamma_n = \gamma_n'$. □

Remark 4.14. The assumption $d_n = nd_1$ is valid for hyperelliptic curves of genus $g \geq n$, trigonal curves of genus $g \geq 3n + 1$ and general tetragonal curves of genus $g \geq 4n + 1$ (see Remark 4.5). For hyperelliptic and trigonal curves we already have that $\gamma_n = \gamma_n' = \gamma_1$ by Proposition 2.6 (a). For tetragonal curves we have $\gamma_n' = \gamma_1 = 2$ by Proposition 3.5; also $\gamma_n = 2$ for $g \geq n + 4$ by Lemma 2.2 and Theorem 3.6 (e). The corollary also applies to general k-gonal curves of genus $g \geq 2(k-2)n + 4$.

Theorem 4.15. *Let E be a semistable bundle of rank n and degree d_n.*

(a) *If $\frac{d_p}{p} \geq \frac{d_{p+1}}{p+1}$ for all $p < n$ and $d_n \neq nd_1$, then*

$$h^0(E) \leq n + 1$$

and there exist semistable bundles of rank n and degree d_n with $h^0 = n + 1$.

(b) *If $d_n = nd_1$, then*

$$h^0(E) \leq 2n$$

and there exist semistable bundles of rank n and degree d_n with $h^0 = 2n$.

(c) *If $\frac{d_p}{p} \geq \frac{d_n}{n}$ for all $p < n$ and E is stable, then*

$$h^0(E) \leq n + 1.$$

Proof. (a) Suppose $h^0(E) = n + s$ with $s \geq 2$. If E possesses no subbundle N of rank $p < n$ with $h^0(N) \geq p + 1$, then by Lemma 4.8,

$$\deg E \geq d_{ns} > d_n, \tag{4.2}$$

which is a contradiction.

Now let N be a subbundle of minimal rank p such that $h^0(N) \geq p + 1$. Then Lemma 4.8 implies that

$$\deg N \geq d_{p(h^0(N)-p)}.$$

Hence

$$\frac{\deg N}{p} \geq \frac{d_{p(h^0(N)-p)}}{p} \geq \frac{d_p}{p} \geq \frac{d_n}{n}, \tag{4.3}$$

which contradicts the semistability of E unless all these inequalities are equalities. So $h^0(N) = p + 1$, $\deg N = d_p$ and $\frac{d_p}{p} = \frac{d_n}{n}$, i.e., $\mu(N) = \mu(E)$.

It follows that E/N is semistable of rank $n - p$ and degree $d_n - d_p$. Now $d_n - d_p < d_{n-p}$ by Lemma 4.3. So $h^0(E/N) \leq n - p$ by Corollary 4.12 and $h^0(E) \leq n + 1$.

To prove existence, let L be a line bundle of degree d_n with $h^0(L) = n + 1$. Then by Proposition 4.9, E_L is a semistable bundle of rank n and degree d_n with $h^0(E_L) \geq n + 1$. This completes the proof of (a).

We prove (b) by induction, the case $n = 1$ being obvious. If $h^0(E) \leq n + 1$ the result is clear. So suppose $h^0(E) = n + s$ with $s \geq 2$. Arguing as in the proof of (a) we obtain a proper subbundle N of E of rank p and degree d_p with $h^0(N) = p + 1$. Moreover, E/N is semistable of rank $n - p$ and degree $d_n - d_p$, where now $d_n - d_p = d_{n-p}$. By inductive hypothesis, we have $h^0(E/N) \leq 2(n - p)$ and hence

$$h^0(E) \leq 2(n - p) + p + 1 = 2n - p + 1 \leq 2n. \tag{4.4}$$

To prove existence, choose a line bundle L of degree d_1 with $h^0(L) = 2$ and take $E = \bigoplus_{i=1}^{n} L$. In fact, E is the dual span of the line bundle L^n.

(c) If E is stable, then (4.2) and (4.3) give contradictions. Hence $h^0(E) \leq n + 1$. $\qquad\square$

Corollary 4.16. *Suppose $\frac{d_p}{p} \geq \frac{d_{p+1}}{p+1}$ for all $p < n$ and let E be a semistable bundle of rank n and degree $d \leq d_n$. Then Conjecture 3.1 holds for E.*

Proof. If $d < d_n$, then $h^0(E) \leq n$ by Corollary 4.12 in accordance with Conjecture 3.1 (ii).

If $d = d_n = nd_1$, then, since $d_1 \geq \gamma_1 + 2$, we have by Theorem 4.15,

$$h^0(E) \leq 2n \leq \frac{d - \gamma_1 n}{2} + n$$

in accordance with Conjecture 3.1 (i).

If $d = d_n \neq nd_1$, then $h^0(E) \leq n + 1$ by Theorem 4.15. Now

$$\frac{d_n - n}{\gamma_1 + 1} \geq \frac{d_1 - 1}{\gamma_1 + 1} \geq 1,$$

since $d_1 \geq \gamma_1 + 2$. So

$$h^0(E) \leq \frac{d_n - n}{\gamma_1 + 1} + n$$

in accordance with Conjecture 3.1 (ii). □

For general C, we can prove a precise result on the stability of E_L. This can be deduced from [12, Theorem 2]; for the sake of completeness and since [12] is unpublished, we include a proof.

Proposition 4.17. *Suppose C is general and L is a line bundle computing d_n. Then E_L is semistable and $h^0(E_L) = n + 1$. Moreover, E_L is stable unless $n = g$ and $\det E_L$ is isomorphic to $K_C(p_1 + p_2)$ for some $p_1, p_2 \in C$.*

Proof. For $n > g$ stability is proved in Proposition 4.9 (c). So suppose $n \leq g$. According to Proposition 4.9 (d) and Remark 4.4 (c), in order to prove semistability it suffices to show that

$$\frac{1}{p}\left(g - \left[\frac{g}{p+1}\right]\right) \geq \frac{1}{n}\left(g - \left[\frac{g}{n+1}\right]\right).$$

This is satisfied if

$$\frac{1}{p}\left(g - \frac{g}{p+1}\right) \geq \frac{1}{n}\left(g - \frac{g-n}{n+1}\right)$$

which is equivalent to

$$\frac{g}{p+1} \geq \frac{g+1}{n+1}. \tag{4.5}$$

This is true for $n \leq g$. For $n < g$ the same proof shows that (4.5) is true with strict inequality. Since $d_n \neq nd_1$, Theorem 4.15 shows that $h^0(E_L) = n + 1$.

When $n = g$ and $\det E_L \not\simeq K_C(p_1 + p_2)$, it follows directly from [11, Theorem 1.2] that E_L is stable. □

Remark 4.18. Suppose C is hyperelliptic and $n \leq g - 1$. Then by Remark 4.5 (a), $d_n = 2n$. If H is the hyperelliptic line bundle, then

$$E = \bigoplus_{i=1}^{n} H$$

is semistable of degree d_n with $h^0(E) = 2n$. Moreover, E is generated, so we can choose a subspace V of $H^0(E)$ of dimension $n + 1$ which generates E, giving an exact sequence

$$0 \to \det E^* \to V \otimes \mathcal{O}_C \to E \to 0.$$

Noting that $\det E \simeq H^n$ and dualizing, we obtain

$$0 \to E^* \to V^* \otimes \mathcal{O}_C \to H^n \to 0. \tag{4.6}$$

Now $\deg H^n = d_n$. So $h^0(H^n) = n + 1$ and (4.6) is the evaluation sequence of H^n. Thus

$$E \simeq E_{H^n}.$$

Moreover H^n is the unique line bundle of degree d_n with $h^0 \geq n+1$. So if $2 \leq n \leq g-1$, the bundle E_L constructed in (4.1) can never be stable.

Remark 4.19. If C is a trigonal curve and $n \leq \left[\frac{g-1}{3}\right]$, then $d_n = 3n$ by Remark 4.5 (b). So we can use a similar argument to that of Remark 4.18, by replacing H by the unique line bundle T of degree 3 computing d_1. Then

$$E = \bigoplus_{i=1}^{n} T$$

is semistable of degree d_n with $h^0(E) = 2n$. Moreover, if $n < \frac{g-1}{3}$, then T^n is the unique line bundle of degree d_n with $h^0 \geq n+1$ by [31, Proposition 1]. So if $2 \leq n < \frac{g-1}{3}$, then the bundle E_L constructed in (4.1) can never be stable. Note that we need $g \geq 8$ in order to allow $n \geq 2$.

Remark 4.20. Similarly, if C is a general k-gonal curve ($k \geq 4$) of genus $g \geq 4k-4$ ($g \geq 9$ if $k = 4$) and $n \leq \frac{1}{k-2}\left[\frac{g-4}{2}\right]$ ($n \leq \left[\frac{g-1}{4}\right]$ if $k = 4$), then by Remark 4.5 (c) we have $d_n = kn$. Let Q be a line bundle of degree k with $h^0(Q) = 2$. Then

$$E = \bigoplus_{i=1}^{n} Q$$

is semistable of degree d_n with $h^0(E) = 2n$. For $n \geq 2$, E is strictly semistable.

Note that Q is unique for $g \geq (k-1)^2 + 1$. This follows from the fact that a curve of type (a, b) on a smooth quadric surface is of arithmetic genus $(a-1)(b-1)$.

When $k = 4$ and $g \geq 11$, Q^2 is the unique line bundle of degree 8 with $h^0 \geq 3$ by [18, Theorem 3.2]. So for a general tetragonal curve of genus $g \geq 11$, there do not exist stable bundles of the form E_L with L of degree 8 and $h^0(L) = 3$. We do not know whether in other cases Q^n is the unique line bundle of degree d_n with $h^0 \geq n+1$. So it is possible that there could exist stable bundles of the form E_L.

The next theorem improves the results of Theorem 3.6.

Theorem 4.21. *Let C be a curve with Clifford index $\gamma_1 \geq 2$.*
(a) *If $n = g$, then*

$$\gamma_n = 2 - \frac{2}{g};$$

(b) *if $n = g-2$, then*

$$\gamma_n = 2 - \frac{1}{g-2};$$

(c) *if $n = g-3$, then*

$$\gamma_n = 2.$$

Proof. Note first that $d_g = 2g$, $d_{g-1} = 2g-2$, $d_{g-2} = 2g-3$ for $\gamma_1 \geq 1$ and $d_{g-3} = 2g-4$ for $\gamma_1 \geq 2$.

(a) According to Theorem 3.6 (b) we need only show that $\gamma_n \geq \frac{8}{5}$ if $n = g = 5$. This will follow as in the proof of Theorem 3.6 (b) if we can show that there is no semistable bundle E of rank 5 and degree $2n+1 = 11$ with $h^0(E) \geq 7$.

Any such bundle is necessarily stable and if F is an elementary transformation of E, then $\deg F = 10$ and F is semistable. Now $d_5 = 10 \neq 5d_1$ and one can easily check that the conditions of Theorem 4.15 (a) hold. So $h^0(F) \leq 6$. Since this holds for any elementary transformation of E, it follows that E is generated with $h^0(E) = 7$. Now, following through the proof of Theorem 4.15, we see that (4.2) gives $\deg E \geq d_{10}$, a contradiction. So there must exist a subbundle N of E of rank $p < n$ which is minimal with respect to the condition $h^0(N) > \mathrm{rk}N$. In this case we have $\deg N \geq d_p$ and, by stability of E, $\frac{\deg N}{p} < \mu(E) = \frac{11}{5}$. The only possibility for this is $p = 4$, $\deg N = d_4 = 8$. It is easy to check that N must be semistable. So Theorem 4.15 (a) gives $h^0(N) \leq 5$. It follows that $h^0(E/N) \geq 2$. So $\deg(E/N) \geq d_1 = 4$. Hence $\deg E \geq 12$, a contradiction.

(b) For $n = g - 2$, we need to show that there exists a semistable bundle E of rank $g - 2$ and degree $2g - 3$ with $h^0(E) \geq g - 1$ (see the proof of Theorem 3.6 and in particular the inequality (3.5)). Since $\frac{d_{g-2}}{g-2} = \frac{2g-3}{g-2} = 2 + \frac{1}{g-2}$ and $\frac{d_p}{p} > 2$ for all $p < g - 2$ by Lemma 4.6, the hypotheses of Proposition 4.9 (d) hold and we can take $E = E_L$ with L a line bundle of degree d_{g-2} with $h^0(L) = g - 1$.

(c) For $n = g - 3$, we again consider the proof of Theorem 3.6. We need to show that there is no semistable bundle E of rank $g - 3$ and degree $2n + 1 = 2g - 5$ with $h^0(E) \geq g - 2$. Since $\gamma_1 \geq 2$, Lemma 4.6 implies that $d_p \geq 2p + 2$ for all $p < g - 3$. So $\frac{d_p}{p} > \frac{d_{g-3}}{g-3} = 2 + \frac{2}{g-3}$. Hence the conditions of Corollary 4.12 apply, giving $h^0(E) \leq g - 3$. $\qquad\square$

Corollary 4.22. *If $\gamma_1 = 2$, then for all $n \geq 1$,*

$$\gamma_n = \begin{cases} 1 + \frac{g-2}{n} & n > g, \\ 2 - \frac{2}{g} & n = g, \\ 2 - \frac{2}{g-1} & \text{if} \quad n = g - 1, \\ 2 - \frac{1}{g-2} & n = g - 2, \\ 2 & n \leq g - 3. \end{cases}$$

In particular $\gamma_2 = 2$.

Proof. This follows from Theorems 3.6 and 4.21 and Lemma 2.2. For the last part, note that $\gamma_1 = 2$ implies that $g \geq 5$; so $2 \leq g - 3$. $\qquad\square$

Remark 4.23. Corollary 4.22 applies in particular to any tetragonal curve. If C is bielliptic, rather more is known. By [3] (see also [34, Theorem 3.1 and Lemma 3.2]), there exist semistable bundles E of any rank n and degree $d \geq 2$ with $h^0(E) = [\frac{d}{2}]$; in other words,

$$\gamma(E) = \frac{1}{n}\left(d - 2\left[\frac{d}{2}\right] + 2n\right) = \begin{cases} 2 & \text{for} \quad d \text{ even}, \\ 2 + \frac{1}{n} & \text{for} \quad d \text{ odd}. \end{cases}$$

If $2(n+1) \leq d \leq n(g-1)$, these bundles contribute to γ_n and, if $4n \leq d \leq n(g-1)$, they contribute to γ'_n. Since $\gamma'_n = 2$ by Lemma 2.2 and Proposition 3.5, there are many bundles on C which compute γ'_n.

The remaining results of this section will be useful in estimating the value of γ_n.

Lemma 4.24. *Suppose* $p|n$ *and* $\frac{d_q}{q} \geq \frac{d_p}{p}$ *for* $q < p$. *Then*

$$\gamma_n \leq \frac{1}{p}(d_p - 2).$$

Proof. Let F be a semistable bundle of rank p and degree d_p with $h^0(F) \geq p + 1$, which exists by Proposition 4.9. Then, if

$$E = \bigoplus_{i=1}^{\frac{n}{p}} F,$$

we have

$$\gamma(E) = \gamma(F) \leq \frac{1}{p}(d_p - 2).$$

The bundle E contributes to γ_n, which gives the result. $\qquad\square$

Proposition 4.25. *Suppose* E *is a semistable bundle of rank* $n \geq 2$.

(a) *If* $\frac{d_p}{p} \geq \frac{d_n}{n}$ *for all* $p < n$ *and* $h^0(E) = n + 1$, *then*

$$\gamma(E) \geq \frac{1}{n}(d_n - 2).$$

(b) *If* $h^0(E) \geq n + 2$ *and there exists no proper subbundle* N *of* E *with* $h^0(N) > \mathrm{rk}N$, *then*

$$\gamma(E) \geq \frac{1}{n}(d_n - 2).$$

If $n \geq 3$, *this is a strict inequality.*

(c) *If* $h^0(E) \leq 2n - 1$ *and there exists a line subbundle* N *of* E *with* $h^0(N) \geq 2$, *then*

$$\gamma(E) > \gamma_1.$$

(d) *Suppose* $h^0(E) = n + s$ *with* $s \geq 1$ *and there exists a subbundle* N *of* E *of rank* $p \geq 2$ *with* $h^0(N) = p + t$, $t \geq 1 + \frac{2s}{n} - \frac{2}{p}$, *and no subbundle* N' *of rank* $< p$ *with* $h^0(N') > \mathrm{rk}N'$. *Then*

$$\gamma(E) \geq \frac{1}{p}(d_p - 2).$$

If further $\frac{d_p}{p} \geq \frac{d_n}{n}$ *and* $t \geq 1 + \frac{2}{n}(s - 1)$, *then*

$$\gamma(E) \geq \frac{1}{n}(d_n - 2).$$

Proof. (a) If $h^0(E) = n + 1$, then by Corollary 4.12, $d \geq d_n$ which implies the assertion.

(b) Suppose $h^0(E) = n + s$ with $s \geq 2$. By Lemma 4.8, $\deg E \geq d_{ns}$ and

$$\gamma(E) \geq \frac{1}{n}(d_{ns} - 2s)$$

$$\geq \frac{1}{n}(d_n + ns - n - 2s)$$

$$= \frac{1}{n}\big(d_n + (n-2)s - n\big) \geq \frac{1}{n}(d_n - 2)$$

and the last inequality is strict if $n \geq 3$.

(c) By definition of d_1, $\deg N \geq d_1$. So by semistability, $\deg E \geq nd_1$ and hence

$$\gamma(E) \geq \frac{1}{n}\big(nd_1 - 2(h^0(E) - n)\big) = d_1 - \frac{2(h^0(E) - n)}{n} > d_1 - 2 \geq \gamma_1.$$

(d) Lemma 4.8 gives $\deg N \geq d_{pt}$. By semistability,

$$\deg E \geq \frac{n}{p}d_{pt} \geq \frac{n}{p}(d_p + pt - p).$$

So

$$\gamma(E) \geq \frac{1}{p}(d_p + pt - p) - \frac{2s}{n}.$$

The inequality $t \geq 1 + \frac{2s}{n} - \frac{2}{p}$ now gives

$$\gamma(E) \geq \frac{1}{p}(d_p - 2).$$

If $\frac{d_p}{p} \geq \frac{d_n}{n}$ and $t \geq 1 + \frac{2}{n}(s - 1)$, we get

$$\gamma(E) \geq \frac{d_p}{p} - \frac{2}{n} \geq \frac{d_n}{n} - \frac{2}{n},$$

proving the second assertion. $\qquad\qquad\qquad\qquad\qquad\qquad\qquad\qquad\square$

Proposition 4.26. *Suppose E is a semistable bundle of rank $n \geq 2$. If $\frac{d_{n-1}}{n-1} \geq \frac{d_n}{n}$, $h^0(E) = n + 2$ and there exists a subbundle N of rank $n - 1$ of E with $h^0(N) = n$ and no subbundle N' of E of smaller rank with $h^0(N') > \mathrm{rk}N'$, then*

$$\gamma(E) \geq \min\left\{\gamma_1, \frac{1}{n}(d_n - 2)\right\}.$$

Proof. By Lemma 4.8, $\deg N \geq d_{n-1}$. On the other hand, the definition of d_1 implies that $\deg(E/N) \geq d_1$. Hence

$$\deg E \geq d_{n-1} + d_1.$$

If $d_{n-1} \geq (n-1)d_1 - 2(n-2)$, then $\deg E \geq nd_1 - 2(n-2)$ and hence

$$\gamma(E) \geq \frac{1}{n}\big(nd_1 - 2(n-2) - 4\big) = d_1 - 2 \geq \gamma_1.$$

This covers in particular the case $n = 2$. If $n \geq 3$ and

$$d_{n-1} \leq (n-1)d_1 - 2(n-2) - 1,$$

then
$$d_n \le \frac{n}{n-1}d_{n-1} = d_{n-1} + \frac{1}{n-1}d_{n-1} \le d_{n-1} + d_1 - \frac{2n-3}{n-1},$$
implying $d_n \le d_{n-1} + d_1 - 2$. So $\deg E \ge d_n + 2$ and hence
$$\gamma(E) \ge \frac{1}{n}(d_n - 2). \qquad \square$$

5. Rank two

The results of Section 4 allow us to obtain a precise formula for γ_2 and to improve the inequality for γ_2' obtained in Proposition 3.8.

Theorem 5.1. $\gamma_2 = \min\left\{\gamma_2', \frac{d_2}{2} - 1\right\}$.

Proof. The inequality $\gamma_2 \ge \min\left\{\gamma_2', \frac{d_2}{2} - 1\right\}$ is an immediate consequence of Proposition 4.25 (a). Moreover, by the definition of γ_2' and Proposition 4.9, both inequalities can be equalities. $\qquad \square$

Theorem 5.2. $\gamma_2' \ge \min\left\{\gamma_1, \frac{d_4}{2} - 2\right\}$.

Proof. Suppose E contributes to γ_2'. Since E is semistable, any line subbundle L of E has $\deg L \le g - 1$. If $h^0(L) \ge 2$, then $\gamma(L) \ge \gamma_1$.

Now consider the quotient $L' = E/L$. By semistability of E, $\deg L' \ge \deg L$. If $h^0(L') \le h^0(L)$ then $\gamma(L') \ge \gamma(L) \ge \gamma_1$. Note also that
$$\deg(K_C \otimes L'^*) = 2g - 2 - \deg E + \deg L \ge \deg L.$$
So if $h^0(K_C \otimes L'^*) \le h^0(L)$ we have again $\gamma(K_C \otimes L'^*) \ge \gamma_1$, i.e., $\gamma(L') \ge \gamma_1$. Otherwise we have $h^0(L') \ge 2$ and $h^1(L') \ge 2$. Hence also $\gamma(L') \ge \gamma_1$. This implies
$$\gamma(E) \ge \frac{1}{2}\bigl(\gamma(L) + \gamma(L')\bigr) \ge \gamma_1. \qquad (5.1)$$

Now suppose E has no line subbundle L with $h^0(L) \ge 2$. Write $h^0(E) = 2 + s$ with $s \ge 2$. Then by Lemma 4.8,
$$\deg E \ge d_{2s}.$$
By Lemma 4.2 (a) this implies
$$\deg E \ge d_4 + 2s - 4.$$
So
$$\gamma(E) \ge \frac{1}{2}(d_4 + 2s - 4 - 2s) = \frac{d_4}{2} - 2. \qquad (5.2)$$
The proposition follows from (5.1) and (5.2). $\qquad \square$

Corollary 5.3. $\gamma_2 = \min\left\{\gamma_1, \frac{d_2}{2} - 1\right\} \ge \min\left\{\gamma_1, \frac{\gamma_1}{2} + 1\right\}$. *In particular, if* $\gamma_1 \ge 2$ *and* d_2 *computes the Clifford index, then*
$$\gamma_2 = \frac{\gamma_1}{2} + 1.$$

Proof. By Lemma 4.2 (a), $d_4 \geq d_2 + 2$. Hence the first equality follows from Theorems 5.2, 5.1 and Lemma 2.2. The inequality follows from the fact that $d_2 \geq \gamma_1 + 4$. Moreover, if d_2 computes the Clifford index, then $d_2 = \gamma_1 + 4$. □

Corollary 5.4. *Let C be a general curve of genus g.*

(a) *If $g = 4$, then*

$$\gamma_2 = \gamma_2' = \gamma_1 = 1 < \frac{d_2}{2} - 1.$$

(b) *If $g \geq 5$, then*

$$\gamma_2 = \frac{d_2}{2} - 1 = \frac{1}{2}\left(g - \left[\frac{g}{3}\right]\right).$$

(c) *If $g \geq 7$, $g \neq 8$, then*

$$\gamma_2 < \gamma_2'.$$

Proof. When $g = 4$, we note that $\gamma_1 = 1$ and $d_2 = 5$ which proves (a).

(b) By Corollary 5.3 we have $\gamma_2 = \frac{d_2}{2} - 1$ whenever

$$\frac{d_2}{2} - 1 \leq \gamma_1 = \left[\frac{g-1}{2}\right].$$

By Remark 4.4 (c) this inequality is equivalent to

$$\left[\frac{g-1}{2}\right] \geq \frac{1}{2}\left(g - \left[\frac{g}{3}\right]\right).$$

It is easy to see that this is true for $g \geq 5$.

(c) By (b), $\gamma_2 = \frac{1}{2}(g - [\frac{g}{3}])$. By Theorem 5.2,

$$\gamma_2' \geq \min\left\{\gamma_1, \frac{d_4}{2} - 2\right\} = \min\left\{\left[\frac{g-1}{2}\right], \frac{1}{2}\left(g - \left[\frac{g}{5}\right]\right)\right\}.$$

A simple computation gives the assertion. □

Remark 5.5. From Theorem 5.2, $\gamma_2' \geq \min\{\gamma_1, \frac{d_4}{2} - 2\}$. For a general curve this implies

$$\gamma_2' \geq \min\left\{\left[\frac{g-1}{2}\right], \frac{1}{2}\left(g - \left[\frac{g}{5}\right]\right)\right\}.$$

From this and Lemma 2.2 we obtain $\gamma_2' = \gamma_1$ provided

$$\left[\frac{g-1}{2}\right] \leq \frac{1}{2}\left(g - \left[\frac{g}{5}\right]\right).$$

This holds for $g \leq 10$, $g = 12$ and $g = 14$. For $g \leq 10$ the fact that $\gamma_2' = \gamma_1$ follows already from Proposition 3.8. These results and Corollary 5.4 can be deduced also from [34, Lemma 4.5].

Following on from this remark, we have

Proposition 5.6. *Suppose that $\gamma_1 \geq 2$ and there is no semistable bundle on C of rank 2 and degree d with $h^0 \geq 2 + s$, $s \geq 2$ and $d_{2s} \leq d < 2\gamma_1 + 2s$. Then Conjecture 3.1 holds for $n = 2$.*

Proof. If E contributes to γ_2' and $\gamma(E) < \gamma_1$, the proof of Theorem 5.2 (see in particular (5.1)) shows that $d = \deg E \geq d_{2s}$ where $s = h^0(E) - 2$. The hypotheses now imply that $d \geq 2\gamma_1 + 2s$. So $\gamma(E) \geq \frac{1}{2}(2\gamma_1 + 2s - 2s) = \gamma_1$, a contradiction. It follows that $\gamma_2' = \gamma_1$. Moreover, E is in the range of Conjecture 3.1 (i).

Suppose now that E contributes to γ_2, but not to γ_2'. Then $h^0(E) = 3$ and by Corollary 4.12, $d \geq d_2$. Since $d_2 \geq \gamma_1 + 4$ by Lemma 4.6, this gives

$$\frac{1}{\gamma_1 + 1}(d - 2) + 2 \geq \frac{\gamma_1 + 2}{\gamma_1 + 1} + 2 > h^0(E),$$

verifying Conjecture 3.1 (ii). $\qquad\square$

Remark 5.7. The hypotheses of Proposition 5.6 certainly hold if $d_4 \geq 2\gamma_1 + 4$. In any case, a semistable bundle E on C of rank 2 and degree $d \leq 2g - 2$ with $h^0(E) \geq 4$ and $\gamma(E) < \gamma_1$ has no line subbundle with $h^0 \geq 2$ (compare again the proof of Theorem 5.2). We could consider extensions of the form

$$0 \to L \to E \to M \to 0,$$

where L and M are line bundles with $h^0(L) = 1$ and $h^0(M) \geq s + 1$. The problem in constructing E in this way is that one needs to lift a large number of sections of M to E; this is a difficult problem and is likely to require geometric information about C beyond that provided by the gonality sequence (compare [45]).

6. Ranks three and four

Theorem 6.1. *Suppose $\frac{d_2}{2} \geq \frac{d_3}{3}$. Then*

$$\gamma_3 = \min\left\{ \gamma_3', \frac{1}{3}(d_3 - 2) \right\}.$$

Proof. Suppose E contributes to γ_3. If $h^0(E) \geq 6$, then $\gamma(E) \geq \gamma_3'$ by definition of γ_3'. All other possibilities are covered by Propositions 4.25 and 4.26. Hence $\gamma_3 \geq \min\{\gamma_3', \frac{1}{3}(d_3 - 2)\}$.

Moreover, by the definition of γ_3' there exists E with $\gamma(E) = \gamma_3'$. By Proposition 4.9 (d), there exists a semistable bundle E of rank 3 and degree d_3 with $h^0(E) \geq 4$. Since $d_3 \leq 3g - 3$ by Lemma 4.2 (d), this gives $\gamma_3 \leq \gamma(E) \leq \frac{1}{3}(d_3 - 2)$. This completes the proof. $\qquad\square$

Theorem 6.2. *If $\frac{d_3}{3} \geq \frac{d_4}{4}$, then*

$$\gamma_4 = \min\left\{ \gamma_4', \frac{1}{4}(d_4 - 2), \frac{1}{2}(d_2 - 2) \right\}.$$

Proof. The result holds for $\gamma_1 = 0$ and 1 by Proposition 2.6 (a) and the fact that $d_2 \geq 4$ and $d_4 \geq 6$. So suppose $\gamma_1 \geq 2$.

Suppose that E contributes to γ_4. In order to prove the inequality

$$\gamma_4 \geq \min\left\{ \gamma_4', \frac{1}{4}(d_4 - 2), \frac{1}{2}(d_2 - 2) \right\}$$

we may assume by Propositions 4.25 and 4.26 that $h^0(E) = 7$ and E admits a subbundle N of rank p with $2 \leq p \leq 3$ and $h^0(N) = p + 1$ and such that E does not admit a subbundle of smaller rank with $h^0 > \text{rk}$.

Case 1: $p = 2$. We have $h^0(N) = 3$ and $h^0(E) = 7$, so $h^0(E/N) \geq 4$. If E/N has no line subbundle with $h^0 \geq 2$, then Lemma 4.8 gives

$$\deg E \geq \deg N + \deg(E/N) \geq d_2 + d_4 \geq d_4 + 4.$$

This implies

$$\gamma(E) \geq \frac{1}{4}(d_4 + 4 - 6) = \frac{1}{4}(d_4 - 2). \tag{6.1}$$

If E/N has a line subbundle M with $h^0(M) \geq 3$, then $\deg M \geq d_2$ by definition of d_2 and $\deg((E/N)/M) \geq \frac{d}{4} \geq \frac{d_2}{2}$, since E is semistable. So

$$\deg E \geq d_2 + d_2 + \frac{d_2}{2} \geq 2d_2 + 2,$$

since $d_2 \geq 4$. Hence

$$\gamma(E) \geq \frac{1}{4}(2d_2 + 2 - 6) = \frac{1}{2}(d_2 - 2). \tag{6.2}$$

If E/N has a line subbundle M with $h^0(M) = 2$, then $\deg M \geq d_1$ and $\deg((E/N)/M) \geq d_1$. So

$$\deg E \geq d_2 + 2d_1.$$

If $d_2 \leq 2d_1 - 2$, then $\deg E \geq 2d_2 + 2$, so (6.2) holds. If $d_2 \geq 2d_1 - 1$, then $\deg E \geq 4d_1 - 1$ and hence

$$\gamma(E) \geq \frac{1}{4}(4d_1 - 1 - 6) = d_1 - \frac{7}{4} > d_1 - 2 \geq \gamma_1 \geq \gamma_4'. \tag{6.3}$$

Case 2: $p = 3$. We have $h^0(N) = 4$ and $h^0(E) = 7$. So $h^0(E/N) \geq 3$ and hence $\deg(E/N) \geq d_2$ by definition of d_2. If $d_3 \geq d_2 + 2$, then $\deg E \geq 2d_2 + 2$ and hence again (6.2) holds. If $d_3 = d_2 + 1$, then using the hypothesis,

$$\deg E \geq d_2 + d_3 = 2d_3 - 1 \geq \frac{3}{2}d_4 - 1.$$

If $d_4 \geq 10$, this gives $\deg E \geq d_4 + 4$, so (6.1) holds. Otherwise, by Lemma 4.6, we have $g = 5$ or $g = 6$. In either case, using Remark 4.4 (c) and Lemma 4.6, we get $d_2 = 6$ and $d_4 \leq 9$. So

$$\deg E \geq d_2 + d_3 \geq d_4 + 4$$

and again (6.1) holds.

The inequality $\gamma_4 \geq \min\left\{ \gamma_4', \frac{1}{4}(d_4 - 2), \frac{1}{2}(d_2 - 2) \right\}$ follows from the inequalities (6.1), (6.2) and (6.3).

The proof of the equality is similar to the last part of the proof of Theorem 6.1. To obtain $\gamma(E) = \frac{1}{2}(d_2 - 2)$, we define E to be $E = E_L \oplus E_L$ where L is a line bundle of degree d_2 with $h^0(L) = 3$. $\qquad\square$

Remark 6.3. The conditions $\frac{d_2}{2} \geq \frac{d_3}{3}$ and $\frac{d_3}{3} \geq \frac{d_4}{4}$ are satisfied for general curves (see proof of Proposition 4.17) and also for hyperelliptic, trigonal, general tetragonal and bielliptic curves (see Remark 4.5).

For $\gamma_1 \leq 2$, we already know that $\gamma_n' = \gamma_1$ and we have precise values for the γ_n from Corollary 4.22, so Theorems 6.1 and 6.2 do not add anything to our knowledge in these cases. For a general curve of genus $g \geq 7$, we can show that $\frac{1}{2}(d_2 - 2)$, $\frac{1}{3}(d_3 - 2)$ and $\frac{1}{4}(d_4 - 2)$ are all ≥ 2. So, in the absence of any good lower bound for γ_n', Theorems 6.1 and 6.2 tell us that

$$2 \leq \gamma_n \leq \gamma_n' \leq \gamma_1$$

for $n = 3, 4$.

7. Rank five

In this section we obtain partial results for γ_5.

Lemma 7.1. *Suppose* $\frac{d_4}{4} \geq \frac{d_5}{5}$. *Let* E *be a semistable bundle of rank* 5 *with* $h^0(E) \leq 9$ *and* N *a subbundle of rank* p, $2 \leq p \leq 4$, *with* $h^0(N) \geq p + 2$. *Suppose further that* E *has no subbundle of rank* $< p$ *with* $h^0 > \mathrm{rk}$. *Then*

$$\gamma(E) \geq \frac{1}{5}(d_5 - 2).$$

Proof. Lemma 4.8 implies that $\deg N \geq d_{2p}$. Semistability gives

$$\deg E \geq \frac{5 d_{2p}}{p}.$$

If $p = 2$, the hypothesis implies that

$$\deg E \geq \frac{5 d_4}{2} \geq \frac{5}{2}\frac{4}{5}d_5 = 2d_5 \geq d_5 + 8,$$

since $d_5 \geq 8$ by Lemma 4.6. If $p = 3$ or 4, then $2p > 5$, so

$$\deg E \geq \frac{5 d_{2p}}{p} \geq \frac{5}{p}(d_5 + 2p - 5) = d_5 + \frac{5 - p}{p}d_5 + 10 - \frac{25}{p} \geq d_5 + 2 + \frac{15}{p}.$$

So $\deg E \geq d_5 + 6$. Hence in any case

$$\gamma(E) \geq \frac{1}{5}(d_5 - 2). \qquad \square$$

Lemma 7.2. *Let* F *be a vector bundle of rank* 2 *with* $h^0(F) \geq 2 + t$, $t \geq 1$. *Suppose* F *is a quotient of a semistable bundle* E *of rank* n *and degree* $d > 0$. *Then*

$$\deg F \geq \min_{1 \leq u \leq t-1}\left\{ d_{2t}, d_t + \frac{d}{n}, d_u + d_{t-u} \right\}.$$

Proof. If F has no line subbundle with $h^0 \geq 2$, then Lemma 4.8 implies that $\deg F \geq d_{2t}$.

Otherwise let N be a line subbundle with $h^0(N) = 1 + u$, $u \geq 1$. If $u \geq t$, then $\deg N \geq d_t$ and $\deg(F/N) \geq \frac{d}{n}$ by semistability of E. If $u \leq t - 1$, then $\deg N \geq d_u$ and $\deg(F/N) \geq d_{t-u}$, since $h^0(F/N) \geq t - u + 1 \geq 2$. $\quad\square$

Theorem 7.3. *If $\frac{d_p}{p} \geq \frac{d_{p+1}}{p+1}$ for $1 \leq p \leq 4$, then*

$$\gamma_5 \geq \min\left\{\gamma_5', \frac{1}{2}(d_2 - 2), \frac{1}{5}(d_5 - 2), \frac{1}{5}(d_1 + 2d_2 - 6), \frac{1}{5}(d_1 + d_4 - 4),\right.$$
$$\left.\frac{1}{5}(2d_1 + d_3 - 6), \frac{1}{5}(3d_1 + d_2 - 8), \frac{1}{5}(d_2 + d_3 - 5)\right\}.$$

Proof. Let E be a semistable vector bundle of rank 5 and degree d. By Proposition 4.25 and Lemma 7.1 we may assume that $h^0(E) = 5 + s$ with $2 \leq s \leq 4$ and E admits a subbundle N of rank p with $2 \leq p \leq 4$ and $h^0(N) = p + 1$ and such that E does not admit a subbundle of smaller rank with $h^0 > \mathrm{rk}$.

If $\gamma_1 \leq 2$ or $g \leq 8$, we have precise values for all γ_n by Proposition 2.6 (a) and Theorems 3.6 and 4.21. One can check that these values satisfy the required inequality. So we may assume that $\gamma_1 \geq 3$ and $g \geq 9$, implying by Lemma 4.6 that

$$d_2 \geq 7, \; d_3 \geq 9, \; d_4 \geq 11 \text{ and } d_5 \geq 13.$$

In fact, in the proof we use only $\gamma_1 \geq 2$, $d_2 \geq 6$ and $d_3 \geq 9$.

Case 1: $p = 2$. We have $h^0(N) = 3$. So Lemma 4.8 implies $\deg N \geq d_2$ and hence $\frac{d}{5} \geq \frac{d_2}{2} \geq 3$ by semistability of E. Moreover, E/N has rank 3 and $h^0(E/N) \geq s+2$.

Case 1 a: Suppose E/N has no proper subbundle with $h^0 > \mathrm{rk}$. Then by Lemma 4.8, $\deg(E/N) \geq d_{3(s-1)}$. So

$$\deg E \geq d_2 + d_{3(s-1)} \geq d_2 + d_3 + 3(s - 2)$$

and

$$\gamma(E) \geq \frac{1}{5}(d_2 + d_3 + 3(s - 2) - 2s) \geq \frac{1}{5}(d_2 + d_3 - 4). \tag{7.1}$$

Case 1 b: Suppose E/N has a line subbundle M with $h^0(M) \geq 2$. If $h^0(M) \geq 3$, then $\deg M \geq d_2$. So by semistability of E,

$$\frac{d}{5} \leq \frac{\deg((E/N)/M)}{2} \leq \frac{d - 2d_2}{2}.$$

Hence

$$d \geq \frac{10}{3}d_2$$

and

$$\gamma(E) \geq \frac{1}{5}\left(\frac{10}{3}d_2 - 8\right) = \frac{d_2}{2} + \frac{d_2}{6} - \frac{8}{5} \geq \frac{d_2}{2} - \frac{3}{5}. \tag{7.2}$$

Now suppose $h^0(M) = 2$. Then $\deg M \geq d_1$ and $\deg((E/N)/M) \geq \frac{2d}{5}$ by semistability of E. So

$$\deg E \geq d_2 + d_1 + \frac{2d}{5} \geq 2d_2 + d_1 \geq d_2 + d_3.$$

If $s = 2$, this gives

$$\gamma(E) \geq \frac{1}{5}(d_2 + d_3 - 4). \qquad (7.3)$$

If $s = 3$, then $h^0((E/N)/M) \geq 3$. So $\deg((E/N)/M) \geq \min\{\, d_2, d_1 + \frac{d}{5} \,\}$ by Lemma 7.2 and

$$
\begin{aligned}
\deg E \; &\geq \; \min\left\{\, d_1 + 2d_2, 2d_1 + d_2 + \frac{d}{5} \,\right\} \\
&\geq \; \min\left\{\, d_1 + 2d_2, 2d_1 + \frac{3d_2}{2} \,\right\} \geq \min\{\, d_1 + 2d_2, 2d_1 + d_3 \,\}.
\end{aligned}
$$

Hence

$$\gamma(E) \geq \min\left\{\, \frac{1}{5}(d_1 + 2d_2 - 6), \frac{1}{5}(2d_1 + d_3 - 6) \,\right\}. \qquad (7.4)$$

If $s = 4$, then $\deg M \geq d_1$ and $h^0((E/N)/M) \geq 4$. Using Lemma 7.2, this gives $\deg((E/N)/M) \geq \min\{\, d_4, 2d_1, d_2 + \frac{d}{5} \,\}$. So

$$
\begin{aligned}
\deg E \; &\geq \; \min\left\{\, d_2 + d_1 + d_4, d_2 + 3d_1, 2d_2 + d_1 + \frac{d}{5} \,\right\} \\
&\geq \; \min\{\, d_5 + 6, 3d_1 + d_2, d_1 + 2d_2 + 3 \,\}
\end{aligned}
$$

and hence

$$\gamma(E) \geq \min\left\{\, \frac{1}{5}(d_5 - 2), \frac{1}{5}(3d_1 + d_2 - 8), \frac{1}{5}(d_1 + 2d_2 - 5) \,\right\}. \qquad (7.5)$$

Case 1 c: Suppose E/N has a subbundle M of rank 2 with $h^0(M) \geq 3$ and no line subbundle with $h^0 \geq 2$. Then $\deg M \geq d_2$ and $\deg((E/N)/M) \geq \frac{d}{5}$. If $h^0(M) \geq 4$, then $\deg M \geq d_4$ by Lemma 4.8, so

$$\deg E \geq d_2 + d_4 + \frac{d}{5} \geq d_2 + d_3 + 4$$

and

$$\gamma(E) \geq \frac{1}{5}(d_2 + d_3 - 4). \qquad (7.6)$$

Now suppose $h^0(M) = 3$. Then

$$\deg E \geq 2d_2 + \frac{d}{5} \geq \frac{5}{2}d_2.$$

If $s = 2$, this gives

$$\gamma(E) \geq \frac{1}{5}\left(\frac{5}{2}d_2 - 4\right) = \frac{d_2}{2} - \frac{4}{5}. \qquad (7.7)$$

If $s = 3$, then $h^0((E/N)/M) \geq 2$. So $\deg E \geq 2d_2 + d_1$ and

$$\gamma(E) \geq \frac{1}{5}(d_1 + 2d_2 - 6). \tag{7.8}$$

If $s = 4$, then $h^0((E/N)/M) \geq 3$ and $\deg E \geq 3d_2$. So,

$$\gamma(E) \geq \frac{1}{5}(3d_2 - 8) \geq \frac{1}{5}\left(\frac{5}{2}d_2 + 3 - 8\right) = \frac{1}{2}(d_2 - 2). \tag{7.9}$$

Case 2: $p = 3$. We have $h^0(N) = 4$. So Lemma 4.8 implies $\deg N \geq d_3$ and by semistability, $\frac{d}{5} \geq \frac{d_3}{3} \geq 3$. Then E/N has rank 2 and $h^0(E/N) \geq s+1$. So we can apply Lemma 7.2 with $t = s - 1$.

If $s = 2$, Lemma 7.2 gives

$$\deg E \quad \geq \quad \min\left\{ d_3 + d_2, d_3 + d_1 + \frac{d}{5} \right\}$$

$$\geq \quad \min\left\{ d_2 + d_3, d_1 + \frac{4}{3}d_3 \right\} \geq \min\{ d_2 + d_3, d_1 + d_4 \}$$

and hence

$$\gamma(E) \geq \min\left\{ \frac{1}{5}(d_2 + d_3 - 4), \frac{1}{5}(d_1 + d_4 - 4) \right\}. \tag{7.10}$$

If $s = 3$, Lemma 7.2 gives

$$\deg E \geq \min\left\{ d_3 + d_4, d_3 + d_2 + \frac{d}{5}, d_3 + 2d_1 \right\} \geq \min\{ d_2 + d_3 + 2, 2d_1 + d_3 \}.$$

So

$$\gamma(E) \geq \min\left\{ \frac{1}{5}(d_2 + d_3 - 4), \frac{1}{5}(2d_1 + d_3 - 6) \right\}. \tag{7.11}$$

If $s = 4$, Lemma 7.2 gives

$$\deg E \geq \min\left\{ d_3 + d_6, 2d_3 + \frac{d}{5}, d_3 + d_1 + d_2 \right\} \geq d_2 + d_3 + 4.$$

So

$$\gamma(E) \geq \frac{1}{5}(d_2 + d_3 - 4). \tag{7.12}$$

Case 3: $p = 4$. We have $h^0(N) = 5$. We can assume by Proposition 4.26 that $h^0(E) \geq 8$. If $h^0(E) = 8$, then $h^0(E/N) \geq 3$ and hence $\deg(E/N) \geq d_2$. So

$$\deg E \geq d_2 + d_4 \geq d_2 + d_3 + 1.$$

Hence

$$\gamma(E) \geq \frac{1}{5}(d_2 + d_3 + 1 - 6) = \frac{1}{5}(d_2 + d_3 - 5). \tag{7.13}$$

If $h^0(E) = 9$, then $h^0(E/N) \geq 4$ and hence $\deg(E/N) \geq d_3$. So

$$\deg E \geq d_4 + d_3.$$

If $d_4 \geq 2d_2 - 2$, then $\deg E \geq d_2 + d_3 + d_2 - 2 \geq d_2 + d_3 + 4$. So

$$\gamma(E) \geq \frac{1}{5}(d_2 + d_3 + 4 - 8) = \frac{1}{5}(d_2 + d_3 - 4). \tag{7.14}$$

If $d_4 \leq 2d_2 - 3$, then

$$d_5 \leq \frac{5}{4}d_4 = d_4 + \frac{1}{4}d_4 \leq d_4 + \frac{1}{2}d_2 - \frac{3}{4}.$$

So $\deg E \geq d_5 - \frac{1}{2}d_2 + \frac{3}{4} + d_3$ and hence

$$\gamma(E) = \frac{1}{5}(\deg E - 8) \geq \frac{1}{5}(d_5 - 2), \tag{7.15}$$

provided $d_3 - \frac{1}{2}d_2 + \frac{3}{4} > 5$, i.e., $d_3 > \frac{1}{2}d_2 + \frac{17}{4}$. Since $d_3 \geq d_2 + 1$ and we are assuming $d_3 \geq 9$, this holds.

The assertion follows from the inequalities (7.1), ..., (7.15). □

Corollary 7.4. *Let C be a general curve. Then*

$$\gamma_5 = \min\left\{ \gamma_5', \frac{1}{5}\left(g - \left[\frac{g}{6}\right] + 3 \right) \right\}.$$

Proof. For $g = 4$ this follows from Proposition 2.6 (a). For $g = 5$ and $g = 6$ we can check it directly from Corollary 4.22. For $g \geq 7$, the inequality $\gamma_5 \geq \min\{ \gamma_5', \frac{1}{5}(g - [\frac{g}{6}] + 3) \}$ follows by evaluating the numbers on the right-hand side of the formula in Theorem 7.3 using Remark 4.4 (c). There exists E computing γ_5' by definition. Moreover, the conditions of Propositions 4.9 (e) and 4.17 imply the existence of a stable bundle E_L of rank 5 and degree $d_5 = g - [\frac{g}{6}] + 5$ with $h^0(E_L) = 6$. □

Remark 7.5. It would be interesting to determine whether $\gamma_5' \geq \frac{1}{5}(g - [\frac{g}{6}] + 3)$.

Remark 7.6. We do not know how many of the terms on the right-hand side of the inequality in Theorem 7.3 are necessary.

Under the hypotheses of the proposition, we do know that there exists a semistable bundle E of degree d_5 with $h^0(E) \geq 6$ (see Proposition 4.9). For this E,

$$\gamma(E) \leq \frac{1}{5}(d_5 - 2).$$

Moreover, there exists by Proposition 4.9 a semistable bundle N of rank 2 and degree d_2 with $h^0(N) \geq 3$. Suppose d_2 is even and let L be a line bundle of degree $\frac{d_2}{2}$ with $h^0(L) \geq 1$. Then

$$E = N \oplus N \oplus L$$

is semistable of degree $\frac{5}{2}d_2$ with $h^0(E) \geq 7$. So

$$\gamma(E) \leq \frac{1}{5}\left(\frac{5}{2}d_2 - 4 \right) = \frac{1}{2}d_2 - \frac{4}{5}.$$

Even in this case we do not know whether there always exist bundles E with $\gamma(E) \leq \frac{1}{2}d_2 - 1$.

If $\frac{d_2}{2} = \frac{d_3}{3}$, we can take $E = N \oplus N'$, where N is semistable of rank 2 and degree d_2 with $h^0(N) \geq 3$ and N' is semistable of rank 3 and degree d_3 with

$h^0(N') \geq 4$. Then E is semistable of rank 5 with $h^0(E) \geq 7$, so

$$\gamma(E) \leq \frac{1}{5}(d_2 + d_3 - 4).$$

In an attempt to construct a semistable bundle E of rank 5 with

$$\gamma(E) = \frac{1}{5}(d_2 + d_3 - 5), \tag{7.16}$$

as in the proof of the proposition, we start with a bundle N of rank 4 and degree d_4 with $h^0(N) = 5$ and a line bundle L of degree d_2 with $h^0(L) = 3$. We consider extensions

$$0 \to N \to E \to L \to 0.$$

If all sections of L lift to E, then $h^0(E) = 8$ and

$$\gamma(E) = \frac{1}{5}(d_2 + d_4 - 6).$$

To achieve (7.16), we need $d_4 = d_3 + 1$. If $d_4 \leq d_2 + 4$, then

$$\deg E \geq 2d_4 - 4 \geq 2\frac{4}{5}d_5 - 4 = d_5 + \frac{3}{5}d_5 - 4,$$

implying that $\deg E \geq d_5 + 4$ if $d_5 \geq 12$. Hence in this case we have

$$\gamma(E) \geq \frac{1}{5}(d_5 + 4 - 6) = \frac{1}{5}(d_5 - 2). \tag{7.17}$$

This is true for $g \geq 8$. There remains the possibility that $d_4 = d_3 + 1$ and $d_3 \geq d_2 + 4$. If one can show that in this case we cannot have E semistable with $h^0(E) = 8$, then we can replace $\frac{1}{5}(d_2 + d_3 - 5)$ by $\frac{1}{5}(d_2 + d_3 - 4)$ which looks more natural.

8. Plane curves

For smooth plane curves the numbers d_r are known by Noether's Theorem (see [14]). For stating it, note that for any positive integer r, there are uniquely determined integers α, β with $\alpha \geq 1, 0 \leq \beta \leq \alpha$ such that

$$r = \frac{\alpha(\alpha + 3)}{2} - \beta.$$

The reason for this is that for any α,

$$\frac{\alpha(\alpha + 3)}{2} - (\alpha + 1) = \frac{(\alpha - 1)(\alpha + 2)}{2}.$$

Noether's Theorem. *Let C be a smooth plane curve of degree δ. For any positive integer r,*

$$d_r = \begin{cases} \alpha\delta - \beta & \text{if} \quad r < g = \frac{(\delta-1)(\delta-2)}{2}, \\ r + \frac{(\delta-1)(\delta-2)}{2} & \text{if} \quad r \geq g. \end{cases}$$

In particular $d_1 = \delta - 1$, $d_2 = \delta$ and d_2 computes $\gamma_1 = \delta - 4$. Note that $r < g = \frac{(\delta-1)(\delta-2)}{2}$ is equivalent to $\alpha \leq \delta - 3$.

Proposition 8.1. *Let C be a smooth plane curve of degree $\delta \geq 5$. Then*

(a)
$$\gamma_2 = \begin{cases} 1 & \text{if } \delta = 5, \\ \frac{\delta}{2} - 1 & \text{if } \delta \geq 6. \end{cases}$$

(b)
$$\gamma_2' = \gamma_1 = \delta - 4.$$

Proof. (a) is a special case of Corollary 5.3. For (b), we have $\gamma_1 = \delta - 4$ and $d_4 = 2\delta - 1$. By Theorem 5.2, $\gamma_2' \geq \min\{\delta - 4, \delta - \frac{5}{2}\} = \delta - 4$. So (b) follows from Lemma 2.2. $\qquad\square$

Remark 8.2. Part (a) of Proposition 8.1 holds also for a curve which admits as a plane model a general nodal curve of degree $\delta \geq 7$ with $\nu \leq \frac{1}{2}(\delta^2 - 7\delta + 14)$ nodes. This follows from the third paragraph on page 90 of [16] stating that $\gamma_1 = \delta - 4$ in this case. Also $d_1 = \delta - 2$ and $d_2 = \delta$.

Hence there exist curves of all genera $g \geq 8$ with

$$\gamma_2 = \frac{\gamma_1}{2} + 1 < \gamma_1 = d_2 - 4.$$

In particular these curves are not general and both d_1 and d_2 compute γ_1.

Theorem 6.1 does not apply for plane curves, since $\frac{d_2}{2} < \frac{d_3}{3}$ for $\delta \geq 5$. Indeed, its statement is wrong for plane curves. Instead we have

Proposition 8.3. *Let C be a smooth plane curve of degree $\delta \geq 5$. Then*

$$\gamma_3 = \min\left\{\gamma_3', \frac{1}{3}\left(\left\lceil\frac{3\delta + 1}{2}\right\rceil - 2\right)\right\}.$$

Proof. If $\delta = 5$, then both sides of the equality are 1 by Proposition 2.6 (a). So assume $\delta \geq 6$. From the definition of γ_3' we may assume that E is a semistable bundle of rank 3 and slope $\leq g - 1$ with $4 \leq h^0(E) \leq 5$.

Suppose first that $h^0(E) = 4$. If E has no proper subbundle N with $h^0(N) > \mathrm{rk}N$, then Lemma 4.8 implies that

$$\deg E \geq d_3 = 2\delta - 2.$$

If E has a line subbundle N with $h^0(N) \geq 2$, then $\deg N \geq d_1$ and so by semistability,

$$\deg E \geq 3d_1 = 3\delta - 3.$$

Suppose E has a subbundle N of rank 2 with $h^0(N) \geq 3$ and no line subbundle with $h^0 \geq 2$. If $h^0(N) = 4$, then $\deg N \geq d_4$ by Lemma 4.8 and hence by semistability,

$$\deg E \geq \frac{3}{2}d_4 = 3\delta - \frac{3}{2}.$$

If $h^0(N) = 3$, then $\deg N \geq d_2 = \delta$ by Lemma 4.8 and so

$$\deg E \geq \frac{3}{2}\delta.$$

Hence, if $h^0(E) = 4$ we get

$$\gamma(E) = \frac{1}{3}(\deg E - 2) \geq \frac{1}{3}\left(\left\lceil\frac{3\delta + 1}{2}\right\rceil - 2\right). \tag{8.1}$$

Now suppose $h^0(E) = 5$. If E has no proper subbundle N with $h^0(N) > \mathrm{rk}N$, then Lemma 4.8 implies that

$$\deg E \geq d_6 = 3\delta - 3.$$

If E has a line subbundle N with $h^0(N) \geq 2$, then $\deg N \geq d_1$ and so by semistability,

$$\deg E \geq 3d_1 = 3\delta - 3.$$

Suppose E has a subbundle N of rank 2 with $h^0(N) \geq 3$ and no line subbundle with $h^0 \geq 2$. If $h^0(N) \geq 4$, then $\deg N \geq d_4 = 2\delta - 1$ by Lemma 4.8 and so

$$\deg E \geq \frac{3}{2}d_4 = 3\delta - \frac{3}{2}.$$

If $h^0(N) = 3$, then $\deg N \geq d_2 = \delta$ and $\deg(E/N) \geq d_1 = \delta - 1$ and so

$$\deg E \geq 2\delta - 1.$$

Hence, if $h^0(E) = 5$ we get

$$\gamma(E) = \frac{1}{3}(\deg E - 4) \geq \frac{1}{3}(2\delta - 5). \tag{8.2}$$

Since $2\delta - 5 \geq \left[\frac{3\delta+1}{2}\right] - 2$ for $\delta \geq 6$, (8.1) and (8.2) imply the inequality $\gamma_3 \geq \min\left\{\gamma_3', \frac{1}{3}\left(\left[\frac{3\delta+1}{2}\right] - 2\right)\right\}$.

To show equality we have to show that the bound $\frac{1}{3}\left(\left[\frac{3\delta+1}{2}\right] - 2\right)$ can be attained. Since $d_1 > \frac{d_2}{2}$, Theorem 4.15 (a) implies the existence of a semistable bundle N of rank 2 and degree $d_2 = \delta$ with $h^0(N) = 3$.

If δ is even take an effective line bundle M of degree $\frac{\delta}{2}$. Then

$$E = N \oplus M$$

is semistable, has degree $\frac{3\delta}{2}$ and $h^0(E) = 4$.

If δ is odd, choose an effective line bundle M of degree $\frac{\delta+1}{2}$. We have $h^0(M) = 1$, since $d_1 = \delta - 1$. Any non-zero section of M induces a map $H^1(M^* \otimes N) \to H^1(\mathcal{O}_C \otimes N)$. Comparing dimensions one checks that this map has a non-trivial kernel. Every non-trivial extension

$$0 \to N \to E \to M \to 0$$

in the kernel of this map defines a bundle E of rank 3 and degree $\frac{3\delta+1}{2}$ with $h^0(E) = 4$. Since any such extension is stable, this completes the proof. $\square$

Proposition 8.4. *Let C be a smooth plane curve of degree $\delta \geq 5$. Then*

$$\gamma_4 = \min\left\{\gamma_4', \frac{\delta}{2} - 1\right\}.$$

Proof. We have $d_3 = 2\delta - 2$ and $d_4 = 2\delta - 1$. So $\frac{d_3}{3} > \frac{d_4}{4}$ and Theorem 6.2 applies to give the assertion. $\square$

Proposition 8.5. *Let C be a smooth plane curve of degree $\delta \geq 5$. Then*

$$\gamma_5 = \min\left\{\gamma_5', \frac{2}{5}(\delta - 1)\right\}.$$

Proof. If $\delta = 5$, then, since $\gamma_1 = 1$, we get from Proposition 2.6 (a) that $\gamma_5 = \gamma_5' = 1$ and the result is obvious. Suppose $\delta \geq 6$. From Noether's Theorem we get

$$d_1 = \delta - 1, \ d_2 = \delta, \ d_3 = 2\delta - 2, \ d_4 = 2\delta - 1 \text{ and } d_5 = 2\delta.$$

It follows that $\frac{d_p}{p} > \frac{d_5}{5}$ for all $p < 5$ and $\frac{d_p}{p} \geq \frac{d_{p+1}}{p+1}$ except when $p = 2$. Hence Proposition 4.25 and Lemma 7.1 are valid. The only place in the proof of Theorem 7.3, where the assumption $\frac{d_2}{2} \geq \frac{d_3}{3}$ is used, is in the proof of inequality (7.4). If we replace $\frac{1}{5}(2d_1 + d_3 - 6)$ by $\frac{1}{5}(2d_1 + \frac{3}{2}d_2 - 6)$, the proof is valid. With this modification, Theorem 7.3 becomes

$$\gamma_5 \geq \min\left\{\gamma_5', \frac{1}{2}(\delta - 2), \frac{1}{5}(2\delta - 2), \frac{1}{5}\left(\frac{7}{2}\delta - 8\right), \frac{1}{5}(4\delta - 11), \frac{1}{5}(3\delta - 7)\right\}$$

$$= \min\left\{\gamma_5', \frac{1}{5}(2\delta - 2)\right\}.$$

By the definition of γ_5' the equality $\gamma(E) = \gamma_5'$ can be attained. The equality $\gamma(E) = \frac{1}{5}(2\delta - 2)$ is attained by the bundle E_L, where L is a line bundle of degree $d_5 = 2\delta$ with $h^0(L) = 6$. Since $\frac{d_p}{p} > \frac{d_5}{5}$ for all $p < 5$, the bundle E_L is stable by Proposition 4.9 (e) and hence $h^0(E_L) = 6$ by Theorem 4.15. □

9. Problems

In this section we present some problems which are related to the contents of this paper.

Problem 9.1. *Find an improved lower bound for γ_5 and good lower bounds for $\gamma_n, n \geq 6$.*

We expect that the term $\frac{1}{p}(d_p - 2)$ for $p|n$ will appear (see Lemma 4.24 and Proposition 4.25). There is some evidence that terms of the form $\frac{1}{n}(d_p + d_{n-p} - 4)$ may appear, but it is possible that a careful argument may eliminate them. See also Remark 7.6.

Problem 9.2. *Find good lower bounds for γ_n' for $n \geq 3$.*

In relation to this we have the conjecture

Conjecture 9.3. $\gamma_n' = \gamma_1$.

The conjecture is the most important point of Mercat's Conjecture 3.1 (see Proposition 3.3). If this is true, a complete proof seems a long way off. In many ways it would be more interesting if the conjecture were false, since this would imply the existence of new semistable bundles reflecting aspects of the geometry of the curve C not detected by classical Brill–Noether theory. A small piece of evidence in favour of the conjecture is presented by Proposition 8.1 which shows that there exist curves of arbitrary γ_1 for which $\gamma_2' = \gamma_1$.

Problem 9.4. *One can define Clifford indices γ_n^s and $\gamma_n^{s\prime}$ by restricting to stable bundles. Of course $\gamma_n \leq \gamma_n^s$ and $\gamma_n' \leq \gamma_n^{s\prime}$. Find examples for which we have strict inequalities or prove there are none.*

Problem 9.5. *Obtain more information about the gonality sequence of a curve C.*

This should contribute to Problem 9.1. For example, the knowledge of the gonality sequence for smooth plane curves enabled us to prove Proposition 8.5 which is a significant improvement on Theorem 7.3.

We have seen that the classical Clifford index alone is not sufficient to determine γ_n. However, the following problem remains open.

Problem 9.6. *Show that γ_n is determined by the gonality sequence or find counter-examples.*

Let E be a semistable bundle on C and V a subspace of $H^0(E)$ which generates E. Let $M_{V,E}$ be defined by (1.2). It has been conjectured by Butler [12] that for general C, E, V, the bundle $M_{V,E}$ is semistable. (Actually this is a slight modification of Butler's original conjecture [12, Conjecture 2] which is set in the context of coherent systems.) In [11, Theorem 1.2] Butler proved for any C and any semistable E that $M_{V,E}$ is semistable if $\mu(E) \geq 2g$. There are many similar results in the literature (a summary of the current state of knowledge for the case where C is general and E is a line bundle may be found in [6, Section 9]). Our Proposition 4.9 is a further example where C is not required to be general.

Problem 9.7. *Give a proof of Butler's conjecture or obtain counter-examples for either general or special curves.*

A solution of this conjecture would be interesting not only in its own right but because the bundles $M_{V,E}$ are related to syzygies and to Picard bundles (see in particular [20, 29], where it is shown that, for sufficiently large degree, the bundle $M_{H^0(E),E}$ is, up to twisting by a line bundle, the restriction of a Picard bundle to the curve C embedded in the relevant moduli space). A further observation is that, if we use (1.2) to map C to the Grassmannian G of n-dimensional quotients of V, then E and $M_{V,E}^*$ are the pullbacks of the tautological quotient bundle and subbundle respectively, so $E \otimes M_{V,E}$ is isomorphic to the pullback of the tangent bundle of G. Thus Butler's conjecture implies semistability of this pullback (see [10, Theorem 4.5] for a recent result in this direction).

We next move on to consider extensions. Given an exact sequence

$$0 \to L \to E \to M \to 0 \tag{9.1}$$

with L and M line bundles, there is a geometric criterion for lifting a section of M to E (see [27]). In our context this leads to several problems.

Problem 9.8. *Extend this to a usable criterion for the case when L and M are vector bundles.*

Problem 9.9. *Try to find a usable criterion for lifting several sections.*

Given vector bundles L and M, the classes of nontrivial extensions (9.1) are parametrized by the projective space $P = P(\text{Ext}^1(M, L))$. The extensions with E semistable form an open set U of P, whereas the extensions for which a given number of independent sections of M are liftable to E form a closed subset V of P.

Problem 9.10. *Determine conditions under which $U \cap V$ is non-empty.*

If $\dim V > \dim(P \setminus U)$, the intersection $U \cap V$ is clearly non-empty. This has been used in several papers, however there are many situations in which the dimensional condition does not hold.

Problem 9.11. *Improve the lemma of Paranjape and Ramanan (Lemma 4.8) and determine conditions under which the converse is true.*

This would be very useful for improving some of the bounds for γ_n and constructing bundles E with low values for $\gamma(E)$.

We finish with one very specific problem.

Problem 9.12. *What is the minimal value of d for which there exists a stable (semistable) bundle of rank 2 with $h^0 \geq 4$?*

By using Lemma 4.8 one can show that, in the semistable case,

$$\min\{2d_1, d_4\} \leq d \leq 2d_1$$

(see the proof of Theorem 5.2), but no information beyond this seems to be available at the moment.

References

[1] D. Arcara: *A lower bound for the dimension of the base locus of the generalized theta divisor.* C. R. Math. Acad. Sci. Paris 340 (2005), no. 2, 131–134.

[2] E. Ballico: *Spanned vector bundles on algebraic curves and linear series.* Rend. Istit. Mat. Univ. Trieste 27 (1995), no. 1-2, 137–156 (1996).

[3] E. Ballico: *Brill–Noether theory for vector bundles on projective curves.* Math. Proc. Camb. Phil. Soc. 128 (1998), 483–499.

[4] P. Belkale: *The strange duality conjecture for generic curves.* J. Amer. Math. Soc. 21 (2008), no. 1, 235–258.

[5] P. Belkale: *Strange duality and the Hitchin/WZW connection.* J. Diff. Geom. 82 (2009), 445–465.

[6] U.N. Bhosle, L. Brambila-Paz, P.E. Newstead: *On coherent systems of type $(n, d, n+1)$ on Petri curves.* Manuscr. Math. 126 (2008), 409–441.

[7] L. Brambila-Paz: *Non-emptiness of moduli spaces of coherent systems.* Int. J. Math. 19 (2008), 777–799.

[8] L. Brambila-Paz, I. Grzegorczyk, P.E. Newstead: *Geography of Brill–Noether loci for small slopes.* J. Alg. Geom. 6 (1997), 645–669.

[9] L. Brambila-Paz, V. Mercat, P.E. Newstead, F. Ongay: *Nonemptiness of Brill–Noether loci.* Int. J. Math. 11 (2000), 737–760.

[10] L. Brambila-Paz, A. Ortega: *Brill–Noether bundles and coherent systems on special curves.* Moduli spaces and vector bundles, London Math. Soc. Lecture Notes Ser., 359, Cambridge Univ. Press, Cambridge, 2009, pp. 456–472.

[11] D.C. Butler: *Normal generation of vector bundles over a curve.* J. Diff. Geom. 39 (1994), 1–34.

[12] D.C. Butler: *Birational maps of moduli of Brill–Noether pairs.* arXiv:alg-geom/9705009v1.

[13] S. Casalaina-Martin, T. Gwena, M. Teixidor i Bigas: *Some examples of vector bundles in the base locus of the generalized theta divisor.* arXiv:0707.2326v1.

[14] C. Ciliberto: *Alcuni applicazioni di un classico procedimento di Castelnuovo.* Sem. di Geom., Dipart. di Matem., Univ. di Bologna, (1982-83), 17–43.

[15] J. Cilleruelo, I. Sols: *The Severi bound on sections of rank two semistable bundles on a Riemann surface.* Ann. Math. 154 (2001), 739–758.

[16] M. Coppens: *The gonality of general smooth curves with a prescribed plane nodal model.* Math. Ann. 289 (1991), 89–93.

[17] M. Coppens, C. Keem, G. Martens: *Primitive linear series on curves.* Manuscr. Math. 77 (1992), 237–264.

[18] M. Coppens, G. Martens: *Linear series on 4-gonal curves.* Math. Nachr. 213 (2000), 35–55.

[19] M. Coppens, G. Martens: *Linear series on a general k-gonal curve.* Abh. Math. Sem. Univ. Hamburg 69 (1999), 347–371.

[20] L. Ein, R. Lazarsfeld: *Stability and restrictions of Picard bundles, with an application to the normal bundles of elliptic curves.* London Math. Soc. Lecture Notes Ser., 179, Cambridge Univ. Press, Cambridge, 1992, pp. 149–156.

[21] D. Eisenbud: *Linear sections of determinantal varieties.* Amer. J. Math. 110 (1988), 541–575.

[22] D. Eisenbud, H. Lange, G. Martens, F.-O. Schreyer: *The Clifford dimension of a projective curve.* Comp. Math. 72 (1989), 173–204.

[23] M.L. Green: *Koszul cohomology and the geometry of projective varieties.* J. Differential Geom. 19 (1984), no. 1, 125–171.

[24] M.L. Green, R. Lazarsfeld: *On the projective normality of complete linear series on an algebraic curve.* Invent. Math. 83 (1985), no. 1, 73–90.

[25] T. Gwena, M. Teixidor i Bigas: *Maps between moduli spaces of vector bundles and the base locus of the theta divisor.* Proc. Amer. Math. Soc. 137 (2009), no. 3, 853–861.

[26] S. Kim: *On the Clifford sequence of a general k-gonal curve.* Indag. Mathem. 8 (1997), 209–216.

[27] H. Lange, M.S. Narasimhan: *Maximal subbundles of rank two vector bundles on curves.* Math. Ann. 266 (1983), 55–72.

[28] H. Lange, P.E. Newstead: *On Clifford's theorem for rank-3 bundles.* Rev. Mat. Iberoamericana 22 (2006), 287–304.

[29] Y. Li: *Spectral curves, theta divisors and Picard bundles.* Int. J. Math. 2 (1991), 525–550.

[30] A. Marian, D. Oprea: *The level-rank duality for non-abelian theta functions.* Invent. Math. 168 (2007), no. 2, 225–247.

[31] G. Martens, F.-O. Schreyer: *Line bundles and syzygies of trigonal curves.* Abh. Math. Sem. Univ. Hamburg 56 (1986), 169–189.

[32] V. Mercat: *Le problème de Brill–Noether pour des fibrés stables de petite pente.* J. reine angew. Math. 506 (1999), 1–41.

[33] V. Mercat: *Fibrés stables de pente 2.* Bull. London Math. Soc. 33 (2001), 535–542.

[34] V. Mercat: *Clifford's theorem and higher rank vector bundles.* Int. J. Math. 13 (2002), 785–796.

[35] K. Paranjape, S. Ramanan: *On the canonical ring of a curve.* Algebraic Geometry and Commutative Algebra in Honor of Masayoshi Nagata (1987), 503–516.

[36] M. Popa: *On the base locus of the generalized theta divisor.* C. R. Acad. Sci. Paris Sér. I Math. 329 (1999), no. 6, 507–512.

[37] M. Popa, M. Roth: *Stable maps and Quot schemes.* Invent. Math. 152 (2003), no. 3, 625–663.

[38] R. Re: *Multiplication of sections and Clifford bounds for stable vector bundles on curves.* Comm. in Alg. 26 (1998), 1931–1944.

[39] O. Schneider: *Sur la dimension de l'ensemble des points base du fibré déterminant sur $SU_C(r)$.* Ann. Inst. Fourier (Grenoble) 57 (2007), no. 2, 481–490.

[40] X.-J. Tan: *Clifford's theorems for vector bundles.* Acta Math. Sin. 31 (1988), 710–720.

[41] M. Teixidor i Bigas: *Brill–Noether theory for stable vector bundles.* Duke Math. J. 62 (1991), 385–400.

[42] M. Teixidor i Bigas: *Rank two vector bundles with canonical determinant.* Math. Nachr. 265 (2004), 100–106.

[43] M. Teixidor i Bigas: *Petri map for rank two vector bundles with canonical determinant.* Compos. Math. 144 (2008), 705–720.

[44] M. Teixidor i Bigas: *Syzygies using vector bundles.* Trans. Amer. Math. Soc. 359 (2007), no. 2, 897–908.

[45] C. Voisin: *Sur l'application de Wahl des courbes satisfaisant la condition de Brill–Noether–Petri.* Acta Math. 168 (1992), 249–272.

Herbert Lange
Department Mathematik
Universität Erlangen–Nürnberg
Bismarckstraße $1\frac{1}{2}$
D-91054 Erlangen, Germany
e-mail: lange@mi.uni-erlangen.de

Peter E. Newstead
Department of Mathematical Sciences
University of Liverpool
Peach Street
Liverpool L69 7ZL, UK
e-mail: newstead@liv.ac.uk

Affine Flag Manifolds and Principal Bundles
Trends in Mathematics, 203–217
© 2010 Springer Basel AG

Division Algebras and Unit Groups on Surfaces

Fabian Reede and Ulrich Stuhler

Abstract. Various types of finiteness questions over algebraic curves and algebraic surfaces over finite fields are considered and studied in the context of the theory of vector bundles over curves resp. surfaces. Two different kinds of modifications of bundles and concepts of connectivity between such bundles are introduced. There are various computations of Chern classes of the bundles involved.

Mathematics Subject Classification (2000). 11G25, 11R52, 11R58, 14D20, 14H05, 14J60.

Keywords. Division algebra, unit group, building, vector bundle, modification, Chern class.

Introduction

In the first part of this paper we reprove several basic results from the arithmetic of function fields in the context of algebraic geometry. All of these classical results will turn out as immediate consequences of the Riemann–Roch theorem for curves. Additionally we use this approach to give some basic finiteness results concerning unit groups of central division algebras of function fields considering their action on the relevant product of Bruhat–Tits buildings associated with the situation. Nothing in this chapter will be really new. In the second part of this paper we study the approach outlined in the first part in the situation of unit groups of central division algebras over function fields of algebraic surfaces over finite fields. Our main interest here is again in finiteness results and we try to understand the methods from part one in this new situation to a certain extent.

1. Classical finiteness results: The case of a curve

Suppose X is a smooth projective curve of genus g over the finite field $k = \mathbb{F}_q$ of q elements, $S = \{x_1, \ldots, x_s\}$ is a nonempty finite set of closed points in X, $U := X \backslash S$ is an affine scheme over k, say $U \cong \mathrm{Spec}(A)$ with a k-algebra A. The function field of X is denoted by $F := k(X)$, and $\eta := \mathrm{Spec}(k(X)) \hookrightarrow X$ is the embedding of the generic point η into X.

We remind the reader of the Riemann–Roch theorem for the curve X: Suppose for this that we have a locally free sheaf of $\mathcal{O}_X$-modules $\mathcal{E}$ of rank n. Then we have

$$\chi(X, \mathcal{E}) := h^0(X, \mathcal{E}) - h^1(X, \mathcal{E}) = \deg(\mathcal{E}) + n(1 - g)$$

where as usual $h^i(X, \mathcal{E}) = \dim_k H^i(X, \mathcal{E})$. We can conclude for $h^0(X, \mathcal{E})$, the dimension of the k-vector space of sections,

$$h^0(X, \mathcal{E}) \geq \deg(\mathcal{E}) + n(1 - g).$$

Here $\deg(\mathcal{E})$ is the degree of $\mathcal{E}$, or equivalently its first Chern class $c_1(\mathcal{E})$.

We give the first application, namely finiteness of the class number of the Dedekind ring A:

Proposition 1.1. *The class number of A, $|\mathrm{Pic}(A)|$, is finite.*

Proof. Given any divisor D of X with support contained in $U = \mathrm{Spec}(A)$, we choose the point $x_1 \in S$ and consider $m \in \mathbb{Z}$, such that for the degree of the divisor $\widetilde{D}_1 := D + m(x_1)$

$$g \leq \deg(\widetilde{D}_1) = \deg(D) + m \cdot \deg(x_1) < g + \deg(x_1)$$

holds, where we have $\deg(x_1) = [k(x_1) : k]$, the degree of the residue field extension. By Riemann–Roch, we find a nonzero section $f \in \Gamma(X, \mathcal{O}_X(\widetilde{D}_1))$. We obtain the isomorphism of invertible sheaves $\mathcal{O}_X(\widetilde{D}_1) \xrightarrow{\sim} \mathcal{O}_X(\widetilde{D}_1 + (f))$ and induced from this the isomorphism

$$\mathcal{O}_X(D)|_U \xrightarrow{\;\sim\;} \mathcal{O}_X(\widetilde{D}_1)|_U \xrightarrow{\;\sim\;} \mathcal{O}_X(\widetilde{D}_1 + (f))|_U$$

in the Picard group $\mathrm{Pic}(U)$.

Now, $D_1 := \widetilde{D}_1 + (f)$ is an effective divisor of the form $D_1 = \sum_{x \in X^{(0)}} v_x(D_1)(x)$. We have $v_x(D_1) \geq 0$ for all closed points $x \in X$. Furthermore

$$\deg(D_1) = \sum_{x \in X^{(0)}} v_x(D_1)\deg(x) < g + \deg(x_1)$$

by construction of D_1. As there are for a given $N \in \mathbb{Z}$ only finitely many $x \in X^{(1)}$ with $\deg(x) = [k(x) : k] < N$ (Exercise), it follows that there are only finitely many possibilities for such a divisor D_1. Therefore $\mathrm{Pic}(U)$ is finite for any U. $\square$

We study next the structure of the group of units $A^\times = \mathcal{O}_X(U)^\times$. We consider divisors of the form $\sum_{i=1}^s n_i(x_i) \in \mathrm{Div}(X)$. We obtain a homomorphism

$$\mathrm{div}\colon A^\times \longrightarrow \bigoplus_{i=1}^s \mathbb{Z}(x_i), \quad f \mapsto \mathrm{div}(f),$$

where we have $\mathrm{div}(f) = \sum_{x \in X^{(0)}} v_x(f)(x) \in \bigoplus_{i=1}^s \mathbb{Z}(x_i)$ because $f \in A^\times$.

Obviously $\deg(\operatorname{div}(f)) = \deg((f)) = 0$, where

$$\deg((f)) = \sum_{x \in X^{(0)}} v_x(f)\deg(x) = \sum_{i=1}^{s} v_{x_i}(f)\deg(x_i).$$

Therefore $\operatorname{div}(A^\times)$ is a free abelian group of rank at most $s-1$. If $f \in A^\times$ satisfies $\operatorname{div}(f) = 0$, then $f \in \Gamma(X, \mathcal{O}_X)^\times = k^\times$. As $k^\times$ is a finite group, we obtain for the free rank $\operatorname{rk}(A^\times)$ of $A^\times$ the equality $\operatorname{rk}(A^\times) = \operatorname{rk}(\operatorname{div}(A^\times)) \le s-1$. On the other hand we can conclude easily $\operatorname{rk}(\operatorname{div}(A^\times)) \ge s-1$. Namely, we consider the divisors $\widetilde{D}_i := (\deg(x_1)(x_i) - \deg(x_i)(x_1))$ for $i = 2,\ldots,s$. Because $\operatorname{Pic}(U)$ is finite, there is $m_i \in \mathbb{N}$, such that $D_i := m_i\widetilde{D}_i$ is a principal divisor in $\operatorname{Div}(U)$ for $i = 2,\ldots,s$. Denoting $D_i = (f_i)$, $f_i \in A^\times$ for $i = 2,\ldots,s$, one immediately sees that $\operatorname{rk}(\mathbb{Z}(f_2) + \cdots + \mathbb{Z}(f_s)) = \operatorname{rk}(\mathbb{Z}D_2 + \cdots + \mathbb{Z}D_s) = s-1$. Therefore, we can conclude that $\operatorname{rk}_\mathbb{Z}(A^\times) \ge s-1$. Putting everything together, we obtain $\operatorname{rk}_\mathbb{Z}(A^\times) = s-1$.

Proposition 1.2 (Dirichlet's unit theorem, function field case). *The group of units,* $(A^\times, \cdot)$, *is a finitely generated abelian group with a free part of rank* $s-1$.

Remarks 1.3. As is well known, both results generalize to finitely generated rings over a finite field. Proposition 1.1 corresponds to the Mordell–Weil theorem for the involved Picard variety. Finite generation of the group of units $A^\times$ of such a ring has been shown by Roquette (see [5]).

Next we consider the following situation: X as above is a smooth projective curve, $U = X \backslash S$ is affine (so S is not empty) and $S = \{x_1, \ldots, x_s\}$ is the set of the (closed) points "at infinity". $F = k(X)$ is the field of rational functions on X and D is a central division algebra over F (with center F). We consider a sheaf of orders $\mathcal{A}$ over X, such that for the generic point $\eta \colon \operatorname{Spec}(F) \hookrightarrow X$ the corresponding stalk $\mathcal{A}_\eta$ is identified with the central division algebra D. So in particular, $\mathcal{A}$ is locally free as a sheaf of $\mathcal{O}_X$-modules and for almost all $x \in X$ the stalks $\mathcal{A}_{X,x}$ are maximal orders over $\mathcal{O}_{X,x}$ in $D = \mathcal{A}_\eta$. We will study the finiteness properties of groups $\Gamma \subset D^\times$, commensurable with the unit groups $\mathcal{A}(U)^\times$. It is immediate to see that we can assume right from the beginning that $\mathcal{A}$ actually is a sheaf of maximal orders in $D = \mathcal{A}_\eta$, that is, for all $x \in X$ the stalk $\mathcal{A}_{X,x}$ is a maximal $\mathcal{O}_{X,x}$-order in D. The reason for this is that all the groups above are in fact commensurable with each other (Exercise). Associated with the maximal orders $\mathcal{A}_{X,x}$ are the corresponding orders A_x in $D \otimes_F F_x$, where $A_x := \mathcal{A}_{X,x} \otimes_{\mathcal{O}_{X,x}} \mathcal{O}_x$, $\mathcal{O}_x$ the completion of $\mathcal{O}_{X,x}$ at the point x. We consider sheaves of locally free left $\mathcal{A}$-modules $\mathcal{E}$ with an identification of the generic fiber $\mathcal{E}_\eta \xrightarrow{\sim} D$. So these sheaves are in particular locally free $\mathcal{O}_X$-modules of rank $\operatorname{rk}(\mathcal{E}) = n^2 = [D : F]$. We consider in particular $\mathcal{A}$-modules $\mathcal{E}$ satisfying $\mathcal{E}(U) = \mathcal{A}(U)$ (making use of the identification $\mathcal{E}_\eta \xrightarrow{\sim} D$). Given $\mathcal{E}$ as above, we then have additionally embeddings $\mathcal{E}_{x_i} \hookrightarrow \mathcal{E}_\eta = D$ and induced from this: $\mathcal{E}_i := \mathcal{E}_{x_i} \otimes_{\mathcal{O}_{X,x_i}} \mathcal{O}_{x_i} \hookrightarrow D \otimes_F F_{x_i} =: D_{x_i}$ $(i = 1, \ldots, s)$. Using flat descent, it is immediate to see that the category of all $\mathcal{E}$

as above (with morphisms only isomorphisms) can be identified with the category of all tuples $(\mathcal{E}(U) = \mathcal{A}(U); \mathcal{E}_1, \ldots, \mathcal{E}_s)$, where the $\mathcal{E}_i$ are free $\mathcal{A}_{x_i}$-submodules of $D \otimes_F F_{x_i}$ $(i = 1, \ldots, s)$. An isomorphism $\phi \colon \mathcal{E} \xrightarrow{\sim} \mathcal{E}'$ of such $\mathcal{A}$-modules induces an isomorphism

$$
\begin{array}{ccc}
\mathcal{E}(U) & \xrightarrow{\ \sim\ } & \mathcal{E}'(U) \\
\| & & \| \\
\mathcal{A}(U) & \xrightarrow{\ \sim\ } & \mathcal{A}(U)
\end{array}
$$

of left $\mathcal{A}(U)$-modules and is therefore given by right multiplication with an element $g \in \mathcal{A}(U)^\times$ which additionally has to satisfy $\mathcal{E}_i \cdot g = \mathcal{E}'_i$ for $i = 1, \ldots, s$, using the embedding $\mathcal{A}(U) \hookrightarrow D_{x_i}$. Suppose we have $D_{x_i} = D \otimes_F F_{x_i} \cong \mathbb{M}(r_i, D_i)$, where D_i is a central division algebra over $F_i := F_{x_i}$ for $i = 1, \ldots, s$. Denoting $\mathcal{O}_i \subseteq D_i$ the (unique!) maximal local order in D_i, we have an induced isomorphism $\mathcal{A}_{x_i} \cong \mathbb{M}(r_i, \mathcal{O}_i)$. The $\mathcal{E}_i$ considered above are given in the form $(\mathcal{A}_{x_i} u_i)$, where u_i is an invertible element in D_{x_i}. We consider the following equivalence relation for submodules $\mathcal{E}_i, \mathcal{E}'_i \subseteq D_{x_i}$, namely

$$
\mathcal{E}_i \sim \mathcal{E}'_i \quad :\Longleftrightarrow \quad \exists m : \mathcal{E}'_i = \pi_i^m \mathcal{E}_i
$$

for $\pi_i \in \mathcal{O}_i$ a uniformizing element in $\mathcal{O}_i$. As it is well known such a set of equivalence classes forms a simplicial complex $\mathfrak{X}_i$, the Bruhat–Tits building associated with the group $\mathrm{GL}(r_i, D_i)$ which can be described as follows: The r-simplices $\langle [\mathcal{E}_i^{(0)}], \ldots, [\mathcal{E}_i^{(r)}] \rangle$ are formed by vertices $\mathcal{E}_i^{(0)} \subset \cdots \subset \mathcal{E}_i^{(r)} \subset \pi_i^{-1} \mathcal{E}_i^{(0)}$ (of course taking into account the equivalence relation). The following facts hold:

1. The $\mathfrak{X}_i$ $(i = 1, \ldots, s)$ are contractible simplicial complexes of dimension $r_i - 1$.
2. As a consequence, the polysimplicial complex $\prod_{i=1}^s \mathfrak{X}_i$ is contractible.
3. There is an action of $\mathcal{A}(U)^\times$ and related groups on $\prod_{i=1}^s \mathfrak{X}_i$. This group acts via the quotient $\mathcal{A}(U)^\times / \mathcal{O}(U)^\times$. There are subgroups of finite index $\Gamma \subset \mathcal{A}(U)^\times / \mathcal{O}(U)^\times$ which act fixpoint free. Here the group action on $\mathfrak{X}$ is given as follows: for an r-simplex $\langle [\mathcal{E}_0], \ldots, [\mathcal{E}_r] \rangle$ and $g \in \mathcal{A}(U)^\times$ one has

$$
\langle [\mathcal{E}_0], \ldots, [\mathcal{E}_r] \rangle \cdot g := \langle [\mathcal{E}_0 g], \ldots, [\mathcal{E}_r g] \rangle.
$$

It is immediate to check that the action is well defined and has the properties indicated above. 1., 2., and 3. are now standard facts which can be proved easily.

4. The quotient complex $\mathfrak{X}/\Gamma$ is finite (or equivalently: the topological space $|\mathfrak{X}/\Gamma|$, the realization of $\mathfrak{X}/\Gamma$, is compact).

Proof of 4. It suffices to show 4. for the case $\Gamma = \mathcal{A}(U)^\times / \mathcal{O}(U)^\times$. Furthermore it is enough to show that there are only finitely many vertices in the quotient $\mathfrak{X}/\Gamma$ (Exercise). This is equivalent to proving that there are only finitely many equivalence classes $[\mathcal{E}]$, where $\mathcal{E}$ is a sheaf of $\mathcal{A}$-modules with $\mathcal{E}(U) = \mathcal{A}(U)$, locally free of rank one as a sheaf of $\mathcal{A}$-modules. Making use of the equivalence relation, we can replace the sheaf $\mathcal{E}$ by any $\mathcal{E}' := \mathcal{E} \otimes \mathcal{O}_X(D)$ with a divisor D with support

$\mathrm{supp}(D) \subset \{x_1, \ldots, x_s\}$. It is immediate to see that $[\mathcal{E}_{x_i}] = [\mathcal{E}'_{x_i}]$. Furthermore there is a constant $c > 0$, such that we can assume by replacing $\mathcal{E}$ by an appropriate $\mathcal{E}'$ as above that we have

- an embedding $0 \hookrightarrow \mathcal{A} \hookrightarrow \mathcal{E}$
- $\deg(\mathcal{E}) \leq c$.

Then we can conclude as follows: because we have an exact sequence

$$0 \longrightarrow H^0(X, \mathcal{A}) \longrightarrow H^0(X, \mathcal{E}) \longrightarrow H^0(X, \mathcal{E}/\mathcal{A})$$

we have $h^0(X, \mathcal{E}) \leq h^0(X, \mathcal{A}) + h^0(X, \mathcal{E}/\mathcal{A})$. As $\mathcal{A}_\eta = D$ is a division algebra and $H^0(X, \mathcal{A})$ is a finite-dimensional algebra over $k = \mathbb{F}_q$, $H^0(X, \mathcal{A})$ is a commutative algebra over k of finite dimension without zero divisors. Therefore $H^0(X, \mathcal{A})$ is a field extension of k. Furthermore its dimension over k is bounded by $\dim_F(\mathcal{A}_\eta)^{\frac{1}{2}} = \dim_F(D)^{\frac{1}{2}} = n$, because any commutative field extension F' of F in D has dimension $\dim_F(F') \leq n$ and $F' = H^0(X, \mathcal{A}) \cdot F$ would be such an extension. Because $\deg(\mathcal{A})$ is fixed and $\deg(\mathcal{A}) \leq \deg(\mathcal{E}) \leq c$ holds, we have $h^0(X, \mathcal{E}/\mathcal{A}) = \lg_k(\mathcal{E}/\mathcal{A}) \leq c - \deg(\mathcal{A})$. In toto we see that we can find a representative $\mathcal{E}$ with $h^0(X, \mathcal{E}) \leq c'$, where $c' = h^0(X, \mathcal{A}) + (c - \deg(\mathcal{A}))$ which is an a priori bound. Making use of the Harder–Narasimhan filtration, it follows that we have only finitely many possibilities for $\mathcal{E}$ considered as a sheaf of $\mathcal{O}_X$-modules. But having fixed $\mathcal{E}$ as a sheaf of $\mathcal{O}_X$-modules, there are additionally only finitely many possibilities to give on $\mathcal{E}$ an action of $\mathcal{A}$ such that $\mathcal{E}$ will be additionally a sheaf of $\mathcal{A}$-modules. The reason for this is as follows: The action of $\mathcal{A}$ on $\mathcal{E}$ induces a homomorphism $\varphi \colon \mathcal{A} \to \mathcal{E}nd_{\mathcal{O}_X}(\mathcal{E}, \mathcal{E})$ into the sheaf of endomorphisms of $\mathcal{E}$ over $\mathcal{O}_X$. φ is a homomorphism of algebras, but in particular a homomorphism of $\mathcal{O}_X$-modules. But of course $\mathrm{Hom}_{\mathcal{O}_X}(\mathcal{A}, \mathcal{E}nd_{\mathcal{O}_X}(\mathcal{E}, \mathcal{E}))$ is a finite-dimensional k-vector space, so in particular finite itself. Therefore the quotient $\mathfrak{X}/\Gamma$ is finite. $\qquad \square$

Remarks 1.4.

i) This compactness result is in fact a special but typical case of the Godement criterion from the theory of arithmetic groups (see [2]). From 1.–4. one obtains that all the groups Γ are FP-groups, with finite cohomological dimension. The cohomological dimension of Γ equals

$$\dim(\mathfrak{X}) = \sum_{i=1}^{s} \dim(\mathfrak{X}_i) = \sum_{i=1}^{s} (r_i - 1)$$

where we had $D_{x_i} = D \otimes F_{x_i} \cong \mathbb{M}(r_i, D_i)$.

ii) There is an additional beautiful fact concerning the groups Γ:
- $H^i(\Gamma, \mathbb{Q}) = 0$ for $i \neq 0, \dim(\mathfrak{X})$
- $H^0(\Gamma, \mathbb{Q}) = \mathbb{Q}$

- For the Euler characteristic $\chi(\Gamma, \mathbb{Q})$ one has

$$\chi(\Gamma, \mathbb{Q}) = \sum_{i=1}^{\dim(\mathfrak{X})} (-1)^i \dim_{\mathbb{Q}}(H^i(\Gamma, \mathbb{Q}))$$

$$= 1 + (-1)^r \dim_{\mathbb{Q}}(H^r(\Gamma, \mathbb{Q})) \text{ where } r := \dim(\mathfrak{X})$$

$$= \int_{\Gamma \backslash G} \omega$$

where $G = \prod_{i=1}^s \mathrm{PGL}(r_i, D_i)$, ω is the so-called Euler–Poincaré measure. Using known results on the Tamagawa measure, one can compute the integral explicitly.

2. Locally free sheaves of modules over Azumaya algebras: The case of a surface

To get started with studying unit groups of central division algebras in the case of an algebraic surface, we recall some of the relevant results which we will use in the next section. The reader can consult for proofs [3], [7] or [9]. There is a slight problem coming from the fact that these papers require the underlying field k to be algebraically closed. Actually this assumption is not necessary and we will make some comments on this later on. So, we consider an integral projective scheme X over the field k. $\mathcal{A}$ is a sheaf of associative $\mathcal{O}_X$-algebras satisfying the following properties:

- As a sheaf of $\mathcal{O}_X$-modules, $\mathcal{A}$ is coherent and torsion free.
- The stalk $\mathcal{A}_\eta$ over the generic point $\eta \in X$ is a central simple algebra over the function field $F = k(X) = \mathcal{O}_{X,\eta}$.

Remarks 2.1. Of course, later on this chapter, we will restrict to the surface case, but at the moment this is not necessary.

Definition 2.2. A family of generically simple torsion free $\mathcal{A}$-modules over a k-scheme S is a sheaf $\mathcal{E}$ of left modules under the pullback $\mathcal{A}_S$ of $\mathcal{A}$ to $X \times_k S$ with the following properties:

- $\mathcal{E}$ is coherent over $\mathcal{O}_{X \times_k S}$ and flat over S.
- For every point $s \in S$, the fiber $\mathcal{E}_s$ is a generically simple torsion free $\mathcal{A}_{k(s)}$-module.

Here $k(s)$ is the residue field of S at s, and the fiber $\mathcal{E}_s$ is by definition the pullback to $X \times_k \mathrm{Spec}(k(s))$ induced by $\mathrm{Spec}(k(s)) \hookrightarrow X$ over k. The corresponding moduli functor is denoted $\mathcal{M} = \mathcal{M}_{\mathcal{A}/X} \colon \mathrm{Sch}_k \to \mathrm{Sets}$, it sends a k-scheme S to the set of isomorphism classes of families $\mathcal{E}$ of generically simple torsion free $\mathcal{A}$-modules over S.

To formulate the relevant results, we choose an ample line bundle $\mathcal{O}_X(1)$ on X and put $d := \dim(X)$. The Hilbert polynomial $P(E)$ of a coherent sheaf E on X with respect to this choice is given by

$$P(E; m) := \chi(E(m)) = \sum_{i=0}^{d} (-1)^i \dim_k H^i(X, E(m))$$

where, as usual, $E(m) := E \otimes \mathcal{O}_X(1)^{\otimes m}$. As is well known, the Hilbert polynomial is locally constant in flat families. We denote by $\mathcal{M}_{\mathcal{A}/X;P} \colon \mathrm{Sch}_k \to \mathrm{Sets}$ the subfunctor of $\mathcal{M}_{\mathcal{A}/X}$ that parameterizes families $\mathcal{E}$ all of whose fibers $\mathcal{E}_s$ have Hilbert polynomial P. We have the following result:

Theorem 2.3.

i) *The moduli functor $\mathcal{M}_{\mathcal{A}/X;P}$ is representable by a coarse moduli scheme $M_{\mathcal{A}/X;P}$ which is a separated scheme of finite type over k*

ii) *The moduli scheme $M_{\mathcal{A}/X;P}$ is obtained as a geometric quotient in the form $M_{\mathcal{A}/X;P} = \mathrm{GL}(N)\backslash R$.*

iii) *The quotient morphism $R \to M_{\mathcal{A}/X;P}$ is a principal $\mathrm{PGL}(N)$-bundle (for N appropriate), locally trivial in the fppf-topology.*

iv) *$M_{\mathcal{A}/X;P}$ is a projective scheme over k.*

Remarks 2.4.

i) In particular, the moduli functor $\mathcal{M}_{\mathcal{A}/X}$ of all generically simple torsion free $\mathcal{A}$-modules has a coarse moduli scheme $M_{\mathcal{A}/X} = \coprod_P M_{\mathcal{A}/X;P}$, a disjoint sum of projective schemes over k

ii) If X is smooth of dimension d, there is another such decomposition $M_{\mathcal{A}/X} = \coprod_{c_1,\ldots,c_d} M_{\mathcal{A}/X;c_1,\ldots,c_d}$, given by fixing the Chern classes $c_i \in \mathrm{CH}^i(X)$ in the Chow group of cycles modulo algebraic equivalence. Indeed, each part $M_{\mathcal{A}/X;c_1,\ldots,c_d}$ is open and closed in some $M_{\mathcal{A}/X;P}$, where P is given by Hirzebruch–Riemann–Roch.

iii) In case X is a smooth projective surface, the decomposition above comes down to $M_{\mathcal{A}/X} = \coprod_{\substack{c_1 \in \mathrm{NS}(X) \\ c_2 \in \mathbb{Z}}} M_{\mathcal{A}/X;c_1,c_2}$. We are mainly interested when the generic stalk $\mathcal{A}_\eta$ is a central division algebra over $F = k(X)$.

iv) $M_{\mathcal{A}/X}(\overline{k})$ classifies the isomorphism classes of the above-mentioned $\mathcal{A}$-sheaves on $X \times_k \overline{k}$. The sets of points $M_{\mathcal{A}/X;c_1,c_2}(k)$ are all finite for a finite field $k = \mathbb{F}_q$. On the other hand a sheaf of $\mathcal{A}$-modules $\mathcal{E}$ over X defines a point $[\mathcal{E} \otimes_k \overline{k}] \in M_{\mathcal{A}/X}(\overline{k})$ which is invariant under the Galois action of $\mathrm{Gal}(\overline{k}/k)$. So it defines a point in $M_{\mathcal{A}/X}(k)$. If two different sheaves of $\mathcal{A}$-modules $\mathcal{E}, \mathcal{E}'$ of the type above, both defined over k, get isomorphic over $\overline{k}$, then we obtain a 1-cocycle in $H^1(\mathrm{Gal}(\overline{k}/k), \mathrm{Aut}_{\mathcal{A}}(\mathcal{E} \otimes_k \overline{k}))$ in the usual way of Galois cohomology. But $\mathrm{Aut}_{\mathcal{A} \otimes_k \overline{k}}(\mathcal{E} \otimes_k \overline{k})$ obviously is the group of units in the endomorphism ring (of global endomorphisms) $\mathrm{End}_{\mathcal{A} \otimes_k \overline{k}}(\mathcal{E} \otimes_k \overline{k})$. The letter algebra will be a finite-dimensional associative $\overline{k}$-algebra, because $X \times_k \overline{k}$ is projective over $\overline{k}$. Furthermore we have an embedding $\mathrm{End}_{\mathcal{A} \otimes_k \overline{k}}(\mathcal{E} \otimes_k \overline{k}) \hookrightarrow \mathrm{End}_{\mathcal{A}_\eta \otimes_k \overline{k}}(\mathcal{A}_\eta \otimes_k \overline{k})$, where $\mathrm{End}_{A_\eta \otimes_k \overline{k}}(A_\eta \otimes_k \overline{k}) \cong (A_\eta \otimes_k \overline{k})^{\mathrm{op}}$ is acting by right multiplication. If $A_\eta \otimes_k \overline{k}$ is again a division algebra, then $\mathrm{End}_{A \otimes_k \overline{k}}(\mathcal{E} \otimes_k \overline{k})$ has no zero divisors which implies, because it is a finite-dimensional $\overline{k}$-algebra, that it is a division algebra over $\overline{k}$ of finite dimension. It follows immediately that $\mathrm{End}_{A \otimes_k \overline{k}}(\mathcal{E} \otimes_k \overline{k}) \cong \overline{k}$ and $\mathrm{Aut}_{A \otimes_k \overline{k}}(\mathcal{E} \otimes_k \overline{k}) \cong \overline{k}^\times$. But then by Hilbert 90, we have $H^1(\mathrm{Gal}(\overline{k}/k), \mathrm{Aut}_{A \otimes_k \overline{k}}(\mathcal{E} \otimes_k \overline{k})) = (1)$. Therefore the isomorphism classes over k are really classified by $M_{A/X;c_1,c_2}$ in this case, so they are in particular finite! If $A_\eta \otimes_k \overline{k}$ is not a division algebra, it is nevertheless true that the Galois cohomology $H^1(\mathrm{Gal}(\overline{k}/k), \mathrm{Aut}_{A \otimes_k \overline{k}}(\mathcal{E} \otimes_k \overline{k}))$ is finite. So we obtain the necessary finiteness results again.

In the last part of this section we collect formulas for Chern classes which are useful for our purposes. We consider only the case of surfaces. Then given a coherent sheaf $\mathcal{E}$ over X, we have the Chern classes $c_1(\mathcal{E})$ and $c_2(\mathcal{E})$. Suppose additionally that we have an invertible sheaf $\mathcal{L}$. Then we have the following formulas for Chern classes:

Lemma 2.5. *Denote $r = \mathrm{rk}(\mathcal{E})$, where $\mathcal{E}$ is a sheaf of locally free $\mathcal{O}_X$-modules. Then one has*

   i) $c_1(\mathcal{E} \otimes \mathcal{L}) = c_1(\mathcal{E}) + rc_1(\mathcal{L})$.

  ii) $c_2(\mathcal{E} \otimes \mathcal{L}) = c_2(\mathcal{E}) + (r-1)c_1(\mathcal{E})c_1(\mathcal{L}) + \binom{r}{2}c_1(\mathcal{L})^2$.

 iii) *Suppose $j \colon Z \hookrightarrow X$ is a closed, reduced and irreducible subscheme of X of codimension r. Then as a consequence of the Grothendieck–Riemann–Roch theorem one has:*

     • $c_i(j_*(\mathcal{O}_Z)) = 0$ *for $i < r$*

     • $c_r(j_*(\mathcal{O}_Z)) = (-1)^{(r-1)}(r-1)![Z]$

    where $[Z] \in \mathrm{CH}^r(X)$ is the cycle defined by Z in the Chow group.

 iv) *For a vector bundle $\mathcal{F}$ on Z of rank n, one has*

     • $c_i(j_*(\mathcal{F})) = 0$ *for $i < r$*

     • $c_r(j_*(\mathcal{F})) = n(-1)^{(r-1)}(r-1)![Z]$

    in the notation from above.

  v) *For the convenience of the reader we also recall the Grothendieck–Riemann–Roch theorem: If $\mathrm{rk}(\mathcal{E}) = r$, one has*

$$\chi(X, \mathcal{E}) = \frac{1}{2}(c_1(\mathcal{E})(c_1(\mathcal{E}) - K_X)) - c_2(\mathcal{E}) + r\chi(\mathcal{O}_X).$$

 vi) *The discriminant of a locally free sheaf $\mathcal{E}$ over X is defined by*

$$\Delta(\mathcal{E}) = 2rc_2(\mathcal{E}) - (r-1)c_1(\mathcal{E})^2.$$

 vii) *One has $\Delta(\mathcal{E} \otimes \mathcal{L}) = \Delta(\mathcal{E})$ upon tensoring with an invertible sheaf $\mathcal{L}$ on X.*

For completeness we quote the following result of Bogomolov [1, Chapter 9, Theorem 1; p. 245]: Suppose X is an algebraic surface and H is an ample divisor. Suppose $\mathcal{E}$ is an H-stable rank r vector bundle on X. Then one has $\Delta(\mathcal{E}) \geq 0$.

Finally to complete these remarks on Chern classes, let us recall the fundamental property of the total Chern class: If

$$0 \longrightarrow \mathcal{E}_1 \longrightarrow \mathcal{E} \longrightarrow \mathcal{E}_2 \longrightarrow 0$$

is an exact sequence of locally free sheaves, we have $c_\bullet(\mathcal{E}) = c_\bullet(\mathcal{E}_1)c_\bullet(\mathcal{E}_2)$ for the total Chern class $c_\bullet(\mathcal{E}) = \sum_{i \geq 0} c_i(\mathcal{E})$ in $\bigoplus_{i \geq 0} \mathrm{CH}^i(X) =: \mathrm{CH}^\bullet(X)$, the Chow ring of cycles modulo algebraic equivalence.

Remarks 2.6.

i) For the total Chern class $c_\bullet(\mathcal{E}^*)$ of the dual vector bundle one shows easily $c_\bullet(\mathcal{E}^*) = (c_\bullet(\mathcal{E}))^*$, where given a class $\alpha = \sum_{i \geq 0} \alpha_i \in \mathrm{CH}^\bullet(X)$ one has $\alpha^* := \sum_{i \geq 0} (-1)^i \alpha_i$.

ii) Of course one could also consider Chern classes with values in the Chow ring modulo rational equivalence.

iii) Using the multiplicativity of Chern classes described above, one can define a total Chern class $c_\bullet(\mathcal{F})$ for any coherent sheaf $\mathcal{F}$ over X by choosing a finite resolution $0 \to \mathcal{E}_m \to \cdots \to \mathcal{E}_1 \to \mathcal{E}_0 \to \mathcal{F} \to 0$ defining $c_\bullet(\mathcal{F}) := \prod_{i=0}^m c_\bullet(\mathcal{E}_i)^{(-1)^i}$. This is well defined and satisfies multiplicativity for exact sequences again, namely for an exact sequence

$$0 \longrightarrow \mathcal{F}_1 \longrightarrow \mathcal{F} \longrightarrow \mathcal{F}_2 \longrightarrow 0$$

of coherent sheaves one has $c_\bullet(\mathcal{F}) = c_\bullet(\mathcal{F}_1)c_\bullet(\mathcal{F}_2)$ in $\mathrm{CH}^\bullet(X)$.

iv) Finally for completeness we also mention the Chern character. This is a ring homomorphism ch: $K_0(X) \to \mathrm{CH}^\bullet(X)_\mathbb{Q}$. For the case of a surface X we have $\mathrm{ch}(\mathcal{E}) = r + c_1(\mathcal{E}) + \frac{1}{2}(c_1(\mathcal{E})^2 - 2c_2(\mathcal{E}))$.

3. Elementary modifications and connectivity

X denotes a smooth projective algebraic surface over the field $k = \mathbb{F}_q$ of q elements. $\mathcal{A}$ is an Azumaya algebra over X, everywhere unramified, of rank 4 which generically is a division algebra. So $\mathcal{A}_\eta$ is a quaternion division algebra over the field of functions $k(X) = \mathcal{O}_{X,\eta}$. By the Hasse principle any restriction $\mathcal{A}|_C$ of $\mathcal{A}$ to an irreducible algebraic curve $C \hookrightarrow X$ is of the form $\mathcal{E}nd(\mathcal{E})$, that is a sheaf of endomorphisms of a rank two vector bundle $\mathcal{E}$ on C.

Remark 3.1. The assumption that $\mathcal{A}$ is everywhere unramified is actually not necessary and could be replaced by the weaker assumption that $\mathcal{A}$ might ramify in finitely many points of codimension one.

As in Section 2 of this paper we consider torsion free (with respect to $\mathcal{O}_X$) left $\mathcal{A}$-modules $\mathcal{E}$, coherent as $\mathcal{O}_X$-modules which are generically simple $\mathcal{A}_\eta$-modules, that is $\mathcal{E}_\eta$ is a one-dimensional $\mathcal{A}_\eta$-module. Additionally we fix a finite set $\{ C_i \mid i \in \{1, \ldots, n\} \}$ of irreducible curves $C_i \hookrightarrow X$ defined over k.

Denoting $U := X \backslash \bigcup_{i=1}^n C_i$, we assume that U is open affine in X. We consider the set of all sheaves of $\mathcal{A}$-modules $\mathcal{E}$ over X as above, equipped with an

identification $\mathcal{E}(U) \xrightarrow{\sim} \mathcal{A}(U)$ of left $\mathcal{A}(U)$-modules. This induces an identification $\mathcal{E}_\eta \xrightarrow{\sim} \mathcal{A}_\eta = D$ and therefore $\mathcal{E}$ comes with an embedding $\mathcal{E} \hookrightarrow \mathcal{E}_\eta = D$.
We denote the set of pairs

$$\widetilde{\mathfrak{X}} := \left\{ (\mathcal{E}; \mathcal{E}(U) \xrightarrow{\sim} \mathcal{A}(U)) \,|\, \mathcal{E} \text{ is coherent, torsion free and a left } \mathcal{A}\text{-module} \right\}.$$

Additionally, we define for $A \in \mathbb{R}$, a real number,

$$\widetilde{\mathfrak{X}}(A) := \left\{ (\mathcal{E}; \mathcal{E}(U) \xrightarrow{\sim} \mathcal{A}(U)) \in \widetilde{\mathfrak{X}} \,|\, \Delta(\mathcal{E}) \leq A \right\}.$$

We consider now the equivalence relation on $\widetilde{\mathfrak{X}}(A)$ generated by the following two relations:

1. **Isomorphisms of pairs:** Here a pair $(\mathcal{E}; \mathcal{E}(U) \xrightarrow{\sim} \mathcal{A}(U))$ is isomorphic to a pair $(\mathcal{E}'; \mathcal{E}'(U) \xrightarrow{\sim} \mathcal{A}(U))$ if and only if there is an isomorphism of sheaves of $\mathcal{A}$-modules $\phi \colon \mathcal{E} \xrightarrow{\sim} \mathcal{E}'$, such that the following diagram commutes:

$$
\begin{array}{ccc}
\mathcal{E}(U) & \xrightarrow{\sim} & \mathcal{A}(U) \\
{\scriptstyle \phi(U)}\big\downarrow & & \big\| \\
\mathcal{E}'(U) & \xrightarrow{\sim} & \mathcal{A}(U).
\end{array}
$$

2. **Tensorization with U-trivialized invertible sheaves:** Suppose $\mathcal{L}$ is an invertible sheaf on X equipped with a fixed trivialization $\mathcal{L}(U) \xrightarrow{\sim} \mathcal{O}_X(U)$ with the structure sheaf over U. Then we have a canonical operation

$$\mathcal{L} \otimes \left(\mathcal{E}; \mathcal{E}(U) \xrightarrow{\sim} \mathcal{A}(U) \right) = \left(\mathcal{L} \otimes \mathcal{E}; (\mathcal{L} \otimes \mathcal{E})(U) \xrightarrow{\sim} \mathcal{A}(U) \right).$$

We make the following additional assumption: The classes $[C_1], \ldots, [C_n]$ of irreducible components of the closed subset $X \backslash U$ are a generating system of the rational vector space $\mathrm{NS}(X) \otimes \mathbb{Q}$, the Néron–Severi group of X over the rationals.

Lemma 3.2. $\sum_{i=1}^{n} \mathbb{Z}[C_i]$ is an abelian subgroup of finite index in $\mathrm{NS}(X)$.

Proof. This follows immediately from the fact that $\mathrm{NS}(X)$ is finitely generated over the integers and we have assumed that $[C_1], \ldots, [C_n]$ are a generating system over $\mathbb{Q}$. $\qquad\square$

Proposition 3.3. *The set* $(\widetilde{\mathfrak{X}}(A)/\sim)$ *is finite.*

Proof. Using Operation 2. above and using the assumption, we can fix finitely many classes $c_1 \in \mathrm{NS}(X) \otimes \mathbb{Q}$ such that for any pair $(\mathcal{E}; \mathcal{E}(U) \xrightarrow{\sim} \mathcal{A}(U)) \in \widetilde{\mathfrak{X}}(A)$ one can find an equivalent pair $(\mathcal{E}'; \mathcal{E}'(U) \xrightarrow{\sim} \mathcal{A}(U)) \in \widetilde{\mathfrak{X}}(A)$ such that $c_1(\mathcal{E}') = c_1$ is one of the finitely many fixed Chern classes. So there are also only finitely many possibilities for the values of c_1^2. As the discriminant $\Delta(\mathcal{E}')$ is bounded from above and below, it follows that also the values of $c_2(\mathcal{E}')$ are bounded. So there are only finitely many possibilities for c_1 and c_2. But the moduli schemes $M_{\mathcal{A}/X;c_1,c_2}$ are of finite type over k, so the set $M_{\mathcal{A}/X;c_1,c_2}(k)$ is finite itself. It follows from Remark 2.4. iv) that $(\widetilde{\mathfrak{X}}(A)/\sim)$ is then finite. $\qquad\square$

Definition 3.4. We denote $\mathfrak{X} := (\widetilde{\mathfrak{X}}/\sim_2)$, $\mathfrak{X}(A) := (\widetilde{\mathfrak{X}}(A)/\sim_2)$, where $\sim_2$ is the equivalence relation generated by operations of type 2. above, that is by tensorization with line bundles.

Remark 3.5. As $\Delta(\mathcal{E} \otimes \mathcal{L}) = \Delta(\mathcal{E})$ for any coherent sheaf $\mathcal{E}$ and invertible sheaf $\mathcal{L}$, the definition above is reasonable.

Proposition 3.6. *There is a canonical action of the group of units $A(U)^\times$ on the set $\mathfrak{X}(A)$. The equivalence relation induced by this action on $\mathfrak{X}(A)$ is exactly the isomorphism relation 1. from above (up to tensorization).*

Proof. Two pairs $(\mathcal{E}; \mathcal{E}(U) \xrightarrow{\sim} A(U))$ and $(\mathcal{E}'; \mathcal{E}'(U) \xrightarrow{\sim} A(U))$ are isomorphic if and only if there exists $\gamma \in A(U)^\times$, such that upon right multiplication with γ one obtains $\mathcal{E} \cdot \gamma = \mathcal{E}'$ inside $\mathcal{E}_\eta = \mathcal{E}'_\eta = D$. $\qquad\square$

Remark 3.7. It follows from these observations that in case one could introduce some concept of connectivity in $\mathfrak{X}(A)$, similar as in the case of buildings which is natural enough to be compatible with the action of $A(U)^\times$ and if additionally $\mathfrak{X}(A)$ would be a connected graph for example with respect to this concept of connectivity, it follows that the group $A(U)^\times$ is finitely generated! We give the little argument at the end after introducing some concepts of connectivity.

Modifications

In this final part of Section 3 we introduce two kinds of modifications for the class of a pair $[(\mathcal{E}; \mathcal{E}(U) \xrightarrow{\sim} A(U))] \in \mathfrak{X}$. These will lead us to some concept of connectivity. The question is how the discriminant Δ behaves under these connectivities. We do now the necessary Chern class computations for both kinds of modifications.

Modifications of type I. Here we consider a pair $(\mathcal{E}; \mathcal{E}(U) \xrightarrow{\sim} A(U))$ and we have $\mathfrak{m}_{X,x}\mathcal{E} \subsetneq \mathcal{E}' \subsetneq \mathcal{E}$, where $\mathfrak{m}_{X,x}$ is the ideal sheaf of the point $x \in X$ in the structure sheaf $\mathcal{O}_X$. More precisely, we consider only closed points $x \in \bigcup_{i=1}^n C_i$ which additionally should be only on one of the irreducible components C_i, $i = 1, \ldots, n$. We have the isomorphism

$$\mathcal{E}/\mathfrak{m}_{X,x}\mathcal{E} \cong A_{X,x}/\mathfrak{m}_{X,x}A_{X,x} \cong \mathbb{M}(2, k(x))$$

where $k(x)$ is the residue field at x. Therefore it follows that the sheaf of $\mathcal{A}$-modules $\mathcal{E}/\mathcal{E}'$ is a simple nontrivial $\mathcal{A}_x$-module, so is isomorphic as an $\mathcal{O}_{X,x}$-module to $k(x)^2$. From the exact sequence

$$0 \longrightarrow \mathcal{E}' \longrightarrow \mathcal{E} \longrightarrow \mathcal{E}/\mathcal{E}' \longrightarrow 0$$

follows $c_\bullet(\mathcal{E}')c_\bullet(\mathcal{E}/\mathcal{E}') = c_\bullet(\mathcal{E})$ and with the relevant formulas in Section 2 we obtain $c_\bullet(\mathcal{E}/\mathcal{E}') = 1 - 2 \cdot [x]$, where $[x]$ denotes the class of x in the Chow group of X. If $x_0 \in X(k)$ is a point of degree 1 on X, one has furthermore in $\mathrm{CH}^2(X)$:

$$[x] = \deg(x)[x_0] = [k(x) : k][x_0].$$

We obtain

$$c_\bullet(\mathcal{E}) = 1 + c_1(\mathcal{E}) + c_2(\mathcal{E})$$
$$= (1 + c_1(\mathcal{E}') + c_2(\mathcal{E}'))(1 - 2 \cdot \deg(x)[x_0])$$
$$= 1 + c_1(\mathcal{E}') + (c_2(\mathcal{E}') - 2 \cdot \deg(x)[x_0])$$

for the total Chern classes of $\mathcal{E}'$ and $\mathcal{E}$.
We obtain $c_1(\mathcal{E}') = c_1(\mathcal{E})$ and $c_2(\mathcal{E}') = c_2(\mathcal{E}) + 2 \cdot \deg(x)[x_0]$.
For the discriminant of $\mathcal{E}'$ we obtain:

$$\Delta(\mathcal{E}') = 8c_2(\mathcal{E}') - 3c_1(\mathcal{E}')^2$$
$$= 8(c_2(\mathcal{E}) + 2 \cdot \deg(x)[x_0]) - 3c_1(\mathcal{E})^2$$
$$= \Delta(\mathcal{E}) + 16 \cdot \deg(x)[x_0].$$

If we start with a pair $[(\mathcal{E}; \mathcal{E}(U) \xrightarrow{\sim} \mathcal{A}(U))] \in \mathfrak{X}(A)$ fixing a modification $\mathfrak{m}_{X,x}\mathcal{E} \subsetneq \mathcal{E}' \subsetneq \mathcal{E}$ for $\mathcal{E}'$ as a sheaf of $\mathcal{A}$-modules, one obtains an induced identification

$$\mathcal{E}'(U)$$
$$\|$$
$$\mathcal{E}(U) \xrightarrow{\sim} \mathcal{A}(U).$$

From the last computations, one sees that there are only finitely many closed points in $\bigcup_{i=1}^n C_i$, such that $[(\mathcal{E}'; \mathcal{E}'(U) \xrightarrow{\sim} \mathcal{A}(U))]$ is still in $\mathfrak{X}(A)$.

Modifications along a closed curve. We consider one of the irreducible curves $C_i =: C \hookrightarrow X$. We start with $[(\mathcal{E}; \mathcal{E}(U) \xrightarrow{\sim} \mathcal{A}(U))] \in \mathfrak{X}(A)$ and look at

$$0 \longrightarrow \mathcal{E}' \longrightarrow \mathcal{E} \longrightarrow \mathcal{L} \longrightarrow 0$$

where $\mathcal{L}$ is obtained via $\mathcal{E} \to \mathcal{E}|_C \to \mathcal{L}$, such that $\mathcal{L}$ is a quotient of the restriction $\mathcal{E}|_C$ as a sheaf of $\mathcal{A}$-modules. As we do not assume that the sheaf of $\mathcal{A}$-modules is locally free as a sheaf of $\mathcal{O}_X$-modules, we have the following situation:

$$0 \longrightarrow \mathcal{E} \longrightarrow \mathcal{E}^{**} \longrightarrow \mathcal{T} \longrightarrow 0$$

where $\mathcal{E}^{**}$ is the reflexive hull of $\mathcal{E}$, itself a sheaf of $\mathcal{A}$-modules and automatically locally free as a sheaf of $\mathcal{O}_X$-modules. $\mathcal{T}$ is of finite length as a sheaf of $\mathcal{O}_X$-modules. Upon restriction to a closed curve $C \subset X$ (which will be one of the curves C_i, $i = 1, \ldots, n$), we obtain the diagram

$$\begin{array}{ccccccccc}
0 & \longrightarrow & \mathcal{E} & \longrightarrow & \mathcal{E}^{**} & \longrightarrow & \mathcal{T} & \longrightarrow & 0 \\
& & \downarrow & & \downarrow & & \downarrow & & \\
0 & \longrightarrow & \mathcal{T}or_1^{\mathcal{O}_X}(\mathcal{T}, \mathcal{O}_C) & \longrightarrow & \mathcal{E}|_C & \longrightarrow & \mathcal{E}^{**}|_C & \longrightarrow & \mathcal{T}|_C & \longrightarrow & 0
\end{array}$$

of sheaves. (We are a bit sloppy here, as we do no distinguish in our notation between say the sheaf $\mathcal{E}|_C$ on C and its direct image $i_*(\mathcal{E}|_C)$ where $\iota: C \hookrightarrow X$

is the embedding above. But what is meant will be clear from the context.) We obtain from this the exact sequence

$$0 \longrightarrow (\overline{\mathcal{E}|_C}) \longrightarrow \mathcal{E}^{**}|_C \longrightarrow \mathcal{T}|_C \longrightarrow 0 \tag{1}$$

where $(\overline{\mathcal{E}|_C})$ is the image of the morphism $\mathcal{E}|_C \longrightarrow \mathcal{E}^{**}|_C$. So $(\overline{\mathcal{E}|_C})$ is a torsion free subsheaf of the locally free sheaf $\mathcal{E}^{**}|_C$. In particular, $(\overline{\mathcal{E}|_C})$ will be locally free itself, if $\iota\colon C \hookrightarrow X$ is a smooth curve on X. To obtain the modification along C, we consider quotients $\mathcal{E}|_C \longrightarrow \mathcal{F}$. Altogether, we obtain the exact sequence

$$0 \longrightarrow \mathcal{E}' \longrightarrow \mathcal{E} \longrightarrow \mathcal{F} \longrightarrow 0 \tag{2}$$

where $\mathcal{E}'$ is defined as the kernel of the homomorphism $\mathcal{E} \longrightarrow \mathcal{F}$. $\mathcal{E}'$ is the modification we are looking for. As our modification will take place entirely along one of the curves C_i $(i = 1, \ldots, n)$ on X, this construction additionally yields an identification $\mathcal{E}'(U) \xrightarrow{\sim} \mathcal{E}(U) \xrightarrow{\sim} \mathcal{A}(U)$ and therefore an object $[(\mathcal{E}'; \mathcal{E}'(U) \xrightarrow{\sim} \mathcal{A}(U))] \in \mathfrak{X}$. So to compute $\Delta(\mathcal{E}')$ it is necessary to understand the possible quotients $\mathcal{E}|_C \longrightarrow \mathcal{F}$. This we can take apart a little bit more in the following way: We have

$$
\begin{array}{ccccccccc}
0 & \longrightarrow & (\mathcal{E}|_C)^{\mathrm{tors}} & \longrightarrow & \mathcal{E}|_C & \longrightarrow & (\overline{\mathcal{E}|_C}) & \longrightarrow & 0 \\
 & & \downarrow & & \downarrow & & \downarrow & & \\
0 & \longrightarrow & \mathcal{F}^{\mathrm{tors}} & \longrightarrow & \mathcal{F} & \longrightarrow & \overline{\mathcal{F}} & \longrightarrow & 0.
\end{array}
\tag{3}
$$

Therefore the possible quotients $\mathcal{F}$ (of course all the time as sheaves of $\mathcal{A}$-modules or $\mathcal{A}|_C$-modules) are obtained as follows: One has to find the possible quotients

$$(\overline{\mathcal{E}|_C}) \longrightarrow \overline{\mathcal{F}}.$$

Additionally one can take for $\mathcal{F}^{\mathrm{tors}}$ all possible quotients

$$(\mathcal{E}|_C)^{\mathrm{tors}} \longrightarrow \mathcal{F}^{\mathrm{tors}},$$

again taken in the category of $\mathcal{A}|_C$-modules.

Remark 3.8. Before doing the computation of the total Chern classes and discriminants involved, we just remark that the problem above can be further simplified. As $\mathcal{A}|_C \cong \mathcal{E}nd(\mathcal{E}_0)$, the sheaf of endomorphisms of a locally free sheaf $\mathcal{E}_0$ of rank two on C, by Morita equivalence the problem can be reduced immediately to the main problem to find the possible quotients of a rank two vector bundle on C.

We compute the relevant total Chern classes: By Section 2 we have

$$c_\bullet(\mathcal{E})c_\bullet(\mathcal{T}) = c_\bullet(\mathcal{E}^{**});$$

furthermore one has

$$c_\bullet(\mathcal{T}) = 1 - \lg(\mathcal{T})[x],$$

where $[x]$ denotes a cycle of codimension two in X, given by a rational point $x \in X(k)$. Altogether we obtain

$$c_1(\mathcal{E}) = c_1(\mathcal{E}^{**}) \quad \text{and} \quad c_2(\mathcal{E}) - \lg(\mathcal{T})[x] = c_2(\mathcal{E}^{**}).$$

Continuing our computations in the Chow ring $\mathrm{CH}^\bullet(C)$, we have

$$c_\bullet(\mathcal{E}|_C) = c_\bullet(\mathcal{E}^{**}|_C) \text{ and } c_1(\mathcal{E}|_C) = c_1(\mathcal{E}^{**}|_C) = c_1(\mathcal{E})[C].$$

Furthermore one has

$$c_\bullet(\mathcal{T}or_1^{\mathcal{O}_X}(\mathcal{T}, \mathcal{O}_C)) = c_\bullet(\mathcal{T}|_C).$$

Finally $c_\bullet(\mathcal{T}|_C) = 1 + c_1(\mathcal{T}|_C) = 1 + \lg(\mathcal{T}|_C)[x]$, x a k-rational point on C. From the exact sequence (1) we have

$$c_\bullet(\overline{\mathcal{E}|_C})c_\bullet(\mathcal{T}|_C) = c_\bullet(\mathcal{E}^{**}|_C)$$

and therefore

$$c_1(\overline{\mathcal{E}|_C}) = c_1(\mathcal{E}^{**}|_C) - \lg(\mathcal{T}|_C)[x] = c_1(\mathcal{E}|_C) - \lg(\mathcal{T}|_C)[x].$$

From (2) we obtain $c_\bullet(\mathcal{E}')c_\bullet(\mathcal{F}) = c_\bullet(\mathcal{E})$ and more explicitly

$$c_1(\mathcal{E}) = c_1(\mathcal{E}') + c_1(\mathcal{F})$$
$$c_2(\mathcal{E}) = c_2(\mathcal{E}') + c_2(\mathcal{F}) + c_1(\mathcal{E}')c_1(\mathcal{F}).$$

From the exact sequence (3) we obtain again in $\mathrm{CH}^\bullet(X)$:

$$c_\bullet(\mathcal{F}^{\mathrm{tors}})c_\bullet(\overline{\mathcal{F}}) = c_\bullet(\mathcal{F}).$$

Again we have $c_\bullet(\mathcal{F}^{\mathrm{tors}}) = 1 - \lg(\mathcal{F}^{\mathrm{tors}})[x]$, where we have

$$0 \leq \lg(\mathcal{F}^{\mathrm{tors}}) \leq \lg((\mathcal{E}|_C)^{\mathrm{tors}}).$$

As $\overline{\mathcal{F}}$ itself is a locally free sheaf of rank two on the curve C, one obtains easily

$$c_\bullet(\iota_*(\overline{\mathcal{F}})) = 1 + 2[C] + \big(3[C]^2 - j_*(c_1(\overline{\mathcal{F}}))\big).$$

(Here j_* is the obvious map of Chow groups $\mathrm{CH}^1(C) \longrightarrow \mathrm{CH}^2(X)$, considering a divisor on C as a codimension two cycle on X.)
We have everything at hand to write down the total Chern class $c_\bullet(\mathcal{E}')$ and the discriminant $\Delta(\mathcal{E}')$ of the modified sheaf $\mathcal{E}'$ of $\mathcal{E}$. For simplicity we do this only for the case $\mathcal{F} = \overline{\mathcal{F}}$, that is $\mathcal{F}^{\mathrm{tors}} = 0$. We have computed above

$$c_1(\iota_*(\overline{\mathcal{F}})) = 2[C]$$
$$c_2(\iota_*(\overline{\mathcal{F}})) = 3[C]^2 - j_*c_1(\overline{\mathcal{F}}).$$

Therefore we obtain

$$c_1(\mathcal{E}') = c_1(\mathcal{E}) - 2[C]$$
$$c_2(\mathcal{E}') = c_2(\mathcal{E}) + [C]^2 + j_*(c_1(\overline{\mathcal{F}})) - 2c_1(\mathcal{E})[C].$$

For the discriminant one obtains

$$\Delta(\mathcal{E}') = 8c_2(\mathcal{E}') - 3c_1(\mathcal{E}')^2$$
$$= \Delta(\mathcal{E}) - 4[C]^2 + 4\big(2j_*(c_1(\overline{\mathcal{F}})) - c_1(\mathcal{E})[C]\big).$$

The question remains to study the possible modifications which allow to stay inside a set of the form $\mathfrak{X}(A)$ for appropriate A and eventually connect two given vertices of $\mathfrak{X}(A)$.

References

[1] R. Friedmann. *Algebraic surfaces and holomorphic vector bundles*. Springer-Verlag, 1998.

[2] G. Harder. Minkowskische Reduktionstheorie über Funktionenkörpern. *Invent. math.*, 7:33–54, 1969.

[3] N. Hoffmann and U. Stuhler. Moduli schemes of generically simple Azumaya modules. *Doc. Math.*, 10:369–389, 2005.

[4] D. Huybrechts and M. Lehn. *The geometry of moduli spaces of sheaves*. Aspects Math. 31. Vieweg-Verlag, 1997.

[5] S. Lang. *Diophantine geometry*. Wiley-Interscience, 1962.

[6] A. Langer. Semistable sheaves in positive characteristic. *Ann. of Math.*, 159 (1):251–276, 2004.

[7] M. Lieblich. Moduli of twisted sheaves. *Duke Math. J.*, 138:23–118, 2007.

[8] Ch. Hao Liu and Sh.-Tung Yau. Nontrivial Azumaya noncommutative schemes, morphisms therefrom, and their extension by the sheaf of algebras of differential operators: D-branes in a B-field background à la Polchinski–Grothendieck Ansatz. arXiv: 0909.2291 v.1 [math.AG], 2009.

[9] K. Yoshioka. Moduli spaces of twisted sheaves on a projective variety. *Moduli spaces and arithmetic geometry, Adv. Stud. Pure Math.*, 45:1–30, 2006.

Fabian Reede and Ulrich Stuhler
Mathematisches Institut
Bunsenstraße 3–5
D-37073 Göttingen, Germany
e-mail: fabianreede@gmx.net
        stuhler@uni-math.gwdg.de

Affine Flag Manifolds and Principal Bundles
Trends in Mathematics, 219–232
© 2010 Springer Basel AG

A Physics Perspective on Geometric Langlands Duality

Karl-Georg Schlesinger

Abstract. We review the approach to the geometric Langlands program for algebraic curves via S-duality of an $N = 4$ supersymmetric four-dimensional gauge theory, initiated by Kapustin and Witten in 2006. We sketch some of the central further developments. Placing this four-dimensional gauge theory into a six-dimensional framework, as advocated by Witten, holds the promise to lead to a formulation which makes geometric Langlands duality a manifest symmetry (like covariance in differential geometry). Furthermore, it leads to an approach toward geometric Langlands duality for algebraic surfaces, reproducing and extending the recent results of Braverman and Finkelberg.

Mathematics Subject Classification (2000). 81T30, 14D21, 53C07.

Keywords. Langlands duality/program, S-duality, gauge theory, conformal field theory.

1. Introduction

In April 2006, Kapustin and Witten published their pathbreaking work [1] which led to a completely new perspective on geometric Langlands duality for algebraic curves. It starts from a four-dimensional $N = 4$ supersymmetric gauge theory. Assuming that S-duality, a certain symmetry which generalizes the electric-magnetic duality of the Maxwell equations to the case of a nonabelian gauge theory, holds for this theory, it is possible to derive geometric Langlands duality for algebraic curves from this. S-duality is conjectural but very well supported. In this sense, we get a reformulation of geometric Langlands duality for algebraic curves.

In the first part of this contribution, we will review the approach of Kapustin and Witten. We will continue by very briefly sketching some of the new developments which this approach has initiated. Finally, we will place the four-dimensional gauge theory in a six-dimensional string theory framework. This perspective (see [2], [3]) holds the promise to lead to a formulation, making geometric Langlands duality a manifest symmetry (like covariance is manifest in differential geometry and has no longer to be verified by calculations on specific coordinate transformations). The six-dimensional view also leads to an approach toward geometric Langlands duality for algebraic surfaces, reproducing and extending the recent results of Braverman and Finkelberg (see [4], see also [5]).

This article is intended as an introduction to the gauge and string theory approach to the geometric Langlands program for mathematicians. As such, it focuses on a short, non-technical, overview of the central ideas and concepts and does not contain any original research results. Neither do we pretend to give a complete overview of this rapidly developing and highly promising field. To keep the article in a sufficiently focused form, some exciting developments (e.g., the appearance of Arthur's SL (2) in this framework, see [6] and [7]) will be completely left out. For another recently published review on the topic, see [8].

2. $N = 4$ supersymmetric gauge theory

Let G be a compact Lie group. For simplicity (and to keep all the formulae valid in precisely the form used here, without any extra factors), we will assume that G is from the ADE-series. Later, for the six-dimensional framework, this assumption will be essential and no longer a technical assumption for simplicity.

Let X_4 be the four-dimensional space-time (again, we make a technical assumption for simplicity, assuming that the signature of X_4 is Euclidean rather than Minkowskian). We use Greek indices, e.g.,

$$\mu = 0, \ 1, \ 2, \ 3,$$

to label space-time indices and Latin ones

$$i = 1, \ldots, 6,$$

for an internal set of indices. Let A be a connection of a G-bundle over X_4 and F the corresponding curvature form. With

$$D_\mu$$

we denote the covariant derivatives and with

$$\phi_i$$

a set of adjoint-valued scalar fields (i.e., functions valued in the adjoint representation of the Lie algebra of G). The action of the $N = 4$ supersymmetric gauge theory which we want to consider is then given by

$$S_4 = \frac{1}{e^2} \int d^4x \, \mathrm{Tr} \left(\frac{1}{2} \sum_{\mu,\nu=0}^{3} F_{\mu\nu} F^{\mu\nu} + \sum_{\mu=0}^{3} \sum_{i=1}^{6} D_\mu \phi_i D^\mu \phi_i + \frac{1}{2} \sum_{i,j=1}^{6} [\phi_i, \phi_j]^2 \right) + \cdots$$

where the dots indicate the fermionic part of the action and e the coupling constant of the theory (just as Newton's constant in gravity). The fermionic part is necessary for supersymmetry but we will not consider it, here. Though we can explain the essential ideas without considering it explicitly, one should nevertheless keep in mind that the whole construction does only hold with $N = 4$ supersymmetry implemented. Here, $N = 4$ denotes the degree of supersymmetry. The supersymmetry algebra is determined by a choice of representation of the – in this case four-dimensional – spin group, i.e., the double cover of the Lorentz group. The simplest degree of supersymmetry is denoted by $N = 1$ while $N = 4$ means (very roughly speaking) that we have four copies of the simplest representation involved in the definition of the supersymmetry algebra.

If one has higher than $N = 1$ supersymmetry, it is generally possible to derive the theory from an $N = 1$ supersymmetric theory, living in a higher-dimensional space, by dimensional reduction. In this case, the higher-dimensional theory is an $N = 1$ supersymmetric gauge theory in ten dimensions. Dimensional reduction (which will appear over and over again in this article) means that one assumes some of the dimensions (in this case six) to be scrolled up to a compact space of very small volume (in this case, six dimensions are scrolled up to small circles). Sending the volume to zero (i.e., sending the radii of the circles to zero) results in an induced lower-dimensional theory. The fact that we can get the action S_4 from ten dimensions by reducing on six circles is the reason for the appearance of the adjoint-valued scalar fields and the internal indices $i = 1, \ldots, 6$. Indeed, the first term in S_4 is the well-known gauge theory term while the other two arise from the dimensional reduction of the corresponding gauge theory term in ten dimensions.

To get the most general $N = 4$ supersymmetric gauge theory in four dimensions, it is possible to add the so-called *topological term* S_θ to S_4. This is given by

$$S_\theta = -\frac{\theta}{8\pi^2} \int d^4x \, \mathrm{Tr} \left(F \wedge F \right).$$

It is referred to as topological since the integral just gives the second Chern class of the G-bundle. The coupling constant e of S_4 and the parameter θ of the topological term are combined into the complex coupling constant

$$\tau = \frac{\theta}{2\pi} + \frac{4\pi i}{e^2}.$$

Observe that the imaginary part of τ is always positive, i.e., τ is from the upper half-plane $\mathbb{H}$.

3. S-duality

In the quantized theory there is a natural symmetry given by the generator

$$T \colon \tau \mapsto \tau + 1$$

which results from the fact that – roughly speaking – the complex coupling constant appears only in the form

$$e^{2\pi i \tau}$$

in the path integral. On the other hand, there is a natural action of $\mathrm{SL}\,(2, \mathbb{R})$ on $\mathbb{H}$.

The *S-duality conjecture* states the following:

There exists a second symmetry (i.e., generator) S, generating together with T a discrete subgroup of SL $(2, \mathbb{R})$ such that $S_4 + S_\theta$ is invariant under the following combination of operations:

- $\tau \mapsto S(\tau)$
- exchange of electric and magnetic charges
- $G \mapsto {}^L G$.

The latter two operations are not unrelated: In 1977 Goddard, Nuyts and Olive investigated how electric and magnetic charges are classified in a nonabelian gauge theory (see [9]). The result is that one set of charges is given by the weight lattice of the Lie algebra of the gauge group G while the other one is given by the root lattice. Of course, exchanging weight and root lattice is precisely what defines the Langlands dual ${}^L G$ of the group G.

In the same year Montonen and Olive presented S-duality as a conjectural symmetry for nonabelian gauge theory (see [10]). While S-duality does not hold in the non-supersymmetric case, there is strong evidence that it holds with $N = 4$ supersymmetry.

For G from the ADE-series (as we do assume), the generator S has to take the form

$$S: \tau \mapsto -\frac{1}{\tau},$$

i.e., the discrete subgroup of SL $(2, \mathbb{R})$ generated by T and S is the modular group SL $(2, \mathbb{Z})$.

In order to derive geometric Langlands duality for algebraic curves from the S-duality conjecture, one has to perform two essential steps on the four-dimensional gauge theory: First, one has to perform a topological twist and than a dimensional reduction to a two-dimensional theory. We will discuss both steps very briefly, in a non-technical manner, in the next two sections. After that we will introduce the operators of the four-dimensional gauge theory which – after performing the two steps on them – will lead to geometric Langlands duality. For technical details, we refer the reader to [1].

4. Topological twisting

Topological twisting means that one retains only part of the state space of the original theory. For this one introduces the cohomology with respect to a certain differential Q (what physicists call a BRST-operator) with

$$Q^2 = 0.$$

To find a Q suitable for the topological twist, supersymmetry is essential. Passing to the cohomology with respect to Q – i.e., forgetting all Q-exact terms – one retains only part of the information of the original theory. The resulting theory is called topological. For a pure mathematician this nomenclature might be slightly disturbing since the theory is *not* independent of all non-topological information, e.g., we will see that after dimensional reduction it is still dependent on certain holomorphic and symplectic structures. Topological theory in this context means that *on* the Q-cohomology we have independence from the choice of metric.

Concretely, in this case Q is determined by a choice of homomorphism

$$\chi \colon \mathrm{Spin}\,(4) \to \mathrm{Spin}\,(6)$$

which is related to the fact that the gauge theory arises from a ten-dimensional theory by dimensional reduction and a decomposition of $\mathrm{Spin}\,(10)$ into $\mathrm{Spin}\,(4)$ and $\mathrm{Spin}\,(6)$ components. The approach is very similar to the introduction of Donaldson invariants for four-dimensional manifolds by using a topological twist for an $N = 2$ supersymmetric four-dimensional gauge theory.

It turns out that the topological twist is not determined uniquely but there arises a whole family of suitable topological twists, parameterized by the *topological twisting parameter*

$$t \in \mathbb{C}P^1.$$

The complex coupling constant τ and the topological twisting parameter t are then combined into the *canonical parameter*

$$\psi = \frac{\tau + \bar{\tau}}{2} + \frac{\tau - \bar{\tau}}{2} \left(\frac{t - t^{-1}}{t + t^{-1}} \right).$$

The reason for introducing the canonical parameter is that the correlation functions of the observables of the topological theory (i.e., of the Q-cohomology classes) do only depend on ψ and not on the parameters e, θ and t separately.

It is a small lemma to show that on ψ the two generators T and S operate, again, as

$$T \colon \psi \mapsto \psi + 1 \qquad \text{and} \qquad S \colon \psi \mapsto -\frac{1}{\psi}.$$

5. Dimensional reduction

Let now

$$X_4 \cong \Sigma \times C$$

with Σ a (compact or non-compact) Riemann-surface (which will become the two-dimensional space-time after dimensional reduction) and C a compact Riemann surface. As a technical assumption, we will require that

$$\mathrm{genus}\,(C) \geq 2.$$

We now perform the dimensional reduction by assuming that

$$\mathrm{vol}\,(C) \ll \mathrm{vol}\,(\Sigma).$$

In this limit, the four-dimensional $N = 4$ supersymmetric gauge theory induces a two-dimensional field theory on Σ. In two dimensions there do not exist non-trivial gauge theories and the resulting field theory turns out to be a nonlinear sigma model, i.e., a field theory where the (bosonic) fields are given by maps from

Σ to the so-called *target space*. Roughly speaking, the action is given by a minimal area requirement for the image of Σ under these maps into the target space. So, the essential information to determine the nonlinear sigma model is to specify the target space. One shows that in this case the resulting target space is the Hitchin moduli space Hit (G, C) for the gauge group G and the complex curve C (see [11]).

Let A be a G-connection on C and F the corresponding curvature form. Let ϕ be an adjoint-valued 1-form on C. Consider the set of equations

$$F - \phi \wedge \phi = 0 \qquad \text{and} \qquad D\phi = D^*\phi = 0.$$

The solutions to this set of equations, modulo G-gauge transformations, define the Hitchin moduli space Hit (G, C). For those readers with a knowledge of Higgs-bundles one can simply define it as the moduli space of G-Higgs-bundles on C. That we have used the letter ϕ for the adjoint-valued 1-form is not by accident. The topological twist shifts the degree of some fields and we get the adjoint-valued 1-form from the adjoint-valued scalar fields of the original gauge theory. With

$$g = \text{genus}\,(C)$$

one has

$$\dim_{\mathbb{C}} \text{Hit}\,(G, C) = (2g - 2)\dim G.$$

and Hit (G, C) is a Hyperkähler manifold. So, we have a representation of the quaternion algebra on the tangent bundle and complex structures I, J, K with corresponding symplectic structures ω_I, ω_J, ω_K. When we refer to complex or symplectic structures in the sequel, we will have to keep in mind that we have to make precise to which of these structures we refer.

6. Wilson operators

We are now ready to introduce the needed operators in the four-dimensional gauge theory. Usual operators in a quantum field theory, as you remember them from any introductory course on the subject, are attached to points (i.e., they are zero-dimensional objects): They are operators $M(x)$, $M(z)$ attached to points x, z and satisfying the well-known commutation relations (e.g., $M(x)$ and $M(z)$ commute if x and z are space-like separated). Physicists have learned in recent decades that there are other operators, attached to lines (one-dimensional objects), containing essential information in a quantum field theory (in solid state physics or in the study of phase transitions these are prominent operators).

Recall that A is a connection on a G-bundle over X_4. Let S be an oriented loop in X_4, R an irreducible representation of G. With Tr_R we denote taking the trace in the representation R. We define the *Wilson operator* $W_0(R, S)$ as the holonomy of A around S:

$$W_0(R, S) = \text{Tr}_R \exp\left(-\oint_S A\right).$$

Since we want to perform the two steps, topological twisting and dimensional reduction, on the operators, the next question is if these operators induce well-defined operators on Q-cohomology. Unfortunately, the answer is no and this problem cannot be resolved for general values of the topological twisting parameter t. But for the special values $t = i$ and $t = -i$ there exists a solution: For these values there exists a linear combination of A with the adjoint-valued 1-form ϕ, such that the holonomy of the linear combination induces a well-defined operator on cohomology, i.e., we have *topological Wilson operators*. Concretely, the topological Wilson operators are defined by

$$W(R, S) = \text{Tr}_R \exp\left(-\oint_S A + i\phi\right) \quad \text{for } t = i$$

and

$$W(R, S) = \text{Tr}_R \exp\left(-\oint_S A - i\phi\right) \quad \text{for } t = -i.$$

Next, replace the loop S with a line L from p to q. Replace the trace Tr_R with the matrix of parallel transport from the fiber E_p of the G-bundle on X_4 to the fiber E_q, with both fibers considered in the representation R of G. The parallel transport is taken with respect to the connection

$$\mathcal{A} = A + i\phi \qquad \text{and} \qquad \overline{\mathcal{A}} = A - i\phi$$

for $t = \pm i$, respectively. This corresponds to the canonical parameter $\psi = \infty$. In conclusion, for $\psi = \infty$ we have topological Wilson operators, defined by representations R of G.

Assume, now, that S-duality holds for the $N = 4$ supersymmetric gauge theory. This means that there has to exist a second set of topological line operators which exchange with the Wilson operators on lines under S-duality, i.e., for $\psi = 0$ there should exist topological line operators, defined by representations of $^L G$. Indeed, these operators can be constructed in the form of the so-called *'t Hooft operators* which we are not going to discuss explicitly, here.

Finally, we perform the dimensional reduction on the topological line operators. Consider a two-dimensional theory and a line operator $\widehat{L}$ on a line L, close to a boundary with specified boundary condition (i.e., what physicists call a D-brane for a two-dimensional nonlinear sigma model). Imagine L approaching the boundary more and more closely. In the limit, L will be absorbed by the boundary and the operator $\widehat{L}$ disappears, resulting in a change of boundary conditions. Of course, this is a heuristic picture but it can be validated in a calculation. The boundary condition is given by a submanifold (D-brane) of the target space to which the one-dimensional boundary of Σ has to be mapped under the fields, together with a vector bundle W on this submanifold. One can show the operator $\widehat{L}$ to change boundary conditions by changing this vector bundle W. So, we can view the line operators in the two-dimensional theory as abstract operators, operating on boundary conditions.

We call a boundary condition, given by W, an *eigenbrane* of $\widehat{L}$ if there exists a fixed vector space V such that $\widehat{L}$ acts as

$$\widehat{L}\colon W \mapsto V \otimes W.$$

This is similar to eigenfunctions for operators in quantum mechanics, with the function replaced by a vector bundle and the eigenvalue replaced by the fixed vector space V. As in quantum mechanics, we can pose the question if the line operators $\widehat{L}_1$ and $\widehat{L}_2$ on lines L_1 and L_2 can have simultaneous eigenbranes. The answer is that they have simultaneous eigenbranes if and only if

$$\left[\widehat{L}_1, \widehat{L}_2\right] = 0.$$

For the dimensional reduction of the topologically twisted $N = 4$ supersymmetric gauge theory, one can show that there exist simultaneous eigenbranes of *all* topological Wilson operators. These eigenbranes are called *electric eigenbranes*. Similarly, there exist simultaneous eigenbranes of *all* topological 't Hooft operators and these are called *magnetic eigenbranes*.

7. Mirror symmetry

Without defining the three complex structures I, J, K (and corresponding symplectic structures) explicitly for $\mathrm{Hit}\,(G, C)$ (see [1]), we recall that we have to keep them apart when referring to a complex or a symplectic structure. For a nonlinear sigma model on a Ricci-flat Kähler manifold there exist two types of topological twists, called the A- and the B-model (see [12]). The A-model couples only to the symplectic structure of the target space and the B-model only to the holomorphic structure. One proves that

> *Electric eigenbranes are elements of the bounded derived category of coherent sheaves in the complex structure J on $\mathrm{Hit}\,(G, C)$.*

The elements of the bounded derived category of coherent sheaves are the D-branes for the B-model, referred to as B-branes in the physics literature. Including reference to the complex structure J, they are called J_B-branes. Similarly, one shows that

> *Magnetic eigenbranes are elements of the Fukaya category in the symplectic structure ω_K.*

In physics terminology, this means magnetic eigenbranes are K_A-branes. Mirror symmetry exchanges the A- and the B-model. One mathematically rigorous formulation of mirror symmetry, called *homological mirror symmetry* (see [13]) states that Calabi–Yau manifolds X and Y form a mirror pair if there is a suitable equivalence between the Fukaya category of X and the bounded derived category of coherent sheaves of Y and vice versa (to make this technically precise, one does not really work with simple categories but with a triangulated version of A_∞-categories). S-duality of the four-dimensional gauge theory induces homological mirror symmetry for the $\mathrm{Hit}\,(G, C)$ sigma model on Σ or in more physics oriented language, S-duality induces mirror symmetry between the B-model on Hit $(G, C)_J$ (corresponding to $\psi = \infty$) and the A-model on Hit$(^L G, C)_K$ (corresponding to $\psi = 0$). Here, the subscripts refer to the complex, respectively symplectic structure, used on the Hitchin moduli space.

In the geometric Langlands program for algebraic curves C one considers two different moduli spaces: The moduli space $\mathcal{M}$ of flat $^L G_{\mathbb{C}}$-bundles on C and the moduli space $\widetilde{\mathcal{M}}$ of holomorphic G-bundles on C. On $\mathcal{M}$ one considers sheaves with support at a point of $\mathcal{M}$ (skyscraper sheaves) and on $\widetilde{\mathcal{M}}$ one considers the so-called Hecke eigensheaves. It is a central part of [1] to show that the skyscrapers are in one-to-one correspondence to the electric eigenbranes and the Hecke eigensheaves to the magnetic eigenbranes. In consequence, if S-duality holds for the four-dimensional $N = 4$ supersymmetric gauge theory, one can derive geometric Langlands duality for algebraic curves.

At this point, the attentive reader might ask why one needs S-duality of the four-dimensional gauge theory for this and why one does not start directly from the homological mirror symmetry conjecture for the Hit (G, C) sigma model. The answer is that mathematicians know very well that $\widetilde{\mathcal{M}}$ is not a true moduli space (and cannot be for geometric Langlands duality to hold true) but a stack. The nonlinear sigma model treats the target space in a first approach as a proper space. If one takes the stacky nature into account, one rediscovers that one actually derived the model from the four-dimensional gauge theory, i.e., the four-dimensional viewpoint is essential for the geometric Langlands program (see [1]).

There are further examples for the deep interplay between the structures, naturally emerging from physics, and those needed for the mathematics of the geometric Langlands program, in this approach. E.g., the Fukaya category as it is originally defined (see [14], see [15] for an approach to a rigorous treatment), involves the Lagrangian submanifolds of Hit (G, C). But there exist additional A-branes on Hit (G, C) which are only coisotropic submanifolds. A special such A-brane (called the canonical coisotropic brane or c.c. brane, for short), corresponding to a coisotropic submanifold of full dimensionality (i.e., isomorphic to Hit (G, C) itself) and of rank one (i.e., the vector bundle W on the brane is a line bundle) is used in [1] to show that the magnetic eigenbranes satisfy the D-module property which is so important for the Hecke eigensheaves in the geometric Langlands program.

Finally, there exists another physics motivated approach to the geometric Langlands program for algebraic curves, using two-dimensional conformal field theory on C to construct Hecke eigensheaves (see [16] for a beautiful review and the original literature). One might wonder how the two approaches are related, one leading to a two-dimensional nonlinear sigma model on Σ, the other to a conformal field theory on C. A step toward the goal to derive the conformal field theory approach on C also from the four-dimensional gauge theory should be to prove the following property to hold: The dual brane (under S-duality, respectively mirror symmetry) of the c.c. brane should be a brane which has support on the space of opers of [17] ([18]). The dual of the c.c. brane is a coisotropic brane of rank > 1 and is calculated in the gauge theory setting in [19].

After this review of some of the central parts of [1], we are now ready to take a brief look on some of the further developments in 2006–2009.

8. Higher-dimensional operators

As we have seen, beyond the usual zero-dimensional operators (attached to points) there are one-dimensional line operators in a quantum field theory, containing fundamental information. One might ask if there are further even higher-dimensional operators. In a four-dimensional theory, these could be two-dimensional (attached to surfaces) or three-dimensional (attached to volumes). Four-dimensional operators would be trivial.

Two-dimensional operators become important if the gauge connection A has singularities. So far, we have assumed A to be holomorphic but one can allow A to be meromorphic, only, and to have singularities along surfaces. The approach of [1] can be extended to this case and surface operators take a central place, then. When Beilinson and Drinfeld developed the geometric Langlands program, it was intended as an analogue to the classical Langlands program, to get insights from a situation with an additional smoothness property. The case of a meromorphic gauge connection A corresponds to what is called ramification in the classical Langlands program. If A has only simple poles, one has tame ramification, otherwise one has the case of wild ramification (see [20], [21], [22]; see [16] for a discussion how structures in the classical and the geometric Langlands program are analogous). Especially, understanding wild ramification in the geometric case is believed to be important for comparison to the classical Langlands program.

Three-dimensional operators live on volumes and therefore divide the four-dimensional manifold X_4 into two halves. They correspond to what physicists call *domain walls* in a gauge theory. Domain walls allow to change the gauge group. On the one side, we have the gauge theory with gauge group G and on the other side the theory with gauge group $\tilde{G}$. On the domain wall we have the three-dimensional operator, corresponding to specifying a boundary condition which ensures that the two gauge theories join consistently along the domain wall.

In the classical Langlands program, beyond Langlands duality, changing the group G is a central ingredient, giving rise to Langlands functoriality. It was an open question – again of tremendous importance for comparing the geometric to the classical case – what constitutes the counterpart of Langlands functoriality in the geometric Langlands program. Domain walls lead to geometric Langlands functoriality. This is a subject very much in its beginning. From the gauge theory side one has to get knowledge on the three-dimensional boundary conditions which involve data given in the form of three-dimensional quantum field theories (see [23], [19], [24], [3]).

In conclusion, higher-dimensional operators on surfaces and volumes have turned out to be very important for studying analogues of structures which are central for the classical Langlands program.

9. The six-dimensional view

Remember that our four-dimensional gauge theory lives on X_4. There is a conjecture, arising from string theory, which states that there exists a conformally invariant field theory on

$$X_4 \times T^2$$

such that in the limit of small T^2 (dimensional reduction), it induces precisely the $N = 4$ supersymmetric gauge theory on X_4.

On T^2 there is, of course, the natural action of SL $(2, \mathbb{Z})$. We can ask what compensates this action on X_4. It turns out that in this way the SL $(2, \mathbb{Z})$-action on T^2 induces S-duality of the gauge theory on X_4. In consequence, if it would be possible to construct this six-dimensional conformal field theory, one could prove S-duality for the $N = 4$ supersymmetric gauge theory on X_4 and, in consequence, geometric Langlands duality. One should stress that existence of the theory suffices: While for the four-dimensional gauge theory one has to prove something (S-duality) to get geometric Langlands duality, for the six-dimensional conformal field theory one only has to construct the theory since its very existence makes S-duality of the four-dimensional theory (and, hence, geometric Langlands duality) manifest. In this sense, one can view the search for this six-dimensional theory as the search for the geometry behind the Langlands program, making Langlands duality a manifest symmetry. This would be very much like passing from a coordinate description to differential geometry where covariance becomes a manifest symmetry.

The problem is that it is known from string theory that this six-dimensional theory cannot exist consistently on its own. It actually has to be embedded into eleven-dimensional M-theory as the world-volume theory of the $M5$-brane (a five-dimensional extended object with a six-dimensional world volume in M-theory, the central charges, leading to the $M5$-brane, arising as one of the components in the direct sum decomposition of the eleven-dimensional supersymmetry algebra). So, its completion in the UV-regime is related to the so-called six-dimensional micro string theories (see [25], [26] for an easily accessible introduction).

Consider the six-dimensional theory on another manifold X_6, now,

$$X_6 \cong \Sigma \times X_4$$

with Σ a (compact or non-compact) Riemann surface, X_4 a compact Hyperkähler manifold and

$$\operatorname{vol}(X_4) \ll \operatorname{vol}(\Sigma).$$

This is very similar to the situation we considered when reducing the four-dimensional gauge theory to a two-dimensional nonlinear sigma model. This time we get the reduction of the six-dimensional theory to a two-dimensional nonlinear sigma model and the target space turns out to be given by the instanton moduli space $\text{Inst}(X_4)$ on X_4, i.e., the space of all anti-self-dual G-connections on X_4 (remember that now, for the six-dimensional view, G definitely has to be from the ADE-series).

From the side of physics, there are some very nice relations behind this model. The Hitchin–Kobayashi correspondence relates $\text{Inst}(X_4)$ to $\text{Bun}_G(X_4)$, i.e., brings in a relation to Yang–Mills theory on X_4. On the other hand, for

$$G = U(k)$$

the space X_4 turns out to be related to the multi-center Taub-NUT solution TN_k of the Einstein vacuum equations. This gives particular interest to the study of instantons on TN_k (see [27], [28], [29]).

On the mathematical side, this model reproduces and extends – beyond the case $G = U(k)$ the results of Braverman and Finkelberg (see [4], [5]) on geometric Langlands duality for algebraic surfaces X_4 (see [30], [3]). Let us review this in a little bit more detail (basically following [3]).

One can show that the six-dimensional theory cannot have a Lagrangian description, it is a purely quantum field theoretic object. But dimensionally reducing the theory for small S^1 on

$$X_6 \cong X_5 \times S^1$$

one gets in the infrared limit a gauge theory description on X_5. One can now pass to the more complicated case with X_6 not being given as a Cartesian product, as above, but as a $U(1)$-bundle over X_5, i.e., we have a free action of $U(1)$ on X_6. This leads to an additional Chern–Simons like term in the dimensional reduction to X_5. Finally, one can pass to the case of a non-free action of $U(1)$ on X_6 and consider the singular quotient space $X_6/U(1)$. Outside the non-free locus the dimensional reduction works as in the previous case. Consider the special case where the non-free locus has codimension four and consists only of fixed points of $U(1)$. In this case, the Chern–Simons term has an anomaly on the two-dimensional non-free locus W, i.e., on W a third term has to appear in the action of the dimensional reduction to X_5 which cancels this anomaly. This third term arises from a two-dimensional quantum field theory on W, given by the holomorphic part of the WZW-model (at level one and for the group G). The affine Lie algebra of G, which mathematically is behind the WZW-model, naturally explains why the approach to the geometric Langlands program for algebraic surfaces (see [4], [5]) leads to Langlands duality for the affine case.

10. Conclusion

We have seen that the search for a six-dimensional field theory (which has to be a purely quantum field theoretic structure, embedded into eleven-dimensional M-theory) offers a fundamental perspective on the geometric Langlands program: It would lead to Langlands duality as a manifest symmetry, it would unite geometric

Langlands duality for algebraic surfaces and algebraic curves into a single framework, and it would naturally include higher-dimensional operators which are so important for studying the counterparts of ramification and Langlands functoriality on the geometric side. In physics it has strong links to many areas (string- and M-theory, Yang–Mills theory, Taub-NUT solutions of the Einstein equations).

Last not least, though the full six-dimensional theory has not been constructed so far, it is amenable to explicit calculations in dimensional reductions, leading to structures like WZW-models where a lot of results are available from the side of mathematical physics.

Acknowledgement

I would like to thank A. Schmitt for the invitation to contribute this article and to give a talk on a similar topic at the highly stimulating conference VBAC 2009 in Berlin. I would like to thank E. Witten for an e-mail explanation concerning the state of the art in research on the dual of the canonical coisotropic brane.

References

[1] A. Kapustin, E. Witten, *Electric-magnetic duality and the geometric Langlands program*, Commun. Number Theory Phys. **1** (2007), 236 pp.

[2] E. Witten, *Conformal field theory in four and six dimensions*, in U. Tillmann (ed.), *Topology, geometry and quantum field theory*, Proceedings of the 2002 Oxford symposium in honour of the 60th birthday of Graeme Segal, Oxford, UK, June 24–29, 2002, Cambridge University Press, London Mathematical Society Lecture Note Series **308** (2004), 405–419.

[3] E. Witten, *Geometric Langlands from six dimensions*, arXiv:0905.2720.

[4] A. Braverman, M. Finkelberg, *Pursuing the double affine Grassmannian I: Transversal slices via instantons on A_k singularities*, arXiv:0711.2083.

[5] H. Nakajima, *Quiver varieties and branching*, SIGMA, Symmetry Integrability Geom. Methods Appl. **5** (2009), Paper 003, 37 pp. (electronic only).

[6] D. Ben-Zvi, D. Nadler, *The character group of a complex group*, arXiv:0904.1247.

[7] E. Witten, *Geometric Langlands and the equations of Nahm and Bogomolny*, arXiv:0905.4795.

[8] E. Frenkel, *Gauge theory and Langlands duality*, arXiv:0906.2747.

[9] P. Goddard, J. Nuyts, D. Olive, *Gauge theories and magnetic charge*, Nucl. Phys. **B125** (1977), 1–28.

[10] C. Montonen, D. Olive, *Magnetic monopoles as gauge particles?*, Phys. Lett. **B72** (1977), 117–120.

[11] N. Hitchin, *The self-duality equations on a Riemann surface*, Proc. London Math. Soc. (3) **55** (1987), 59–126.

[12] E. Witten, *Mirror manifolds and topological field theory*, in S. T. Yau, *Essays on mirror manifolds*, International Press, Hong Kong, 1991, 120–158.

[13] M. Kontsevich, *Homological algebra of mirror symmetry*, in S.D. Chatterji (ed.), *Proceedings of the international congress of mathematicians*, ICM '94, August 3-11, 1994, Zürich, Switzerland, Vol. I, Birkhäuser, 1995, 120–139.

[14] K. Fukaya, *Morse homotopy, A^∞-category, and Floer homologies*, Proc. of GARC workshop on geometry and topology, Seoul 1993, 1–102.

[15] K. Fukaya, Y.-G. Oh, H. Ohta, K. Ono, *Lagrangian intersection Floer theory – anomaly and obstruction*, AMS/IP Studies in Advanced Mathematics, Vol. 46, 2009, 800 pp.

[16] E. Frenkel, *Lectures on the Langlands program and conformal field theory*, in P. Cartier (ed.) et al., *Frontiers in number theory, physics, and geometry II. On conformal field theories, discrete groups and renormalization*, Papers from the meeting, Les Houches, France, March 9–21, 2003, Springer, 2007, 389–533.

[17] A. Beilinson, V. Drinfeld, *Quantization of Hitchin's integrable system and Hecke eigensheaves*, available under
http://www.math.uchicago.edu/~mitya/langlands/hitchin/BD-hitchin.pdf.

[18] E. Witten, private communication.

[19] D. Gaiotto, E. Witten, *Supersymmetric boundary conditions in $N = 4$ super Yang–Mills theory*, J. Stat. Phys. **135** (2009), 789–855.

[20] S. Gukov, E. Witten, *Gauge theory, ramification, and the geometric Langlands program*, in D. Jerison (ed.) et al., *Current developments in mathematics*, 2006, Sommerville, International Press, 2008, 35–180.

[21] S. Gukov, E. Witten, *Rigid surface operators*, arXiv:0804.1561.

[22] E. Witten, *Gauge theory and wild ramification*, Anal. Appl., Singap., **6** (2008), 429–501.

[23] E. Frenkel, E. Witten, *Geometric endoscopy and mirror symmetry*, Commun. Number Theory Phys. **2** (2008), 113–283.

[24] D. Gaiotto, E. Witten, *S-duality of boundary conditions in $N = 4$ super Yang–Mills theory*, arXiv:0807.3720.

[25] R. Dijkgraaf, *Instanton strings and Hyperkähler geometry*, Nucl. Phys. **B543** (1999), 545–571.

[26] R. Dijkgraaf, E. Verlinde, H. Verlinde, *Notes on matrix and micro strings*, in L. Baulieu (ed.) et al., *Strings, branes and dualities*, Proceedings of the NATO Advanced Study Institute, Cargèse, France, May 26–June 14, 1997, Kluwer Academic Publishers, NATO ASI Ser., Ser. C, Math. Phys. Sci. **520** (1999), 319–356.

[27] S.A. Cherkis, *Moduli spaces of instantons on the Taub-NUT space*, arXiv:0805.1245.

[28] S.A. Cherkis, *Instantons on the Taub-NUT space*, arXiv:0902.4724.

[29] E. Witten, *Branes, instantons, and Taub-NUT space*, arXiv:0902.0948.

[30] M.C. Tan, *Five-branes in M-theory and a two-dimensional geometric Langlands duality*, arXiv:0807.1107.

Karl-Georg Schlesinger
Institute for Theoretical Physics
University of Vienna
Boltzmanngasse 5
A-1090 Vienna, Austria
e-mail: kgschles@esi.ac.at

Affine Flag Manifolds and Principal Bundles
Trends in Mathematics, 233–289
ⓒ 2010 Springer Basel AG

Double Affine Hecke Algebras and Affine Flag Manifolds, I

Michela Varagnolo and Eric Vasserot

Abstract. This is the first of a series of papers which review the geometric construction of the double affine Hecke algebra via affine flag manifolds and explain the main results of the authors on its representation theory. There are also some simplifications of the original arguments and proofs for some well-known results for which there exists no reference.

Mathematics Subject Classification (2000). 20C08, 14M15, 16E20.

Keywords. Ind-scheme, affine flag manifold, K-theory, Hecke algebra.

Introduction

This paper is the first of a series of papers reviewing the geometric construction of the double affine Hecke algebra via affine flag manifolds. The aim of this work is to explain the main results in [33], [32], but also to give a simpler approach to some of them, and to give the proof of some related 'folklore' statements whose proofs are not available in the published literature. This work should therefore be viewed as a companion to loc. cit., and is by no means a logically independent treatment of the theory from the very beginning. In order that the length of each paper remains reasonable, we have split the whole exposition into several parts. This one concerns the most basic facts of the theory: the geometric construction of the double affine Hecke algebra via the equivariant, algebraic K-theory and the classification of the simple modules of the category $\mathcal{O}$ of the double affine Hecke algebra. It is our hope that by providing a detailed explanation of some of the difficult aspects of the foundations, this theory will be better understood by a wider audience.

This paper contains three chapters. The first one is a reminder on $\mathcal{O}$-modules over non Noetherian schemes and over ind-schemes. The second one deals with affine flag manifolds. The last chapter concerns the classification of simple modules in the category $\mathcal{O}$ of the double affine Hecke algebra. Let us review these parts in more details.

In the second chapter of the paper we use two different versions of the affine flag manifold. The first one is an ind-scheme of ind-finite type, while the second one is a pro-smooth, coherent, separated, non Noetherian and non quasi-compact scheme. Thus, in the first chapter we recall some basic facts on $\mathcal{O}$-modules over coherent schemes, pro-schemes and ind-schemes. The first section is a reminder on pro-objects and ind-objects in an arbitrary category. We give the definition of direct and inverse 2-limits of categories. Next we recall the definition of the K-homology of a scheme. We'll use non quasi-compact non Noetherian schemes. Also, it is convenient to consider a quite general setting involving unbounded derived categories, pseudo-coherent complexes and perfect complexes. Fortunately, since all the schemes we'll consider are coherent the definition of the K-theory remains quite close to the usual one. To simplify the exposition it is convenient to introduce the derived direct image of a morphism of non Noetherian schemes, its derived inverse image and the derived tensor product in the unbounded derived categories of $\mathcal{O}$-modules. Finally we consider the special case of pro-schemes (compact schemes, pro-smooth schemes, etc) and of ind-schemes. They are important tools in this work. This section finishes with equivariant $\mathcal{O}$-modules and some basic tools in equivariant K-theory (induction, reduction of the group action, the Thom isomorphism and the Thomason concentration theorem).

The second chapter begins with the definition of the affine flag manifold, which is an ind-scheme of ind-finite type, and with the definition of the Kashiwara affine flag manifold, which is a non quasi-compact coherent scheme. This leads us in Section 2.3.6 to the definition of an associative multiplication on a group of equivariant K-theory $\mathbf{K}^I(\mathfrak{N})$. Here $\mathfrak{N}$ is an ind-scheme which can be regarded as the affine analogue of the Steinberg variety for reductive groups. Then, in Section 2.4.1, we define an affine analogue of the concentration map for convolution rings in K-theory used in [7]. It is a ring homomorphism relating $\mathbf{K}^I(\mathfrak{N})$ to the K-theory of the fixed points subset for a torus action. This concentration map is new, and it simplifies the proofs in [33]. The double affine Hecke algebra is introduced in Section 2.5.1 and its geometric realization is proved in Theorem 2.5.6. We use here an approach similar to the one in [4], where a degenerate version of the double affine Hecke algebra is constructed geometrically. Compare also [10], where the regular representation of the double affine Hecke algebra is constructed geometrically. The proof we give uses a reduction to the fixed points of a torus acting on the affine analogue of the Steinberg variety and the concentration map in K-theory.

The third chapter is a review of the classification of the simple modules in the category $\mathcal{O}$ of the double affine Hecke algebra. The main theorem was proved in [33]. The proof we give here is simpler than in loc. cit. because it uses the concentration map. The first section contains generalities on convolution algebras in the cohomology of schemes of infinite type which are locally of finite type. The proof of the classification is given in the second section.

1. Schemes and ind-schemes

1.1. Categories and Grothendieck groups

1.1.1. Ind-objects and pro-objects in a category. A standard reference for the material in this section is [SGA4, Section 8], [18], [19].

Let $\mathbf{Set}$ be the category of sets. Given a category $\mathcal{C}$ let $\mathcal{C}^\circ$ be the opposite category. The category $\mathcal{C}^\wedge$ of *presheaves over* $\mathcal{C}$ is the category of functors $\mathcal{C}^\circ \to \mathbf{Set}$. We'll abbreviate $\mathcal{C}^\vee$ for the category $((\mathcal{C}^\circ)^\wedge)^\circ$. Yoneda's lemma yields fully faithful functors

$$\mathcal{C} \to \mathcal{C}^\wedge, \ X \mapsto \mathrm{Hom}_{\mathcal{C}}(\cdot, X), \quad \mathcal{C} \to \mathcal{C}^\vee, \ X \mapsto \mathrm{Hom}_{\mathcal{C}}(X, \cdot).$$

Let $\mathcal{A}$ be a category and $\alpha \mapsto X_\alpha$ be a functor $\mathcal{A} \to \mathcal{C}$ or $\mathcal{A}^\circ \to \mathcal{C}$ (also called a *system* in $\mathcal{C}$ indexed by $\mathcal{A}$ or $\mathcal{A}^\circ$). Let $\mathrm{colim}_\alpha X_\alpha$ and $\lim_\alpha X_\alpha$ denote the colimit or the limit of this system whenever it is well defined. If the category $\mathcal{A}$ is small or filtrant the colimit and the limit are said to be *small* or *filtrant*. A poset $\mathcal{A} = (A, \leqslant)$ may be viewed as a category, with A as the set of objects and with a morphism $\alpha \to \beta$ whenever $\alpha \leqslant \beta$. A *direct set* is a poset $\mathcal{A}$ which is filtrant as a category. A *direct system* in $\mathcal{C}$ is a functor $\mathcal{A} \to \mathcal{C}$ and an *inverse system* in $\mathcal{C}$ is a functor $\mathcal{A}^\circ \to \mathcal{C}$, where $\mathcal{A}$ is a direct set. A *direct colimit* (also called *inductive limit*) is the colimit of a direct system. An *inverse limit* (also called *projective limit*) is the limit of an inverse system. Both are small and filtrant.

A *complete* or *cocomplete* category is one that has all small limits or all small colimits. A *Grothendieck category* is a cocomplete Abelian category with a generator such that the small filtrant colimits are exact.

Given a direct system or an inverse system in $\mathcal{C}$ we define the following functors

$$\text{``}\mathrm{colim}_\alpha\text{''} X_\alpha \colon \mathcal{C}^\circ \to \mathbf{Set}, \quad Y \mapsto \mathrm{colim}_\alpha \mathrm{Hom}_{\mathcal{C}}(Y, X_\alpha),$$
$$\text{``}\lim_\alpha\text{''} X_\alpha \colon \mathcal{C} \to \mathbf{Set}, \quad Y \mapsto \mathrm{colim}_\alpha \mathrm{Hom}_{\mathcal{C}}(X_\alpha, Y).$$

The categories of *ind-objects* of $\mathcal{C}$ and the category of *pro-objects* of $\mathcal{C}$ are the full subcategory $\mathbf{Ind}(\mathcal{C})$ of $\mathcal{C}^\wedge$ and the full subcategory $\mathbf{Pro}(\mathcal{C})$ of $\mathcal{C}^\vee$ consisting of objects isomorphic to some $\text{``}\mathrm{colim}_\alpha\text{''} X_\alpha$ and $\text{``}\lim_\alpha\text{''} X_\alpha$ respectively. Note that we have $\mathbf{Pro}(\mathcal{C}) = \mathbf{Ind}(\mathcal{C}^\circ)^\circ$. By the Yoneda functor we may look upon $\mathcal{C}$ as a full subcategory of $\mathbf{Ind}(\mathcal{C})$ or $\mathbf{Pro}(\mathcal{C})$. We'll say that an ind-object or a pro-object is *representable* if it is isomorphic to an object in $\mathcal{C}$. Note that, for each object Y of $\mathcal{C}$ we have the following formulas, see, e.g., [18, Section 1.11], [19, Section 2.6]

$$\mathrm{Hom}_{\mathbf{Ind}(\mathcal{C})}(Y, \text{``}\mathrm{colim}_\alpha\text{''} X_\alpha) = \mathrm{Hom}_{\mathcal{C}^\wedge}(Y, \text{``}\mathrm{colim}_\alpha\text{''} X_\alpha) = \mathrm{colim}_\alpha \mathrm{Hom}_{\mathcal{C}}(Y, X_\alpha),$$
$$\mathrm{Hom}_{\mathbf{Pro}(\mathcal{C})}(\text{``}\lim_\alpha\text{''} X_\alpha, Y) = \mathrm{Hom}_{\mathcal{C}^\vee}(\text{``}\lim_\alpha\text{''} X_\alpha, Y) = \mathrm{colim}_\alpha \mathrm{Hom}_{\mathcal{C}}(X_\alpha, Y).$$

1.1.2. Direct and inverse 2-limits. Let $\mathcal{A} = (A, \leqslant)$ be a directed set. Given a direct system of categories $(\mathcal{C}_\alpha, i_{\alpha\beta} \colon \mathcal{C}_\alpha \to \mathcal{C}_\beta)$ the *2-limit* (also called the *2-colimit*) of this system is the category $\mathcal{C} = 2\,\mathrm{colim}_\alpha \mathcal{C}_\alpha$ whose objects are the pairs (α, X_α)

with X_α an object of $\mathcal{C}_\alpha$. The morphisms are given by

$$\mathrm{Hom}_{\mathcal{C}}((\alpha, X_\alpha), (\beta, X_\beta)) = \operatorname*{colim}_{\gamma \geqslant \alpha, \beta} \mathrm{Hom}_{\mathcal{C}_\gamma}(i_{\alpha\gamma}(X_\alpha), i_{\beta\gamma}(X_\beta)).$$

Given an inverse system of categories $(\mathcal{C}_\alpha, i_{\alpha\beta} \colon \mathcal{C}_\beta \to \mathcal{C}_\alpha)$ the 2-*limit* of this system is the category $\mathcal{C} = 2\lim_\alpha \mathcal{C}_\alpha$ whose objects are the families of objects X_α of $\mathcal{C}_\alpha$ and of isomorphisms $i_{\alpha\beta}(X_\beta) \simeq X_\alpha$ satisfying the obvious composition rules. The morphisms are defined in the obvious way. See [34, Appendix A] for more details on 2-colimits and 2-limits.

1.1.3. Grothendieck groups and derived categories.
Given an Abelian category $\mathcal{A}$ let $\mathcal{C}(\mathcal{A})$ be the category of complexes of objects of $\mathcal{A}$ with differential of degree $+1$ and chain maps as morphisms, let $\mathcal{D}(\mathcal{A})$ be the corresponding (unbounded) derived category, let $\mathcal{D}(\mathcal{A})^-$ be the full subcategory of complexes bounded above, let $\mathcal{D}(\mathcal{A})^b$ be the full subcategory of bounded complexes. Finally let $[\mathcal{A}]$ be the Grothendieck group of $\mathcal{A}$.

The Grothendieck group $[\mathcal{T}]$ of a triangulated category $\mathcal{T}$ is the quotient of the free Abelian group with one generator for each object X of $\mathcal{T}$ modulo the relations $X = X' + X''$ for each distinguished triangle

$$X' \to X \to X'' \to X'[1].$$

Here the symbol $[1]$ stands for the shift functor in the triangulated category $\mathcal{T}$. Throughout, we'll use the same symbol for an object of $\mathcal{T}$ and its class in $[\mathcal{T}]$.

Recall that the Grothendieck group of $\mathcal{D}(\mathcal{A})^b$ is canonically isomorphic to $[\mathcal{A}]$, and that two quasi-isomorphic bounded complexes of $\mathcal{C}(\mathcal{A})$ have the same class in $[\mathcal{A}]$.

1.1.4. Proposition.
Let $(\mathcal{C}_\alpha)$ be a direct system of Abelian categories (resp. of triangulated categories) and exact functors. Then the direct 2-limit $\mathcal{C}$ of $(\mathcal{C}_\alpha)$ is also an Abelian category (resp. a triangulated category) and we have a canonical group isomorphism $[\mathcal{C}] = \operatorname{colim}_\alpha [\mathcal{C}_\alpha]$.

1.2. K-theory of schemes

This section is a recollection of standard results from [SGA6], [31] on the K-theory of schemes, possibly of infinite type.

1.2.1. Background.
For any Abelian category $\mathcal{A}$ a complex in $\mathcal{C}(\mathcal{A})$ is cohomologically bounded if the cohomology sheaves vanish except for a finite number of them. The canonical functor yields an equivalence from $\mathcal{D}(\mathcal{A})^b$ to the full subcategory of $\mathcal{D}(\mathcal{A})$ consisting of cohomologically bounded complexes [18, p. 45].

A quasi-compact scheme is a scheme that has a finite covering by affine open subschemes (e.g., a Noetherian scheme or a scheme of finite type is quasi-compact) and a quasi-separated scheme is a scheme such that the intersection of any two affine open subschemes is quasi-compact (e.g., a separated scheme is quasi-separated). More generally, a scheme homomorphism $f \colon X \to Y$ is said to be quasi-compact, resp. quasi-separated, if for every affine open $U \subset Y$ the inverse image of U is quasi-compact, resp. quasi-separated. Elementary properties of quasi-compact and quasi-separated morphisms can be found in [11, Chapter I, Section 6.1]. For instance quasi-compact and quasi-separated morphisms are stable under composition and pullback, and if $f\colon X \to Y$ is a scheme homomorphism with Y quasi-compact and quasi-separated then X is quasi-compact and separated if and only if f is quasi-compact and quasi-separated. *Throughout, by the word scheme we'll always mean a separated $\mathbb{C}$-scheme and by the word scheme homomorphism we'll always mean a morphism of separated $\mathbb{C}$-schemes.* In particular a scheme homomorphism will always be separated (hence quasi-separated) [11, Chapter I, Section 5.3].

Given a scheme X, the word $\mathcal{O}_X$-module will mean a sheaf on the scheme X which is a sheaf of modules over the sheaf of rings $\mathcal{O}_X$. Unless otherwise stated, modules are left modules. This applies also to $\mathcal{O}_X$-modules. Since $\mathcal{O}_X$ is commutative this specification is indeed irrelevant. Let $\mathcal{O}(X)$ be the Abelian category of all $\mathcal{O}_X$-modules. Given a closed subscheme $Y \subset X$ let $\mathcal{O}(X \text{ on } Y)$ be the full subcategory of $\mathcal{O}_X$-modules supported on Y.

Let $\boldsymbol{Coh}(X)$, $\boldsymbol{Qcoh}(X)$ be the categories of coherent and quasi-coherent $\mathcal{O}_X$-modules. They are Abelian subcategories of $\mathcal{O}(X)$ which are stable under extensions. Quasi-coherent sheaves are preserved by tensor products, by arbitrary colimits and by inverse images [11, Chapter I, Section 2.2]. They are well behaved on quasi-compact (quasi-separated) schemes: under this assumption quasi-coherent $\mathcal{O}_X$-modules are preserved by direct images and any quasi-coherent $\mathcal{O}_X$-module is the limit of a direct system of finitely presented $\mathcal{O}_X$-modules. Further, if X is quasi-compact (quasi-separated) the category $\boldsymbol{Qcoh}(X)$ is a Grothendieck category. In particular for any such X there are enough injective objects in $\boldsymbol{Qcoh}(X)$ [11, Chapter I, Section 6.7, 6.9], [31, Section B.3]. Given a closed subscheme $Y \subset X$ let $\boldsymbol{Coh}(X \text{ on } Y)$, $\boldsymbol{Qcoh}(X \text{ on } Y)$ be the full subcategories of sheaves supported on Y.

We'll abbreviate $\mathcal{C}(X) = \mathcal{C}(\mathcal{O}(X))$ and $\mathcal{D}(X) = \mathcal{D}(\mathcal{O}(X))$. Let $\mathcal{D}(X)_{\mathrm{qc}}$ be the full subcategory of $\mathcal{D}(X)$ of complexes of $\mathcal{O}_X$-modules with quasi-coherent cohomology.

1.2.2. *Remark.* Bökstedt and Neeman proved that if X is quasi-compact (separated) then the canonical functor

$$(1.2.1) \qquad\qquad \mathcal{D}(\boldsymbol{Qcoh}(X)) \to \mathcal{D}(X)_{\mathrm{qc}}$$

is an equivalence [5, Corollary 5.5], [24, Proposition 3.9.6]. Further, the standard derived functors in 1.2.10-12 below, evaluated on quasi-coherent sheaves, are the same taken in $\mathcal{O}(X)$ or in $\boldsymbol{Qcoh}(X)$, see, e.g., [31, Corollary B.9]. So from now on we'll identify the categories $\mathcal{D}(\boldsymbol{Qcoh}(X))$ and $\mathcal{D}(X)_{\mathrm{qc}}$.

A commutative ring R is *coherent* if and only if it is coherent as an R-module, or, equivalently, if every finitely generated ideal of R is finitely presented. For instance a Noetherian ring is coherent, the quotient of a coherent ring by a finitely generated ideal is a coherent ring and the localization of a coherent ring is again coherent.

1.2.3. Definitions. Let X be any scheme. We say that

(a) X is *coherent* if its structure ring $\mathcal{O}_X$ is coherent,
(b) X is *locally of countable type* if the $\mathbb{C}$-algebra $\mathcal{O}_X(U)$ is generated by a countable number of elements for any affine open subset $U \subset X$,
(c) a closed subscheme $Y \subset X$ is *good* if the ideal of Y in $\mathcal{O}_X(U)$ is finitely generated for any affine open subset $U \subset X$.

If the scheme X is coherent then an $\mathcal{O}_X$-module is coherent if and only if it is finitely presented, and we have $f^*(\mathbf{Coh}(Y)) \subset \mathbf{Coh}(X)$ for any morphism $f \colon X \to Y$. If X is quasi-compact and coherent then any quasi-coherent $\mathcal{O}_X$-module is the direct colimit of a system of coherent $\mathcal{O}_X$-modules. Finally a good subscheme Y of a coherent scheme X is again coherent and the direct image of $\mathcal{O}_Y$ is a coherent $\mathcal{O}_X$-module. See [EGAI, Chapter 0, Section 5.3] for details.

1.2.4. K-theory of a quasi-compact coherent scheme. For an arbitrary scheme X the K-homology group (= K-theory) may differ from $[\mathbf{Coh}(X)]$, one reason being that $\mathcal{O}_X$ may not be an object of $\mathbf{Coh}(X)$. Let us recall briefly some relevant definitions and results concerning pseudo-coherence. Details can be found in [SGA6, Chapter I], [31] and [24, Section 4.3]. We'll assume that X is quasi-compact and coherent.

1.2.5. Definition-Lemma.

(a) A complex of quasi-coherent $\mathcal{O}_X$-modules $\mathcal{E}$ is said to be *pseudo-coherent* if it is locally quasi-isomorphic to a bounded above complex of vector bundles. Since X is coherent, this simply means that $\mathcal{E}$ has coherent cohomology sheaves vanishing in all sufficiently large degrees [SGA6, Corollary I.3.5(iii)]. In particular any coherent $\mathcal{O}_X$-module is a pseudo-coherent complex.
(b) Let $\mathbf{Pcoh}(X)$ be the full subcategory of $\mathbf{D}(X)_{\mathrm{qc}}$ consisting of the cohomologically bounded pseudo-coherent complexes. Given a closed subscheme $Y \subset X$ the full subcategory of complexes which are acyclic on $X - Y$ is $\mathbf{Pcoh}(X$ on $Y)$. It is a triangulated category.

Note that for a general scheme the equivalence of categories (1.2.1) does not hold and a pseudo-coherent complex may consist of non quasi-coherent $\mathcal{O}_X$-modules. The K-homology group of the pair (X, Y) is [SGA6, Definition IV.2.2]

$$\mathbf{K}(X \text{ on } Y) = [\mathbf{Pcoh}(X \text{ on } Y)], \quad \mathbf{K}(X) = \mathbf{K}(X \text{ on } X).$$

By 1.2.5 the K-homology groups are well behaved on quasi-compact coherent schemes. More precisely we have the following.

1.2.6. Proposition. *Assume that X is quasi-compact and coherent. For any closed subscheme $Y \subset X$, we have $\mathbf{K}(X$ on $Y) = [\mathbf{Coh}(X$ on $Y)]$. If $Y \subset X$ is good there is a canonical isomorphism $\mathbf{K}(Y) \to \mathbf{K}(X$ on $Y)$.*

If X is coherent but not quasi-compact we define the group $\mathbf{K}(X)$ as follows. Fix a covering $X = \bigcup_w X^w$ by quasi-compact open subsets. The restrictions yield an inverse system of categories with a functor

$$\mathcal{C}oh(X) \to 2\lim_{w} \mathcal{C}oh(X^w).$$

By functoriality of the K-theory we have also an inverse system of Abelian groups. We define

$$\mathbf{K}(X) = \lim_{w} \mathbf{K}(X^w) = \lim_{w} [\mathcal{C}oh(X^w)].$$

The group $\mathbf{K}(X)$ does not depend on the choice of the open covering. It may be regarded as a completion of the K-homology group of X, as defined in [SGA6].

1.2.7. *Remark.* Let X be a quasi-compact scheme. A *perfect complex* over X is a complex of quasi-coherent $\mathcal{O}_X$-modules which is locally quasi-isomorphic to a bounded complex of vector bundles. The K-cohomology group of X is the Grothendieck group of the full subcategory of $\mathcal{D}(X)_{qc}$ of perfect complexes. We'll not use it.

1.2.8. Basic properties of the K-theory of a coherent quasi-compact scheme. Recall that for $\mathcal{O}_X$-modules $\mathcal{E}$, $\mathcal{F}$ the *local hypertor* is the $\mathcal{O}_X$-module $\mathcal{T}or_i^{\mathcal{O}_X}(\mathcal{E}, \mathcal{F})$ whose stalk at a point x is $\mathrm{Tor}_i^{\mathcal{O}_{X,x}}(\mathcal{E}_x, \mathcal{F}_x)$.

1.2.9. Definitions.

(a) An $\mathcal{O}_X$-module $\mathcal{E}$ has a *finite tor-dimension* if there is an integer n such that $\mathcal{T}or_i^{\mathcal{O}_X}(\mathcal{E}, \mathcal{F}) = 0$ for each $i > n$ and each $\mathcal{F} \in \mathcal{Q}coh(X)$.

(b) A scheme homomorphism $f: X \to Y$ has *finite tor-dimension* if there is an integer n such that $\mathcal{T}or_i^{\mathcal{O}_Y}(f_*\mathcal{O}_X, \mathcal{E}) = 0$ for each $i > n$ and each $\mathcal{E} \in \mathcal{Q}coh(Y)$. Equivalently, f has finite tor-dimension if there is an integer n such that for each $x \in X$ there is an exact sequence of $\mathcal{O}_{Y,f(x)}$-modules

$$0 \to P_n \to P_{n-1} \to \cdots \to P_0 \to \mathcal{O}_{X,x} \to 0$$

with P_i flat over $\mathcal{O}_{Y,f(x)}$.

(c) A scheme X *satisfies the Poincaré duality* if any quasi-coherent $\mathcal{O}_X$-module has a finite tor-dimension.

Poincaré duality is a local property. Note that, since taking a local hypertor commutes with direct colimits [EGAIII, Proposition 6.5.6], a coherent scheme satisfies the Poincaré duality if and only if any coherent $\mathcal{O}_X$-module has a finite tor-dimension. Note that if X satisfies the Poincaré duality then any cohomologically bounded pseudo-coherent complex is perfect [31, Theorem 3.21].

Now, let us recall a few basic properties of direct/inverse images of complexes of $\mathcal{O}$-modules. We'll use derived functors in the unbounded derived category of $\mathcal{O}$-modules. This simplifies the exposition. Their definition and properties can be found in [24]. To simplify, in Sections 1.2.10 to 1.2.15 we'll also assume that all schemes are quasi-compact and coherent. Therefore, all morphisms will also be quasi-compact.

1.2.10. Derived inverse image. For any morphism $f\colon Z \to X$ the inverse image functor Lf^* maps $\mathcal{D}(X)_{\mathrm{qc}}$, $\mathcal{D}(X)^-$ into $\mathcal{D}(Z)_{\mathrm{qc}}$, $\mathcal{D}(Z)^-$ respectively. It preserves pseudo-coherent and perfect complexes [24, Proposition 3.9.1], [31, Section 2.5.1]. Further if $\mathcal{E}$ is a cohomologically bounded pseudo-coherent complex then the complex $Lf^*(\mathcal{E})$ is cohomologically bounded if the map f has finite tor-dimension. Under this assumption, for each closed subscheme $Y \subset X$ the functor Lf^* yields a group homomorphism

$$Lf^*\colon \mathbf{K}(X \text{ on } Y) \to \mathbf{K}(Z \text{ on } f^{-1}(Y)).$$

If the schemes X, Z satisfy the Poincaré duality then f has a finite tor-dimension.

1.2.11. Derived tensor product. Let $\otimes_X$ denote the tensor product of $\mathcal{O}$-modules on any scheme X. The standard theory of the derived tensor product of $\mathcal{O}$-modules applies to complexes in $\mathcal{D}(X)^-$, see, e.g., [13, p. 93]. Following Spaltenstein [27] we can extend the theory to arbitrary complexes in $\mathcal{D}(X)$, see also [24, Section 2.5]. This yields a functor

$$\overset{L}{\otimes}_X\colon \mathcal{D}(X) \times \mathcal{D}(X) \to \mathcal{D}(X)$$

which maps $\mathcal{D}(X)_{\mathrm{qc}} \times \mathcal{D}(X)_{\mathrm{qc}}$, $\mathcal{D}(X)^- \times \mathcal{D}(X)^-$ to $\mathcal{D}(X)_{\mathrm{qc}}$, $\mathcal{D}(X)^-$ respectively. It preserves pseudo-coherent complexes [31, Section 2.5.1]. If $\mathcal{E}$, $\mathcal{F}$ are pseudo-coherent complexes their derived tensor product is cohomologically bounded if either $\mathcal{E}$ is perfect or $\mathcal{F}$ is perfect [31, Section 3.15]. Recall that if X satisfies the Poincaré duality then any cohomologically bounded pseudo-coherent complex is perfect. Under this assumption, for each closed subschemes $Y, Z \subset X$ there is a (derived) tensor product

$$\overset{L}{\otimes}_X\colon \mathbf{K}(X \text{ on } Y) \times \mathbf{K}(X \text{ on } Z) \to \mathbf{K}(X \text{ on } Y \cap Z).$$

Given a map f as in 1.2.10 there is a functorial isomorphism in $\mathcal{D}(X)$ [24, Proposition 3.2.4]

$$Lf^*(\mathcal{E} \overset{L}{\otimes}_X \mathcal{F}) = Lf^*(\mathcal{E}) \overset{L}{\otimes}_Z Lf^*(\mathcal{F}).$$

We'll refer to this relation by saying that the derived tensor product commutes with Lf^*.

1.2.12. Derived direct image. For any $f\colon X \to Z$ the direct image functor Rf_* is right adjoint to Lf^* and it maps $\mathcal{D}(X)_{\mathrm{qc}}$, $\mathcal{D}(X)^b$ into $\mathcal{D}(Z)_{\mathrm{qc}}$, $\mathcal{D}(Z)^b$ respectively [24, Proposition 3.9.2]. We say that the map f is *pseudo-coherent* if it factors, locally on X, as $f = p \circ i$ where i is a closed embedding with $i_*\mathcal{O}_X$ coherent and p is smooth. Kiehl's finiteness theorem ensures that if f is proper and pseudo-coherent then Rf_* preserves pseudo-coherent complexes [24, Corollary 4.3.3.2]. Therefore if f is proper and pseudo-coherent, for any closed subscheme $Y \subset X$, the functor Rf_* yields a group homomorphism

$$Rf_*\colon \mathbf{K}(X \text{ on } Y) \to \mathbf{K}(Z \text{ on } f(Y)).$$

1.2.13. *Example.* A good embedding is proper and pseudo-coherent. In this case we have indeed an exact functor $f_* \colon \boldsymbol{Coh}(X) \to \boldsymbol{Coh}(Z)$. It yields the isomorphism $\mathbf{K}(X) \to \mathbf{K}(Z \text{ on } X)$ in 1.2.5. Note that a closed embedding $X \subset Z$ with $Z = \mathbb{A}^{\mathbb{N}}$ and X of finite type is not pseudo-coherent. Here $\mathbb{A}^{\mathbb{N}} = \mathrm{Spec}(\mathbb{C}[x_i; i \in \mathbb{N}])$ is a coherent scheme.

1.2.14. Projection formula. For any $f \colon X \to Z$ there is a canonical isomorphism called the *projection formula* [24, Proposition 3.9.4]

$$Rf_*(\mathcal{E} \overset{L}{\otimes}_X Lf^*(\mathcal{F})) = Rf_*(\mathcal{E}) \overset{L}{\otimes}_Z \mathcal{F}, \quad \forall \mathcal{E} \in \mathcal{D}(X)_{\mathrm{qc}}, \, \mathcal{F} \in \mathcal{D}(Z)_{\mathrm{qc}}.$$

1.2.15. Base change. Consider the following Cartesian square

$$
\begin{array}{ccc}
X' & \xleftarrow{\ f'\ } & Y' \\
{\scriptstyle g}\downarrow & & \downarrow{\scriptstyle g'} \\
X & \xleftarrow{\ f\ } & Y.
\end{array}
$$

Assume that it is tor-*independent*, i.e., assume that we have

$$\mathrm{Tor}_i^{\mathcal{O}_{X,x}}(\mathcal{O}_{X',x'}, \mathcal{O}_{Y,y}) = 0, \quad \forall i > 0, \, \forall x \in X, \forall x' \in X', \forall y \in Y, \, x = g(x') = f(y).$$

Then we have a functorial base-change isomorphism [24, Theorem 3.10.3]

$$Lg^* Rf_*(\mathcal{E}) \simeq Rf'_* L(g')^*(\mathcal{E}), \quad \forall \mathcal{E} \in \mathcal{D}(Y)_{\mathrm{qc}}.$$

1.2.16. Compact schemes. A simple way to produce quasi-compact schemes of infinite type is to use pro-schemes. Let us explain this.

1.2.17. Lemma-Definition. A scheme is *compact* if it is the limit of an inverse system of finite type schemes with affine morphisms. A scheme is compact if and only if it is quasi-compact [31, Theorem C.9].

1.2.18. *Remarks.* Let X be a compact scheme and (X^α) be an inverse system of schemes as above. Then the canonical maps $p_\alpha \colon X \to X^\alpha$ are affine. Further the following hold.

(a) If $\mathcal{F}$ is a coherent $\mathcal{O}_X$-module there is an α and a coherent $\mathcal{O}_{X^\alpha}$-module $\mathcal{F}^\alpha$ such that $\mathcal{F} = (p_\alpha)^*(\mathcal{F}^\alpha)$. Given two coherent $\mathcal{O}_X$-modules $\mathcal{F}$, $\mathcal{G}$ and two coherent $\mathcal{O}_{X^\alpha}$-modules $\mathcal{F}^\alpha$, $\mathcal{G}^\alpha$ as above we set $\mathcal{F}^\beta = (p_{\alpha\beta})^*(\mathcal{F}^\alpha)$ and $\mathcal{G}^\beta = (p_{\alpha\beta})^*(\mathcal{G}^\alpha)$ for each $\beta \geqslant \alpha$. Then we have [31, Section C4], [EGAIV, Section 8.5]

$$\mathrm{Hom}_{\mathcal{O}_X}(\mathcal{F}, \mathcal{G}) = \operatorname*{colim}_{\beta \geqslant \alpha} \mathrm{Hom}_{\mathcal{O}_{X^\beta}}(\mathcal{F}^\beta, \mathcal{G}^\beta).$$

(b) If $f \colon Y \to X$ is a scheme finitely presented over X then there is an $\alpha \in A$ and a finitely presented $f_\alpha \colon Y^\alpha \to X^\alpha$ such that [31, Section C.3]

$$f = f_\alpha \times \mathrm{id}, \quad Y = Y^\alpha \times_{X^\alpha} X = \lim_\beta Y^\beta, \quad Y^\beta = Y^\alpha \times_{X^\alpha} X^\beta, \quad \beta \geqslant \alpha.$$

1.2.19. Definition. A compact scheme $X = \text{``}\lim_\alpha \text{''} X^\alpha$ satisfies the property (S) if $A = \mathbb{N}$ and $(X^\alpha)_{\alpha \in A}$ is an inverse system of smooth schemes of finite type with smooth affine morphisms. A scheme is *pro-smooth* if it is covered by a finite number of open subsets satisfying (S).

1.2.20. Proposition. *A pro-smooth scheme is quasi-compact, coherent, and it satisfies the Poincaré duality.*

Proof. A pro-smooth scheme X is coherent by [17, Proposition 1.1.6]. Let us prove that X satisfies the Poincaré duality. Let $\mathcal{F} \in \boldsymbol{Coh}(X)$. We must prove that $\mathrm{Tor}_i^{\mathcal{O}_{X,x}}(\mathcal{F}_x, \mathcal{E}_x) = 0$ for each $i \gg 0$, each $x \in X$ and each $\mathcal{E} \in \boldsymbol{Qcoh}(X)$. Since the question is local around x we can assume that X is a compact scheme satisfying the property (S). By 1.2.18 (a) there is an $\alpha \in A$ and $\mathcal{F}^\alpha \in \boldsymbol{Coh}(X^\alpha)$ such that $\mathcal{F} = (p_\alpha)^*(\mathcal{F}^\alpha)$. Write again $x = p_\alpha(x)$. Since X^α is smooth of finite type the $\mathcal{O}_{X^\alpha,x}$-module $\mathcal{F}_x^\alpha$ has finite tor-dimension. Since the map p_α is affine and flat we have

$$\mathrm{Tor}_i^{\mathcal{O}_{X,x}}(\mathcal{F}_x, \mathcal{E}_x) = \mathrm{Tor}_i^{\mathcal{O}_{X^\alpha,x}}(\mathcal{F}_x^\alpha, (p_\alpha)_*(\mathcal{E}_x)).$$

Since $(p_\alpha)_*(\mathcal{E})$ is quasi-coherent and the scheme X^α is smooth of finite type, and since taking the Tor's commutes with direct colimits, the right-hand side vanishes for large i.   □

1.2.21. Remarks. (a) Let $\boldsymbol{Sch}$ be the category of schemes and $\boldsymbol{Sch}^{\mathrm{ft}}$ be the full subcategory of schemes of finite type. The category of compact schemes can be identified with a full subcategory in $\boldsymbol{Pro}(\boldsymbol{Sch}^{\mathrm{ft}})$ via the assignment $\lim_\alpha X^\alpha \mapsto \text{``}\lim_\alpha\text{''} X^\alpha$. From now on we'll omit the quotation marks for compact schemes.

(b) Let X be a quasi-compact coherent scheme and $Y \subset X$ be a good subscheme. Since X is a compact scheme we can fix an inverse system of finite type schemes (X^α) with affine morphisms $p_{\alpha\beta} : X^\beta \to X^\alpha$ such that $X = \lim_\alpha X^\alpha$. Since the scheme X is coherent and since Y is a good subscheme, the inclusion $Y \subset X$ is finitely presented. Thus, by 1.2.18 (b) there is an $\alpha \in A$ and a closed subscheme $Y^\alpha \subset X^\alpha$ such that $Y = p_\alpha^{-1}(Y^\alpha)$. Setting $Y^\beta = p_{\alpha\beta}^{-1}(Y^\alpha)$ for each $\beta \geqslant \alpha$ we get a direct system of categories $\boldsymbol{Coh}(X^\alpha \text{ on } Y^\alpha)$ with functors $(p_{\alpha\beta})^*$. The pull-back by the canonical map $p_\alpha : X \to X^\alpha$ yields an equivalence of categories [EGAIV, Theorem 8.5.2]

$$2\operatorname*{colim}_\alpha \boldsymbol{Coh}(X^\alpha \text{ on } Y^\alpha) \to \boldsymbol{Coh}(X \text{ on } Y).$$

Now, assume the maps $p_{\alpha\beta}$ are flat. Then 1.1.4 yields a group isomorphism

$$\operatorname*{colim}_\alpha \mathbf{K}(X^\alpha \text{ on } Y^\alpha) = \mathbf{K}(X \text{ on } Y).$$

1.2.22. Pro-finite-dimensional vector bundles. An important particular case of compact schemes is given by pro-finite-dimensional vector bundles.

1.2.23. Definition. A *pro-finite-dimensional vector bundle* $\pi : X \to Y$ is a scheme homomorphism which is represented as the inverse limit of a system of vector bundles $\pi_n : X^n \to Y$ (of finite rank) with n an integer $\geqslant 0$, such that the morphism $X^m \to X^n$ is a vector bundle homomorphism for each $m \geqslant n$.

1.2.24. Proposition. *A pro-finite-dimensional vector bundle* $\pi\colon X \to Y$ *is flat. If* Y *is compact then* X *is compact. If* Y *is pro-smooth then* X *is pro-smooth. If* Y *is coherent then* X *is coherent.*

Proof. The first claim is obvious. By 1.2.18 (b) any vector bundle over Y is, locally over Y, pulled-back from a vector bundle over some Y^α where (Y^α) is an inverse system as in 1.2.19. This implies the second and the third claim. The last one follows from [17, Proposition 1.1.6]. $\quad\square$

1.3. K-theory of ind-coherent ind-schemes

1.3.1. Spaces and ind-schemes. Let $\boldsymbol{Alg}$ be the category of associative, commutative $\mathbb{C}$-algebras with 1. The category of *spaces* is the category $\boldsymbol{Space}$ of functors $\boldsymbol{Alg} \to \boldsymbol{Set}$. By Yoneda's lemma $\boldsymbol{Sch}$ can be considered as a full subcategory in the category $\boldsymbol{Sch}^\wedge$ of presheaves on $\boldsymbol{Sch}$. It can be as well realized as a full subcategory in $\boldsymbol{Space}$ via the functor

$$\boldsymbol{Sch} \to \boldsymbol{Space}, \quad X \mapsto \mathrm{Hom}_{\boldsymbol{Sch}}(\mathrm{Spec}(\cdot), X).$$

By a *subspace* we mean a subfunctor. A subspace $Y \subset X$ is said to be *closed, open* if for every scheme Z and every $Z \to X$ the subspace $Z \times_X Y \subset Z$ is a closed, open subscheme.

1.3.2. Definitions.

(a) An *ind-scheme* is an ind-object X of $\boldsymbol{Sch}$ represented as $X = \text{``colim}_\alpha\text{''} X_\alpha$ where $A = \mathbb{N}$ and $(X_\alpha)_{\alpha \in A}$ is a direct system of quasi-compact schemes with closed embeddings $i_{\alpha\beta}\colon X_\alpha \to X_\beta$ for each $\alpha \leqslant \beta$.

(b) A *closed ind-subscheme* Y of the ind-scheme X is a closed subspace of X. An *open ind-subscheme* Y of the ind-scheme X is an ind-scheme which is an open subspace of X.

Let $\boldsymbol{Isch}$ be the category of ind-schemes. Since direct colimits exist in the category $\boldsymbol{Space}$ we may regard $\boldsymbol{Isch}$ as a full subcategory of $\boldsymbol{Space}$. Hence, to unburden the notation we'll omit the quotation marks for ind-schemes.

1.3.3. *Remarks.*

(a) A closed subscheme of an ind-scheme is always quasi-compact.

(b) We may consider ind-objects of $\boldsymbol{Sch}$ which are represented by a direct system of non quasi-compact schemes X_α with closed embeddings. To avoid any confusion we'll call them ind$'$-schemes.

(c) Given a closed ind-subscheme $Y \subset X$, for each $\alpha \in A$ the closed immersion $X_\alpha \subset X$ yields a closed subscheme $Y_\alpha = X_\alpha \times_X Y \subset X_\alpha$. Further the closed immersion $X_\alpha \subset X_\beta$, $\alpha \leqslant \beta$, factors to a closed immersion $Y_\alpha \subset Y_\beta$. The ind-scheme Y is represented as $Y = \mathrm{colim}_\alpha Y_\alpha$.

(d) For each ind-scheme X and each quasi-compact scheme Y we have

$$\mathrm{Hom}_{\boldsymbol{Isch}}(Y, X) = \operatorname*{colim}_\alpha \mathrm{Hom}_{\boldsymbol{Sch}}(Y, X_\alpha).$$

1.3.4. Definitions.

(a) An ind-scheme X is *ind-proper* or of *ind-finite type* if it can be represented as the direct colimit of a system of proper schemes or of finite type schemes respectively with closed embeddings.

(b) An ind-scheme X is *ind-coherent* if it can be represented as the direct colimit of a system of coherent quasi-compact schemes with good embeddings.

1.3.5. Coherent and quasi-coherent $\mathcal{O}$-modules over ind-coherent ind-schemes.
Let X be an ind-coherent ind-scheme and $Y \subset X$ be a closed ind-subscheme. Given X_α, Y_α as in 1.3.3 (c) we have a direct system of Abelian categories $\mathbf{Coh}(X_\alpha \text{ on } Y_\alpha)$ with exact functors $(i_{\alpha\beta})_*$. We define the following Abelian categories

$$\mathbf{Coh}(X \text{ on } Y) = 2\operatorname*{colim}_\alpha \mathbf{Coh}(X_\alpha \text{ on } Y_\alpha), \quad \mathbf{Coh}(X) = \mathbf{Coh}(X \text{ on } X).$$

These categories do not depend on the direct system (X_α) up to canonical equivalences. An object of $\mathbf{Coh}(X)$ is called a coherent $\mathcal{O}_X$-module.

We define also quasi-coherent $\mathcal{O}_X$-modules in the following way [2, Section 7.11.3], [8, Section 6.3.2]. We have an inverse system of categories $\mathbf{Qcoh}(X_\alpha)$ with functors $(i_{\alpha\beta})^*$. We set

$$\mathbf{Qcoh}(X) = 2\lim_\alpha \mathbf{Qcoh}(X_\alpha).$$

The category $\mathbf{Qcoh}(X)$ is a tensor category, but it need not be Abelian. It is independent of the choice of the system (X_α) up to canonical equivalences of categories. A quasi-coherent $\mathcal{O}_X$-module can be regarded as a rule that assigns to each scheme Z with a morphism $Z \to X$ a quasi-coherent $\mathcal{O}_Z$-module $\mathcal{E}_Z$, and to each scheme homomorphism $f \colon W \to Z$ an isomorphism $f^*\mathcal{E}_Z \simeq \mathcal{E}_W$ satisfying the obvious composition rules.

Finally we define the Grothendieck group of the pair (X, Y) by

$$\mathbf{K}(X \text{ on } Y) = [\mathbf{Coh}(X \text{ on } Y)].$$

Note that we have $\mathbf{K}(X \text{ on } Y) = \operatorname{colim}_\alpha \mathbf{K}(X_\alpha \text{ on } Y_\alpha)$ where $\mathbf{K}(X_\alpha \text{ on } Y_\alpha) = [\mathbf{Coh}(X_\alpha \text{ on } Y_\alpha)]$ for each α.

1.3.6. Remarks.

(a) There is another notion of quasi-coherent $\mathcal{O}_X$-modules on an ind-scheme, called $\mathcal{O}_X^!$-modules in [2, Section 7.11.4]. They form an Abelian category. We'll not need this.

(b) Any morphism of ind-coherent ind-schemes $f \colon X \to Y$ yields a functor $f^* \colon \mathbf{Qcoh}(Y) \to \mathbf{Qcoh}(X)$. If f is an open embedding the base change yields an exact functor $f^* \colon \mathbf{Coh}(Y) \to \mathbf{Coh}(X)$ and a group homomorphism $f^* \colon \mathbf{K}(Y) \to \mathbf{K}(X)$.

1.3.7. Definition. Let X be an ind-scheme. A closed ind-subscheme $Y \subset X$ is *good* if for every scheme $Z \to X$ the closed subscheme $Z \times_X Y \subset Z$ is good.

Note that if X is ind-coherent and $Y \subset X$ is a good ind-subscheme then Y is again an ind-coherent ind-scheme. If $f : Y \to X$ is an ind-proper homomorphism of ind-schemes of ind-finite type, or a good ind-subscheme of an ind-coherent ind-scheme, then there is a functor $f_* : \mathcal{C}oh(Y) \to \mathcal{C}oh(X)$ and a group homomorphism $Rf_* : \mathbf{K}(Y) \to \mathbf{K}(X)$.

1.4. Group actions on ind-schemes

1.4.1. Ind-groups and group-schemes. Let $\mathcal{G}rp$ be the category of groups. A group-scheme is a scheme representing a functor $\mathcal{A}lg \to \mathcal{G}rp$. An ind-group is an ind-scheme representing a functor $\mathcal{A}lg \to \mathcal{G}rp$.

1.4.2. Definition. We abbreviate *linear group* for linear algebraic group. A *pro-linear group* G is a compact, affine, group-scheme which is represented as the inverse limit of a system of linear groups $G = \lim_n G^n$ with n any integer $\geqslant 0$, such that the morphism $G^m \to G^n$ is a group-scheme homomorphism for each $m \geqslant n$.

1.4.3. *Examples.* Let G be a linear group. For each $\mathbb{C}$-algebra R the set of R-points of G is $G(R) = \mathrm{Hom}_{\mathcal{S}ch}(\mathrm{Spec}(R), G)$.

(a) The algebraic group $G(\mathbb{C}[\varpi]/(\varpi^n))$ represents the functor $R \mapsto G(R[\varpi]/(\varpi^n))$. The functor $R \mapsto G(R[\![\varpi]\!])$ is represented by a group-scheme, denoted by $K = G(\mathbb{C}[\![\varpi]\!])$. The group-scheme K is a pro-linear group, since it is the limit of the inverse system of linear groups $G(\mathbb{C}[\varpi]/(\varpi^n))$ with $n \geqslant 0$.

(b) The functor $R \mapsto G(R[\varpi^{-1}])$ is represented by an ind-group, denoted by $G(\mathbb{C}[\varpi^{-1}])$.

(c) The functor $R \mapsto G(R((\varpi)))$ is represented by an ind-group, denoted by $G(\mathbb{C}((\varpi)))$.

Throughout we'll use the same symbol for an ind-scheme X and the set of $\mathbb{C}$-points $X(\mathbb{C})$. For instance the symbol K will denote both the functor above and the group of $\mathbb{C}[\![\varpi]\!]$-points of the linear group G.

1.4.4. Group actions on an ind-scheme. Let G be an ind-group and X be an ind-scheme. We'll say that G acts on X if there is a morphism of functors $G \times X \to X$ satisfying the obvious composition rules. A *G-equivariant ind-scheme* is an ind-scheme with a (given) G-action. We'll abbreviate *G-ind-scheme* for G-equivariant ind-scheme. We'll also call ind-G-scheme a G-ind-scheme which is represented as the direct colimit of a system of quasi-compact G-schemes (X_α) as in 1.3.2.

1.4.5. Definitions. Let $G = \lim_n G^n$ be a pro-linear group.

(a) A (compact) G-scheme X is *admissible* if it is represented as the inverse limit of a system of G-schemes of finite type with flat affine morphisms (X^α) such that, for each α, the G-action on X^α factors through a G^n-action if $n \geqslant n_\alpha$ for some integer n_α.

(b) A morphism of admissible G-schemes $f \colon X \to Y$ is *admissible* if there are inverse systems of (X^α), (Y^α) as above such that f is the limit of a morphism of systems of G-schemes $(f^\alpha) \colon (X^\alpha) \to (Y^\alpha)$ and the following square is Cartesian for each $\alpha \leqslant \beta$

$$
\begin{array}{ccc}
X^\beta & \xrightarrow{\ f_\beta\ } & Y^\beta \\
\downarrow & & \downarrow \\
X^\alpha & \xrightarrow{\ f_\alpha\ } & Y^\alpha.
\end{array}
$$

(c) An ind-G-scheme X is *admissible* if it is the direct colimit of a system of compact admissible G-schemes with admissible closed embeddings.

1.4.6. Remarks.

(a) Let X be a G-torsor, with $G = \lim_n G^n$ a pro-linear group. For each n let G_n be the kernel of the canonical morphism $G \to G^n$. If the quotient scheme X/G is of finite type then the G-scheme X is admissible. Indeed X is the inverse limit of the system of G-schemes (X/G_n), and the G-action on X/G_n factors through a G^n-action.

(b) If $f \colon Y \to X$ is a finitely presented morphism of G-schemes with X admissible then Y and f are also admissible [31, Section C.3], [EGAIV, Proposition 2.1.4].

1.5. Equivariant K-theory of ind-schemes

To simplify, in this section we'll assume that all schemes are quasi-compact.

1.5.1. Equivariant quasi-coherent $\mathcal{O}$-modules over a scheme. Let G be a group-scheme and X be a G-scheme. Let $a, p \colon G \times X \to X$ be the action and the obvious projection.

1.5.2. Definition. A *G-equivariant quasi-coherent $\mathcal{O}_X$-module* is a quasi-coherent $\mathcal{O}_X$-module $\mathcal{E}$ with an isomorphism $\theta \colon a^*(\mathcal{E}) \to p^*(\mathcal{E})$. The obvious cocycle condition is to hold. Let $\boldsymbol{Qcoh}^G(X)$ be the category of G-equivariant quasi-coherent $\mathcal{O}_X$-modules. Given a closed subset $Y \subset X$ we define the category $\boldsymbol{Qcoh}^G(X \text{ on } Y)$ of G-equivariant quasi-coherent $\mathcal{O}_X$-modules supported on Y in the obvious way.

The category $\boldsymbol{Qcoh}^G(X)$ is Abelian. The forgetful functor

$$
\mathrm{for} \colon \boldsymbol{Qcoh}^G(X) \to \boldsymbol{Qcoh}(X)
$$

is exact and it *reflects exactness*, i.e., whenever a sequence in $\boldsymbol{Qcoh}^G(X)$ is exact in $\boldsymbol{Qcoh}(X)$ it is also exact in $\boldsymbol{Qcoh}^G(X)$. A G-equivariant quasi-coherent $\mathcal{O}_X$-module is said to be *coherent, of finite type* or *finitely presented* if it is coherent, of finite type or finitely presented as an $\mathcal{O}_X$-module. We define the categories $\boldsymbol{Coh}^G(X \text{ on } Y)$ and $\boldsymbol{Coh}^G(X)$ in the obvious way.

Let $\mathcal{C}^G(X)_{\mathrm{qc}}$ be the category of complexes of G-equivariant quasi-coherent $\mathcal{O}_X$-modules, and let $\mathcal{D}^G(X)_{\mathrm{qc}}$ be the derived category of G-equivariant quasi-coherent $\mathcal{O}_X$-modules. Note that this notation may be confusing. We do not claim that $\mathcal{D}^G(X)_{\mathrm{qc}}$ is the same as the derived category of G-equivariant $\mathcal{O}_X$-modules with quasi-coherent cohomology. For a coherent quasi-compact G-scheme X we set

$$\mathbf{K}^G(X \text{ on } Y) = [\mathcal{C}oh^G(X \text{ on } Y)], \quad \mathbf{K}^G(X) = \mathbf{K}^G(X \text{ on } X).$$

The representation ring of G is defined by $\mathbf{R}^G = \mathbf{K}^G(\mathrm{pt})$. It acts on the group $\mathbf{K}^G(X \text{ on } Y)$ by tensor product.

To define the standard derived functors for equivariant sheaves we need more material. There are a number of foundational issues to be addressed in translating the theory of derived functors of quasi-coherent sheaves from the non equivariant setting to the equivariant one. Here we briefly consider the issues that are relevant to the present paper.

1.5.3. Definitions.

(a) An *ample family of line bundles* on X is a family of line bundles $\{\mathcal{L}_i\}$ such that for every quasi-coherent $\mathcal{O}_X$-module $\mathcal{E}$ the evaluation map yields an epimorphism

$$\bigoplus_i \bigoplus_{n>0} \Gamma(X, \mathcal{E} \otimes_X \mathcal{L}_i^{\otimes n}) \otimes \mathcal{L}_i^{\otimes(-n)} \to \mathcal{E}.$$

We'll say that X *satisfies the property* (A_G) if it has an ample family of G-equivariant line bundles.

(b) We say that X *satisfies the (resolution) property* (R_G) if for every G-equivariant quasi-coherent $\mathcal{O}_X$-module $\mathcal{E}$ there is a G-equivariant, flat, quasi-coherent $\mathcal{O}_X$-module $\mathcal{P}$ and a surjection of G-equivariant $\mathcal{O}_X$-modules $f \colon \mathcal{P} \to \mathcal{E}$. We'll also demand that we can choose $\mathcal{P}$ and f in a functorial way with respect to $\mathcal{E}$.

(c) We say that a G-equivariant complex of quasi-coherent $\mathcal{O}_X$-modules $\mathcal{E}$ admits a *K-flat resolution* if there is a G-equivariant quasi-isomorphism $\mathcal{P} \to \mathcal{E}$ with $\mathcal{P}$ a G-equivariant complex of quasi-coherent $\mathcal{O}_X$-modules such that $\mathcal{P} \otimes_X \mathcal{F}$ is acyclic for every acyclic complex $\mathcal{F}$ in $\mathcal{C}^G(X)_{\mathrm{qc}}$, see [27, Definition 5.1].

(d) We say that a G-equivariant complex of quasi-coherent $\mathcal{O}_X$-modules $\mathcal{E}$ admits a *K-injective resolution* if there is a G-equivariant quasi-isomorphism $\mathcal{E} \to \mathcal{I}$ with $\mathcal{I}$ a G-equivariant complex of quasi-coherent $\mathcal{O}_X$-modules such that the complex of chain homomorphisms $\mathcal{F} \to \mathcal{I}$ in $\mathcal{C}^G(X)_{\mathrm{qc}}$ is acyclic for every acyclic complex $\mathcal{F}$ in $\mathcal{C}^G(X)_{\mathrm{qc}}$, see [27, Definition 1.1].

If G is the trivial group we'll abbreviate $(A) = (A_G)$ and $(R) = (R_G)$.

1.5.4. *Remarks.*

(a) The property (A_G) implies the property (R_G). It implies also that any G-equivariant quasi-coherent $\mathcal{O}_X$-module of finite type is the quotient of a

G-equivariant vector bundle, because X is quasi-compact [11, Chapter 0, (5.2.3)].

(b) If G is linear and X is Noetherian, normal, and satisfies the property (A), then X satisfies also the property (A_G) [30, Lemma 2.10 and Section 2.2]. Since any quasi-projective scheme satisfies (A), we recover the well-known fact that X satisfies the property (R_G) if it is quasi-projective and normal and if G is linear.

(c) If G is linear and X is Noetherian and regular, then X satisfies (A_G) by part (b), because it satisfies (A) [SGA6, II.2.2.7.1].

(d) Let X be an admissible G-scheme represented as the inverse limit of a system of G-schemes (X^α) as in 1.4.5 (a). If X^α satisfies (A_G) for some α then X satisfies also (A_G), as well as X^β for each $\beta \geqslant \alpha$ [31, Example 2.1.2(g)]. Thus if X is an admissible G-scheme which satisfies the property (S) in 1.2.19 then it satisfies also the property (A_G) (as well as X^α for each α) by part (c) above.

The G-equivariant quasi-coherent sheaves are well behaved on quasi-compact schemes satisfying the property (A_G). *In the rest of Section 1.5 all G-schemes are assumed to be quasi-compact and to satisfy (A_G).*

1.5.5. Lemma. *Assume that X is coherent. Then any G-equivariant quasi-coherent $\mathcal{O}_X$-module is the direct colimit of a system of G-equivariant coherent $\mathcal{O}_X$-modules.*

Proof. For any G-equivariant quasi-coherent $\mathcal{O}_X$-module $\mathcal{E}$ the property (A_G) yields an epimorphism in $\boldsymbol{\mathcal{Q}coh}^G(X)$

$$\mathcal{F} = \bigoplus_i \bigoplus_{n>0} \Gamma(X, \mathcal{E} \otimes_X \mathcal{L}_i^{\otimes n}) \otimes \mathcal{L}_i^{-\otimes n} \to \mathcal{E}.$$

Any (rational) G-module is locally finite, see, e.g., [14, Section I.2.13]. Choose a finite number of i's and n's and a finite-dimensional G-submodule of $\Gamma(X, \mathcal{E} \otimes_X \mathcal{L}_i^{\otimes n})$ for each i and each n in these finite sets. Then $\mathcal{F}$ is represented as the union of a system of G-equivariant locally free $\mathcal{O}_X$-submodules of finite type $\mathcal{F}_\alpha \subset \mathcal{F}$. Taking the image under the epimorphism above we can represent $\mathcal{E}$ as the direct colimit of a system of G-equivariant $\mathcal{O}_X$-submodules of finite type $\mathcal{E}_\alpha$ with surjective maps $\phi_\alpha \colon \mathcal{F}_\alpha \to \mathcal{E}_\alpha$. The kernel of ϕ_α is again a G-equivariant quasi-coherent $\mathcal{O}_X$-module. Considering its finitely generated G-equivariant quasi-coherent $\mathcal{O}_X$-submodules, we prove as in [11, Chapter I, Corollary 6.9.12] that $\mathcal{E}$ is the direct colimit of a system of G-equivariant finitely presented $\mathcal{O}_X$-modules. Since X is a coherent scheme, any finitely presented $\mathcal{O}_X$-module is coherent. □

1.5.6. Proposition.

(a) *Any complex in $\boldsymbol{\mathcal{C}}^G(X)_{\mathrm{qc}}$ admits a K-flat resolution.*

(b) *There is a left derived tensor product $\boldsymbol{\mathcal{D}}^G(X)_{\mathrm{qc}} \times \boldsymbol{\mathcal{D}}^G(X)_{\mathrm{qc}} \to \boldsymbol{\mathcal{D}}^G(X)_{\mathrm{qc}}$.*

(c) *If $f \colon X \to Y$ is a morphism of G-schemes there is a left derived functor $Lf^* \colon \boldsymbol{\mathcal{D}}^G(Y)_{\mathrm{qc}} \to \boldsymbol{\mathcal{D}}^G(X)_{\mathrm{qc}}$.*

Proof. The non equivariant case is treated in [27]. The equivariant case is very similar and is left to the reader. For instance, part (a) is proved as in [24, Section 2.5], while parts (b), (c) follow from (a) and the general theory of derived functors [24, Section 2.5, 3.1], [18], [19].   □

1.5.7. Proposition.

(a) *The category* $\mathbfcal{Qcoh}^G(X)$ *is a Grothendieck category. It has enough injective objects. Any complex of* $\mathbfcal{C}^G(X)_{qc}$ *has a K-injective resolution.*

(b) *If* $f\colon X \to Y$ *is a morphism of G-schemes there is a right derived functor* $Rf_*\colon \mathbfcal{D}^G(X)_{qc} \to \mathbfcal{D}^G(Y)_{qc}.$

Proof. Part (b) follows from (a) and the general theory of derived functors [24], [18], [19]. Let us concentrate on part (a). The second claim is a well-known consequence of the first one. The third claim follows also from the first one by [26]. See also [1, Theorem 5.4], [19, Section 14]. So we must check that $\mathbfcal{Qcoh}^G(X)$ is a Grothendieck category. To do so we must prove that it has a generator, that it is cocomplete, and that direct colimits are exact. Fix a small category $\mathbfcal{A}$ and a functor $\mathbfcal{A} \to \mathbfcal{Qcoh}^G(X)$, $\alpha \mapsto \mathcal{E}_\alpha$. Composing it with the forgetful functor we get a functor $\mathbfcal{A} \to \mathbfcal{Qcoh}(X)$ with a colimit

$$\mathcal{E} = \operatorname*{colim}_{\alpha} \mathrm{for}(\mathcal{E}_\alpha),$$

because the category $\mathbfcal{Qcoh}(X)$ is cocomplete. For the same reason we have also the following colimits

$$\operatorname*{colim}_{\alpha} a^*(\mathrm{for}(\mathcal{E}_\alpha)), \quad \operatorname*{colim}_{\alpha} p^*(\mathrm{for}(\mathcal{E}_\alpha)).$$

Since the functors a^*, p^* have right adjoints, a general result yields

$$a^*(\mathcal{E}) = \operatorname*{colim}_{\alpha} a^*(\mathrm{for}(\mathcal{E}_\alpha)), \quad p^*(\mathcal{E}) = \operatorname*{colim}_{\alpha} p^*(\mathrm{for}(\mathcal{E}_\alpha)).$$

Next, since $(\mathcal{E}_\alpha)$ is a system of G-equivariant quasi-coherent sheaves we have an isomorphism of systems $a^*(\mathrm{for}(\mathcal{E}_\alpha)) \to p^*(\mathrm{for}(\mathcal{E}_\alpha))$. Taking the colimit we get an isomorphism of quasi-coherent sheaves $a^*(\mathcal{E}) \to p^*(\mathcal{E})$. This isomorphism yields a G-equivariant structure on $\mathcal{E}$. The resulting G-equivariant sheaf is a colimit in $\mathbfcal{Qcoh}^G(X)$. Thus the category $\mathbfcal{Qcoh}^G(X)$ is cocomplete and the functor for preserves colimits. Since for reflects exactness and $\mathbfcal{Qcoh}(X)$ is a Grothendieck category we obtain that the direct colimit is an exact functor in $\mathbfcal{Qcoh}^G(X)$. Finally we must prove that the Abelian category $\mathbfcal{Qcoh}^G(X)$ has a generator. This is obvious, because the proof of 1.5.5 implies that the tensor powers of the $\mathcal{L}_i$'s generate the category $\mathbfcal{Qcoh}^G(X)$.   □

1.5.8. Compatibility of the derived functors.

Recall that we have assumed that all G-schemes are quasi-compact and satisfy the property (A_G). This insures that the standard derived functors are well-defined. Here we briefly discuss the equivariant analogue of the properties in 1.2.10–1.2.15 of the derived tensor product, the derived pull-back and the derived direct image.

First Rf_* is right adjoint to Lf^*, next Rf_* preserves the cohomologically bounded complexes, and finally Lf^* commutes with the derived tensor product. These three properties are proved as in the non equivariant case, see e.g., [24, Proposition 3.2, 3.9], [19, Sections 14, 18]. For instance the second one follows from the spectral sequence $R^p f_* \circ H^q \Rightarrow R^{p+q} f_*$, where $R^p f_* = H^p \circ Rf_*$, and the third one from the fact that both derived functors can be computed via K-flat resolutions.

Next Lf^* and the derived tensor product both commute with the forgetful functor for because they can be computed via K-flat resolutions in both the equivariant and the non equivariant cases, and because the forgetful functor for takes a K-flat resolution in $\mathcal{C}^G(X)_{\mathrm{qc}}$ to a K-flat resolution in $\mathcal{C}(X)_{\mathrm{qc}}$.

The remaining properties require some work. We'll say that a G-equivariant complex of quasi-coherent $\mathcal{O}_X$-modules $\mathcal{E}$ is f_*-acyclic if the canonical morphism $f_*(\mathcal{E}) \to Rf_*(\mathcal{E})$ is a quasi-isomorphism. By the general theory of derived functors, for any G-equivariant complex $\mathcal{E}$ of quasi-coherent $\mathcal{O}_X$-modules, the complex $Rf_*(\mathcal{E})$ can be computed using a f_*-acyclic G-equivariant resolution of $\mathcal{E}$. The following lemma has been indicated to us by S. Riche.

1.5.9. Lemma. *Let $f : X \to Y$ be a morphism of G-schemes. Any complex $\mathcal{E}$ in $\mathcal{C}^G(X)_{\mathrm{qc}}$ has a K-injective resolution $\mathcal{E} \to \mathcal{I}$ such that $for(\mathcal{E}) \to for(\mathcal{I})$ is a f_*-acyclic resolution in $\mathcal{C}(X)_{\mathrm{qc}}$.*

Proof. The proof of [26, Theorem 3.13] implies that there is a K-injective resolution $\mathcal{E} \to \mathcal{I}$ in $\mathcal{C}^G(X)_{\mathrm{qc}}$ such that the terms $\mathcal{I}^k$, $k \in \mathbb{Z}$, of the complex $\mathcal{I}$ are injective in $\mathbf{Qcoh}^G(X)$. Next, by [24, Corollary 3.9.3.5] a complex in $\mathcal{C}(X)_{\mathrm{qc}}$ consisting of f_*-acyclic $\mathcal{O}_X$-modules is acyclic. Thus we must check that $for(\mathcal{I}^k)$ is f_*-acyclic in $\mathbf{Qcoh}(X)$. Consider the averaging functor

$$A : \mathbf{Qcoh}(X) \to \mathbf{Qcoh}^G(X), \quad \mathcal{F} \mapsto a_* p^*(\mathcal{F}).$$

It takes injective objects to injective ones, because it is right adjoint to the forgetful functor for, which is exact. Further, any injective object in $\mathbf{Qcoh}^G(X)$ is the direct summand of an object of the form $A(I)$ with I an injective object in $\mathbf{Qcoh}(X)$. Indeed, for any $\mathcal{F}$ in $\mathbf{Qcoh}^G(X)$ there is an embedding $for(\mathcal{F}) \subset I$ with I an injective object of $\mathbf{Qcoh}(X)$hus we have $\mathcal{F} \subset A(I)$ in $\mathbf{Qcoh}^G(X)$ by adjunction. Finally, if $\mathcal{F}$ is injective then it is a direct summand of $A(I)$. Therefore it is enough to check that if I is an injective object in $\mathbf{Qcoh}(X)$ then $for(A(I))$ is f_*-acyclic in $\mathbf{Qcoh}(X)$. Since a, p are flat and affine, by flat base change, the definition of the functor A yields

$$Rf_*(for(A(I))) = for(A(Rf_*(I)))$$
$$= for(A(f_*(I)))$$
$$= f_*(for(A(I))). \qquad \square$$

Therefore the general theory of derived functors implies that

$$Rf_*(for(\mathcal{E})) = for(Rf_*(\mathcal{E})), \quad \forall \mathcal{E} \in \mathcal{D}^G(X)_{qc}.$$

Further, the projection formula holds, i.e., we have

$$Rf_*(\mathcal{E}) \overset{L}{\otimes}_Y \mathcal{F} = Rf_*(\mathcal{E} \overset{L}{\otimes}_X Lf^*(\mathcal{F})), \quad \forall \mathcal{E} \in \mathcal{D}^G(X)_{qc}, \ \mathcal{F} \in \mathcal{D}^G(Y)_{qc}.$$

Indeed, by adjunction we have a natural projection map from the left-hand side to the right-hand side. To prove that this map is invertible it is enough to observe that it is invertible in the non equivariant case, because the forgetful functor commutes with Rf_*, Lf^* and the derived tensor product. The details are left to the reader.

1.5.10. Equivariant coherent sheaves over an ind-coherent ind-scheme. Let G be a group-scheme, X be an ind-coherent ind-G-scheme, and $Y \subset X$ be a closed ind-subscheme which is preserved by the G-action. Let (Y_α), (X_α) be systems of quasi-compact G-schemes as in 1.4.4, such that the inclusion $Y \subset X$ is represented by a system of inclusions $Y_\alpha \subset X_\alpha$. Since the maps $i_{\alpha\beta} \colon X_\alpha \to X_\beta$ are good G-subschemes we have a direct system of Abelian categories $\mathcal{C}oh^G(X_\alpha \text{ on } Y_\alpha)$ and exact functors $(i_{\alpha\beta})_*$. We set

(1.5.1)
$$\mathcal{C}oh^G(X \text{ on } Y) = 2\operatorname*{colim}_\alpha \mathcal{C}oh^G(X_\alpha \text{ on } Y_\alpha),$$
$$\mathbf{K}^G(X \text{ on } Y) = [\mathcal{C}oh^G(X \text{ on } Y)] = \operatorname*{colim}_\alpha \mathbf{K}^G(X_\alpha \text{ on } Y_\alpha).$$

1.5.11. Proposition. *The category $\mathcal{C}oh^G(X \text{ on } Y)$ is independent of the choice of the system (X_α), up to canonical equivalence. The group $\mathbf{K}^G(X \text{ on } Y)$ is independent of the choice of the system (X_α), up to canonical isomorphism.*

Proof. Let $(\tilde{X}_{\tilde\alpha})$ be another direct system of closed subschemes of X representing X. So we have $X = \operatorname{colim}_\alpha X_\alpha$ and $X = \operatorname{colim}_{\tilde\alpha} \tilde{X}_{\tilde\alpha}$. The second equality means that each X_α is included into some $\tilde{X}_{\tilde\alpha}$ as a closed subset and vice-versa. Therefore the 2-limits of both systems are identified. $\square$

Once again we write $\mathcal{C}oh^G(X) = \mathcal{C}oh^G(X \text{ on } X)$ and $\mathbf{K}^G(X) = \mathbf{K}^G(X \text{ on } X)$. Note that the tensor product $\otimes_X$ yields an action of the ring $\mathbf{R}^G$ on the Abelian group $\mathbf{K}^G(X \text{ on } Y)$.

1.5.12. Admissible ind-coherent ind-schemes and reduction of the group action. Let G be a pro-linear group. Fix a system (G^n) as in 1.4.2. For each integer $n \geqslant 0$ let G_n be the kernel of the canonical map $G \to G^n$. Let X be an admissible coherent G-scheme. Let X^α, n_α be as in 1.4.5 (a). We have a direct system of categories $\mathcal{C}oh^{G^n}(X^\alpha)$, $n \geqslant n_\alpha$, with exact functors. The pull-back by the canonical map $p_\alpha \colon X \to X^\alpha$ yields a functor

(1.5.2) $$2\operatorname*{colim}_\alpha 2\operatorname*{colim}_{n \geqslant n_\alpha} \mathcal{C}oh^{G^n}(X^\alpha) \to \mathcal{C}oh^G(X).$$

1.5.13. Proposition.

(a) *The functor* (1.5.2) *is an equivalence of Abelian categories, and it yields a group isomorphism*

$$\operatorname*{colim}_{\alpha} \operatorname*{colim}_{n \geqslant n_\alpha} \mathbf{K}^{G^n}(X^\alpha) \to \mathbf{K}^G(X).$$

(b) *If G_0 is pro-unipotent, the canonical map $\mathbf{K}^{G^0}(X) \to \mathbf{K}^G(X)$ is invertible.*

Proof. The proof of (b) is standard, see, e.g., [7]. Let us concentrate on part (a). The functor (1.5.2) is fully faithful by 1.2.18 (a). Let us check that it is essentially surjective. To do so, fix a G-equivariant coherent $\mathcal{O}_X$-module $\mathcal{E}$. By 1.2.18 (a) there is an α and a coherent $\mathcal{O}_{X^\alpha}$-module $\mathcal{E}^\alpha$ such that $\mathcal{E} = p_\alpha^*(\mathcal{E}^\alpha)$. We must check that we can choose $\mathcal{E}^\alpha$ such that the G-action on $\mathcal{E}$ factors to a G^n-action on $\mathcal{E}^\alpha$ for some $n \geqslant n_\alpha$. The unit of the adjoint pair of functors $(p_\alpha^*, (p_\alpha)_*)$ yields an inclusion of quasi-coherent $\mathcal{O}_{X^\alpha}$-modules

$$\mathcal{E}^\alpha \subset (p_\alpha)_* p_\alpha^*(\mathcal{E}^\alpha) = (p_\alpha)_*(\mathcal{E}).$$

Since X^α is a Noetherian G-scheme and $(p_\alpha)_*(\mathcal{E})$ is a quasi-coherent G-equivariant $\mathcal{O}_{X^\alpha}$-module, we know that $(p_\alpha)_*(\mathcal{E})$ is the union of all its G-equivariant coherent subsheaves. Fix a G-equivariant coherent $\mathcal{O}_{X^\alpha}$-module $\mathcal{F}^\alpha$ containing $\mathcal{E}^\alpha$. The G-action on $\mathcal{F}^\alpha$ factors to an action of the linear group G^n for some $n \geqslant n_\alpha$. Let $\mathcal{G}^\alpha \subset \mathcal{F}^\alpha$ be the G^n-equivariant quasi-coherent subsheaf generated by $\mathcal{E}^\alpha$. It is again a coherent $\mathcal{O}_{X^\alpha}$-module, because $\mathcal{F}^\alpha$ is coherent and X^α is Noetherian. Since $\mathcal{E}$ is already G-equivariant the inclusion

$$\mathcal{E} = p_\alpha^*(\mathcal{E}^\alpha) \subset p_\alpha^*(\mathcal{G}^\alpha)$$

is indeed an equality of $\mathcal{O}_X$-modules $\mathcal{E} \simeq p_\alpha^*(\mathcal{G}^\alpha)$.   □

Now, let X be an admissible ind-coherent ind-G-scheme represented as the direct colimit of a system of admissible G-schemes (X_α) as in 1.4.5 (c). By (1.5.1) we have

$$\mathcal{C}oh^G(X) = 2 \operatorname*{colim}_{\alpha} \mathcal{C}oh^G(X_\alpha), \quad \mathbf{K}^G(X) = \operatorname*{colim}_{\alpha} \mathbf{K}^G(X_\alpha).$$

If G_0 is pro-unipotent then 1.5.13 yields isomorphisms

$$\mathbf{K}^{G^0}(X) = \mathbf{K}^G(X), \quad \mathbf{R}^{G^0} = \mathbf{R}^G.$$

This is called the *reduction of the group action.*

1.5.14. Thom isomorphism and pro-finite-dimensional vector bundles over ind-schemes. A *vector bundle* over the ind-scheme Y is an ind-scheme homomorphism $X \to Y$ which is represented as the direct colimit of a system of vector bundles

$X_\alpha \to Y_\alpha$. More precisely, we require that $X = \mathrm{colim}_\alpha X_\alpha$, $Y = \mathrm{colim}_\alpha Y_\alpha$, and for $\alpha \leqslant \beta$ we have a Cartesian square

$$
\begin{array}{ccc}
X_\alpha & \longrightarrow & X_\beta \\
\downarrow & & \downarrow \\
Y_\alpha & \longrightarrow & Y_\beta
\end{array}
$$

such that the vertical maps are vector-bundles and the upper horizontal map is a morphism of vector bundles. With any vector bundle over an ind-scheme X we can associate its sheaf of sections which is a quasi-coherent $\mathcal{O}_X$-module, see 1.3.5.

A *pro-finite-dimensional vector bundle* over the ind-scheme Y is defined in the same way by replacing everywhere vector bundles by pro-finite-dimensional vector bundles, see 1.2.23. In other words, it is an ind-scheme homomorphism which is represented as the "double limit" of a system of vector-bundles $X_\alpha^n \to Y_\alpha$ with n an integer $\geqslant 0$. Further, for each α and each $m \geqslant n$ we have a vector-bundle homomorphism $X_\alpha^m \to X_\alpha^n$ over Y_α, and for each n and each $\alpha \leqslant \beta$ we have an isomorphism of vector-bundles $X_\alpha^n \to X_\beta^n \times_{Y_\beta} Y_\alpha$. We require that these data satisfy the obvious composition rules. In particular for each $m \geqslant n$ and each $\beta \geqslant \alpha$ the following square is Cartesian

$$
\begin{array}{ccc}
X_\alpha^m & \longrightarrow & X_\beta^m \\
\downarrow & & \downarrow \\
X_\alpha^n & \longrightarrow & X_\beta^n.
\end{array}
$$

Note that a pro-finite-dimensional vector bundle over an ind-coherent ind-scheme is again an ind-coherent ind-scheme.

Let $\pi \colon X \to Y$ be an admissible G-equivariant pro-finite-dimensional vector bundle over an admissible ind-coherent ind-G-scheme Y. From 1.5.10 and base change we get an exact functor $\pi^* \colon \boldsymbol{Coh}^G(Y) \to \boldsymbol{Coh}^G(X)$. It factors to a group homomorphism $\pi^* \colon \mathbf{K}^G(Y) \to \mathbf{K}^G(X)$. The *Thom isomorphism* implies that π^* is invertible.

1.5.15. Descent and torsors over ind-schemes. Fix a pro-linear group $G = \lim_n G^n$. For each integer $n \geqslant 0$ let G_n be the kernel of the canonical map $G \to G^n$.

Let Y be a scheme. A G-*torsor* over Y is a scheme homomorphism $P \to Y$ which is represented as the inverse limit of a system consisting of a G^n-torsor $P^n \to Y$ for each integer $n \geqslant 0$ such that the morphism of Y-schemes $P^m \to P^n$, $m \geqslant n$, intertwines the G^m-action on the left-hand side and the G^n-action on the right-hand side, via the canonical group-scheme homomorphism $G^m \to G^n$.

Now, assume that $Y = \mathrm{colim}_\alpha Y_\alpha$ is an ind-scheme. A G-*torsor* over Y is an ind-scheme homomorphism $P \to Y$ which is represented as the direct colimit of a system of G-torsors $P_\alpha \to Y_\alpha$. More precisely, we require that $P = \text{colim}_\alpha P_\alpha$, that for each α we have a system of G^n-torsors $P_\alpha^n \to Y_\alpha$ representing the G-torsor $P_\alpha \to Y_\alpha$, and that for each n and each $\beta \geqslant \alpha$ we have an isomorphism of G^n-torsors $P_\alpha^n \to P_\beta^n \times_{Y_\beta} Y_\alpha$. In particular, for each $m \geqslant n$ and each $\beta \geqslant \alpha$ we have a Cartesian square

$$
\begin{array}{ccc}
P_\alpha^m & \longrightarrow & P_\beta^m \\
\downarrow & & \downarrow \\
P_\alpha^n & \longrightarrow & P_\beta^n.
\end{array}
$$

Note that a G-torsor over a pro-smooth scheme is again a pro-smooth scheme, and that a G-torsor over an ind-coherent ind-scheme is again an ind-coherent ind-scheme.

Now, assume that Y is a scheme and let $P \to Y$ be a G-torsor. Note that for each integer $n \geqslant 0$ we have a G^n-torsor $P/G_n \to Y$. In the rest of this subsection we consider the *induction functors*.

Let X be an admissible G-scheme and let X^α, n_α be as in 1.4.5 (a). For each α and each integer $n \geqslant n_\alpha$ the quotient space $(X^\alpha)_Y = P \times_G X^\alpha$ is equal to the Y-scheme $(P/G_n) \times_{G^n} X^\alpha$. Further, if $\beta \geqslant \alpha$ the canonical map $X^\beta \to X^\alpha$ yields a Y-scheme homomorphism $(X^\beta)_Y \to (X^\alpha)_Y$. Thus the quotient space $X_Y = P \times_G X$ is a Y-scheme which is represented as the inverse limit $X_Y = \lim_\alpha (X^\alpha)_Y$. By (1.5.2) we have an equivalence of categories

$$
2\,\text{colim}_\alpha\, 2\,\text{colim}_{n \geqslant n_\alpha}\, \boldsymbol{Coh}^{G^n}(X^\alpha) \to \boldsymbol{Coh}^G(X).
$$

By faithfully flat descent we have a functor

$$
\boldsymbol{Coh}^{G^n}(X^\alpha) \to \boldsymbol{Coh}((P/G_n) \times_{G^n} X^\alpha) = \boldsymbol{Coh}((X^\alpha)_Y).
$$

This yields a functor (called *induction* functor)

$$
\boldsymbol{Coh}^G(X) \to \boldsymbol{Coh}(X_Y).
$$

Next, let Z be an admissible ind-G-scheme which is represented as the direct colimit of a system of admissible G-schemes Z_α as in 1.4.5 (c). Then the quotient space $Z_Y = P \times_G Z$ is an ind-scheme over Y which is represented as the direct colimit $Z_Y = \lim_\alpha (Z_\alpha)_Y$, and $(Z_\alpha)_Y$ is defined as above for each α. If Z is ind-coherent and Y is coherent then the ind-scheme Z_Y is again ind-coherent and (1.5.1) yields a functor (called *induction* functor)

$$
\boldsymbol{Coh}^G(Z) \to \boldsymbol{Coh}(Z_Y).
$$

The induction functor is defined in a similar way if Y is an ind-scheme. The details of the construction are left to the reader.

1.5.16. Remark. We define in a similar way induction functors for quasi-coherent sheaves. Next, let H be a group-scheme acting on the G-torsor $P \to Y$, i.e., the group H acts on P, Y and the action commutes with the G-action and with the projection $P \to Y$. Then there is an H-action on Z_Y and the induction yields a functor $\mathbf{Coh}^G(Z) \to \mathbf{Coh}^H(Z_Y)$.

1.5.17. Complements on the concentration map. Now, assume that $G = S$ is a diagonalizable linear group. Let $\mathbf{X}^S$ be the group of characters of S. Each $\lambda \in \mathbf{X}^S$ defines a one-dimensional representation θ_λ of S. Let θ_λ denote also its class in $\mathbf{R}^S$. The ring $\mathbf{R}^S$ is spanned by the elements θ_λ with $\lambda \in \mathbf{X}^S$. So any element of $\mathbf{R}^S$ may be viewed as a function on S. For any $\Sigma \subset S$ let $\mathbf{R}_\Sigma^S$ be the ring of quotients of $\mathbf{R}^S$ with respect to the multiplicative set of the functions in $\mathbf{R}^S$ which do not vanish identically on Σ.

Now, let X be an ind-coherent ind-S-scheme. We'll say that Σ is X-*regular* if the fixed points subsets X^Σ, X^S are equal. Write

$$\mathbf{K}^S(X)_\Sigma = \mathbf{R}_\Sigma^S \otimes_{\mathbf{R}^S} \mathbf{K}^S(X).$$

The *Thomason concentration theorem* says that the map

$$\mathbf{K}^S(X^S)_\Sigma \to \mathbf{K}^S(X)_\Sigma$$

given by the direct image by the canonical inclusion $X^S \subset X$ is invertible if X is a scheme of finite type and Σ is X-regular [28], [29]. We'll use some form of the concentration theorem in some more general situation, which we consider below. In each case, the proof of the concentration theorem can be reduced to the original statement of Thomason using the discussion above. It is left to the reader.

1.5.18. Let X be a pro-smooth admissible S-scheme. It is easy to check that the fixed-points subset $X^S \subset X$ is a closed subscheme which is again pro-smooth. Thus the obvious inclusion $j : X^S \to X$ has a finite tor-dimension by 1.2.10, 1.2.20. Hence it yields an $\mathbf{R}^S$-linear map $Lj^* : \mathbf{K}^S(X) \to \mathbf{K}^S(X^S)$. This map can be viewed as follows. Any coherent $\mathcal{O}_X$-module $\mathcal{E}$ has locally a finite resolution by locally free modules of finite rank. Hence the pth left derived functor $L_p j^* \mathcal{E} = H^{-p}(Lj^*\mathcal{E})$ vanishes for $p \gg 0$. We have $Lj^*(\mathcal{E}) = \sum_{p \geqslant 0}(-1)^p L_p j^*(\mathcal{E})$. If Σ is X-regular we get a group isomorphism

$$Lj^* : \mathbf{K}^S(X)_\Sigma \to \mathbf{K}^S(X^S)_\Sigma.$$

1.5.19. Let X be an admissible ind-S-scheme of ind-finite type. The inclusion of the fixed points subset $i : X^S \to X$ is a good ind-subscheme. Thus the direct image yields an $\mathbf{R}^S$-linear map $i_* : \mathbf{K}^S(X^S) \to \mathbf{K}^S(X)$. If Σ is X-regular we get a group isomorphism

$$i_* : \mathbf{K}^S(X^S)_\Sigma \to \mathbf{K}^S(X)_\Sigma.$$

1.5.20. Let X be a pro-smooth admissible S-scheme and $f : Y \to X$ be an admissible ind-S-scheme over X. We'll assume that the map f is locally trivial in the following sense: there is an admissible ind-S-scheme F of ind-finite type and an S-equivariant finite affine open cover $X = \bigcup_w X^w$ such that over each X^w the map f is isomorphic to the obvious projection $X^w \times F \to X^w$, where the group S

acts diagonally on the left-hand side. The ind-scheme Y is ind-coherent by 1.2.20. Further, the fixed points subset X^S is again pro-smooth. Setting $Y' = X^S \times_X Y$ we get the following diagram

$$\begin{array}{ccccc}
Y^S & \xrightarrow{\ i\ } & Y' & \xrightarrow{\ j\ } & Y \\
\downarrow & & \downarrow & & \downarrow f \\
X^S & \xrightarrow{\ =\ } & X^S & \longrightarrow & X.
\end{array}$$

Over the open set X^w the map j is isomorphic to the obvious inclusion

$$(X^w)^S \times F \subset X^w \times F.$$

The inclusion $(X^w)^S \subset X^w$ has a finite tor-dimension by 1.2.10, 1.2.20. Thus the map j has also a finite tor-dimension. By base change we have an $\mathbf{R}^S$-linear map $Lj^* \colon \mathbf{K}^S(Y) \to \mathbf{K}^S(Y')$. Since i is the inclusion of a good ind-subscheme the direct image gives a map $i_* \colon \mathbf{K}^S(Y^S) \to \mathbf{K}^S(Y')$. If Σ is Y-regular we get a group isomorphism

$$(i_*)^{-1} \circ Lj^* \colon \mathbf{K}^S(Y)_\Sigma \to \mathbf{K}^S(Y^S)_\Sigma.$$

2. Affine flag manifolds

2.1. Notation relative to the loop group

2.1.1. Let G be a simple, connected and simply connected linear group over $\mathbb{C}$ with the Lie algebra $\mathfrak{g}$. Let $T \subset G$ be a Cartan subgroup and W be the Weyl group of the pair (G, T). Recall that $\mathbf{X}^T$ is the Abelian group of characters of T and that $\mathbf{Y}^T$ is the Abelian group of cocharacters of T. Let $\mathfrak{t}$ be the Lie algebra of T and $\mathfrak{t}^*$ be the set of linear forms on $\mathfrak{t}$. We'll view $\mathbf{X}^T$, $\mathbf{Y}^T$ as lattices in $\mathfrak{t}^*$, $\mathfrak{t}$ in the usual way. Note that, since G is simply connected, the lattice $\mathbf{X}^T$ is spanned by the fundamental weight and the lattice $\mathbf{Y}^T$ is spanned by the simple coroots. Let $\mathbf{X}^T_+ \subset \mathbf{X}^T$ and $\mathbf{Y}^T_+ \subset \mathbf{Y}^T$ denote the monoids of dominant characters and cocharacters.

Fix a Borel subgroup $B \subset G$. Let Δ be the set of roots of (G, T) and $\Pi \subset \Delta$ be the subset of simple roots associated with B. Let $\Delta^\vee$ be the set of coroots. Let θ be the highest root and $\check\theta$ be the corresponding coroot. Let

$$\langle\ ,\ \rangle \colon \mathbf{X}^T \times \mathbf{Y}^T \to \mathbb{Z}, \quad (\ ,\ ) \colon \mathfrak{t}^* \times \mathfrak{t}^* \to \mathbb{C}$$

be the canonical perfect pairing and the nondegenerate W-invariant bilinear form normalized by $(\theta, \theta) = 2$. We'll denote by κ the corresponding homomorphism $\mathfrak{t} \to \mathfrak{t}^*$ and we'll abbreviate $(\check\lambda, \check\mu) = (\kappa(\check\lambda), \kappa(\check\mu))$ for each $\check\lambda, \check\mu \in \mathfrak{t}$.

Let $\tilde\Delta$ be the set of affine roots, $\tilde\Delta_e$ be the subset of positive affine roots and $\tilde\Pi$ be the subset of simple affine roots. Let $\alpha_0 \in \tilde\Pi$ be the unique simple affine root which does not belong to Π. We have $\tilde\Pi = \{\alpha_0, \alpha_1, \ldots, \alpha_n\}$ where n is the rank of G. Let $\tilde\Delta^\vee$ be the set of affine coroots. We have $\tilde\Pi^\vee = \{\check\alpha_0, \check\alpha_1, \ldots, \check\alpha_n\}$ where $\check\alpha_i$ is the affine coroot associated with the simple affine root α_i for each i.

Let $\tilde{W} = W \ltimes \mathbf{Y}^T$ be the affine Weyl group of G. For any affine real root α let $s_\alpha \in \tilde{W}$ be the corresponding affine reflection. We'll abbreviate $s_i = s_{\alpha_i}$ for each i. Since G is simply connected the group $\tilde{W}$ is a Coxeter group with simple reflections the s_i's.

We'll abbreviate $w = (w, 0)$ and $\xi_{\check{\lambda}} = (e, \check{\lambda})$ for each $(w, \check{\lambda}) \in \tilde{W}$. In particular we'll regard W as a subgroup of $\tilde{W}$ in the obvious way. Here e denotes the unit, both in W and in $\tilde{W}$.

2.1.2. We'll fix a decreasing sequence of subsets $\tilde{\Delta}_l \subset \tilde{\Delta}_e$, with $l \in \mathbb{N}$, such that

$$(\tilde{\Delta}_l + \tilde{\Delta}_e) \cap \tilde{\Delta}_e \subset \tilde{\Delta}_l, \quad \#(\tilde{\Delta}_e \setminus \tilde{\Delta}_l) < \infty, \quad \bigcap_l \tilde{\Delta}_l = \varnothing.$$

For instance we may set $\tilde{\Delta}_l = l\delta + \tilde{\Delta}_e$ where δ is the smallest positive imaginary root. Put also $\tilde{\Delta}_l^\circ = -\tilde{\Delta}_l$.

2.1.3. We'll abbreviate $G((\varpi)) = G(\mathbb{C}((\varpi)))$, $\mathfrak{g}((\varpi)) = \mathfrak{g}(\mathbb{C}((\varpi)))$, etc. Recall that $K = G(\mathbb{C}[\![\varpi]\!])$. Let $I \subset K$ be the standard Iwahori subgroup and $I^\circ \subset K^\circ = G(\mathbb{C}[\![\varpi^{-1}]\!])$ be the opposite Iwahori subgroup. Let N, N° be the pro-unipotent radicals of I, I° respectively. The groups I, I°, N, N° are compact.

2.1.4. Let $\mathfrak{n}$, $\mathfrak{n}^\circ$, $\mathfrak{i}$, $\mathfrak{k}$ be the Lie algebras of N, N°, I, K. For any integer $l \geqslant 0$ let $\mathfrak{n}_l \subset \mathfrak{n}$ and $\mathfrak{n}_l^\circ \subset \mathfrak{n}^\circ$ be the product of all weight subspaces associated with the roots in $\tilde{\Delta}_l$, $\tilde{\Delta}_l^\circ$ respectively. Put also

$$\mathfrak{n}^l = \mathfrak{n}/\mathfrak{n}_l, \quad \mathfrak{n}^{\circ,l} = \mathfrak{n}^\circ/\mathfrak{n}_l^\circ, \quad \mathfrak{n}_w = w(\mathfrak{n}) \cap \mathfrak{n}, \quad \mathfrak{n}_w^\circ = w(\mathfrak{n}^\circ) \cap \mathfrak{n}, \quad w \in \tilde{W}.$$

Let N_l, N_l°, etc, be the groups associated with the Lie algebras $\mathfrak{n}_l$, $\mathfrak{n}_l^\circ$, etc. We'll write $\tilde{\Delta}_w$, $\tilde{\Delta}_w^\circ$ for the set of roots of $\mathfrak{n}_w$, $\mathfrak{n}_w^\circ$. Note that $\mathfrak{n}$, $\mathfrak{n}_l$, $\mathfrak{n}_w$ are compact coherent schemes such that $\mathfrak{n}_l$, $\mathfrak{n}_w$ are good subschemes of $\mathfrak{n}$, and that $\mathfrak{n}^l$, $\mathfrak{n}_w^\circ$ are schemes of finite type. Note also that $\mathfrak{n}$, $\mathfrak{n}_l$ are admissible I-equivariant compact schemes for the adjoint I-action. We'll call *Iwahori Lie subalgebra of* $\mathfrak{g}((\varpi))$ any Lie subalgebra which is $G((\varpi))$-conjugate to $\mathfrak{i}$.

2.1.5. The group $\mathbb{C}^\times$ acts on $\mathbb{C}((\varpi))$ by loop rotations, i.e., a complex number $z \in \mathbb{C}^\times$ takes a formal series $f(\varpi)$ to $f(z\varpi)$. This yields $\mathbb{C}^\times$-actions on $G((\varpi))$, I and $\mathfrak{g}((\varpi))$. Consider the semi-direct products

$$\hat{G} = \mathbb{C}^\times \ltimes G((\varpi)), \quad \hat{I} = \mathbb{C}^\times \ltimes I, \quad \hat{I}^\circ = \mathbb{C}^\times \ltimes I^\circ, \quad \hat{T} = \mathbb{C}^\times \times T.$$

The group $\hat{G}$ acts again on $\mathfrak{g}((\varpi))$.

2.1.6. Let $\tilde{G}$ be the maximal, "simply connected", Kac–Moody group associated with G defined by Garland [9]. It is a group ind-scheme which is a central extension

$$1 \to \mathbb{C}^\times \to \tilde{G} \to \hat{G} \to 1.$$

See [23, Section 13.2] for details. Let $\tilde{I}$, $\tilde{T}$, $\tilde{K}$ be the corresponding Iwahori, Cartan and maximal compact subgroup. Note that $\tilde{K} = \hat{K} \times \mathbb{C}^\times$, $\tilde{I} = \hat{I} \times \mathbb{C}^\times$ and $\tilde{T} = \hat{T} \times \mathbb{C}^\times$, i.e., the central extension splits. We define also the opposite Iwahori group

$\tilde{I}^\circ = \hat{I}^\circ \times \mathbb{C}^\times$. Let $\tilde{\mathfrak{g}}$, $\tilde{\mathfrak{i}}$, $\tilde{\mathfrak{k}}$ be the Lie algebras of $\tilde{G}$, $\tilde{I}$, $\tilde{K}$. The group $\tilde{G}$ acts on $\mathfrak{g}((\varpi))$ and $\tilde{\mathfrak{g}}$ by the adjoint action. By an Iwahori Lie subalgebra of $\tilde{\mathfrak{g}}$ we simply mean a Lie subalgebra which is $\tilde{G}$-conjugate to $\tilde{\mathfrak{i}}$.

2.1.7. We'll also use the groups

$$\boldsymbol{G} = \tilde{G} \times \mathbb{C}^\times, \quad \boldsymbol{I} = \tilde{I} \times \mathbb{C}^\times, \quad \boldsymbol{T} = \tilde{T} \times \mathbb{C}^\times.$$

The group $\boldsymbol{G}$ acts also on $\tilde{\mathfrak{g}}$. We simply require that an element $z \in \mathbb{C}^\times$ acts by multiplication by z ($=$ by dilatations). Note that $\boldsymbol{T} = T \times (\mathbb{C}^\times)^3$. To distinguish the different copies of $\mathbb{C}^\times$ we may use the following notation: $\mathbb{C}^\times_{\mathrm{rot}}$ corresponds to loop rotation, $\mathbb{C}^\times_{\mathrm{cen}}$ to the central extension, and $\mathbb{C}^\times_{\mathrm{qua}}$ to dilatations. Thus we have

$$\hat{T} = T \times \mathbb{C}^\times_{\mathrm{rot}}, \quad \tilde{T} = \hat{T} \times \mathbb{C}^\times_{\mathrm{cen}}, \quad \boldsymbol{T} = \tilde{T} \times \mathbb{C}^\times_{\mathrm{qua}}.$$

We'll also write $T_{\mathrm{cen}} = T \times \mathbb{C}^\times_{\mathrm{cen}}$.

2.1.8. We'll abbreviate $\tilde{\mathbf{X}} = \mathbf{X}^{\tilde{T}}$, $\mathbf{X} = \mathbf{X}^T$, $\tilde{\mathbf{Y}} = \mathbf{Y}^{\tilde{T}}$ and $\mathbf{Y} = \mathbf{Y}^T$. The pairing $\langle\ ,\ \rangle$ extends to the canonical pairing

$$\langle\ ,\ \rangle \colon \tilde{\mathbf{X}} \times \tilde{\mathbf{Y}} \to \mathbb{Z}.$$

Let d, c be the canonical generators of $\mathbf{Y}^{\mathbb{C}^\times_{\mathrm{rot}}}$, $\mathbf{Y}^{\mathbb{C}^\times_{\mathrm{cen}}}$. We have

$$\tilde{\mathbf{Y}} = \mathbf{Y}^T \oplus \mathbb{Z}d \oplus \mathbb{Z}c = \mathbb{Z}d \oplus \bigoplus_{i=0}^n \mathbb{Z}\check{\alpha}_i, \quad \check{\alpha}_0 = c - \check{\theta}.$$

The affine fundamental weights are the unique elements $\omega_i \in \tilde{\mathbf{X}}$, $i = 0, 1, \ldots, n$, such that $\langle \omega_i, \check{\alpha}_j \rangle = \delta_{i,j}$ for each i, j. We have

$$\tilde{\mathbf{X}} = \mathbf{X}^T \oplus \mathbb{Z}\delta \oplus \mathbb{Z}\omega_0 = \mathbb{Z}\delta \oplus \bigoplus_{i=0}^n \mathbb{Z}\omega_i, \quad \mathbf{X} = \tilde{\mathbf{X}} \oplus \mathbb{Z}t,$$

where δ is the smallest positive imaginary root. Recall that $\alpha_0 = \delta - \theta$. Then δ, ω_0, t are the canonical generators of $\mathbf{X}^{\mathbb{C}^\times_{\mathrm{rot}}}$, $\mathbf{X}^{\mathbb{C}^\times_{\mathrm{cen}}}$, and $\mathbf{X}^{\mathbb{C}^\times_{\mathrm{qua}}}$ respectively.

2.1.9. There is a $\tilde{W}$-action on $\tilde{\mathbf{X}}$, $\tilde{\mathbf{Y}}$ such that the natural pairing is $\tilde{W}$-invariant. It is given by:

- W fixes the elements δ, ω_0, d, c and it acts in the usual way on $\mathbf{X}^T$, $\mathbf{Y}^T$,
- the element $\xi_{\check{\lambda}}$, $\check{\lambda} \in \mathbf{Y}^T$, maps $\mu \in \tilde{\mathbf{X}}$ to

$$\mu + \langle \mu, c \rangle \kappa(\check{\lambda}) - \big(\langle \mu, \check{\lambda} \rangle + (\check{\lambda}, \check{\lambda}) \langle \mu, c \rangle / 2 \big) \delta,$$

- the element $\xi_{\check{\lambda}}$, $\check{\lambda} \in \mathbf{Y}^T$, maps $\check{\mu} \in \tilde{\mathbf{Y}}$ to

$$\check{\mu} + \langle \delta, \check{\mu} \rangle \check{\lambda} - \big(\langle \kappa(\check{\lambda}), \check{\mu} \rangle + (\check{\lambda}, \check{\lambda}) \langle \delta, \check{\mu} \rangle / 2 \big) c.$$

This action is denoted by $^w\mu$, $^w\check{\mu}$ for each $w \in \tilde{W}$, $\mu \in \tilde{\mathbf{X}}$ and $\check{\mu} \in \tilde{\mathbf{Y}}$.

2.1.10. There is also a $\tilde{W}$-action on $\tilde{T}$. It is given by:

- W fixes $\mathbb{C}^\times_{\text{rot}}$, $\mathbb{C}^\times_{\text{cen}}$ and it acts in the usual way on T,
- the element $\xi_{\bar{\lambda}}$, $\bar{\lambda} \in \mathbf{Y}^T$, maps the pair (s, τ) with $s \in T_{\text{cen}}$ and $\tau \in \mathbb{C}^\times_{\text{rot}}$ to the pair

$$(s\check{\lambda}(\tau)c(\kappa(\check{\lambda})(sh))^{-1}, \tau) \text{ with } h^2 = \check{\lambda}(\tau).$$

Here we regard $\check{\lambda}, c$ as group homomorphisms $\mathbb{C}^\times \to T_{\text{cen}}$ and $\kappa(\check{\lambda})$ as a group homomorphism $T_{\text{cen}} \to \mathbb{C}^\times$.

2.1.11. Since I is a group-scheme the ring $\mathbf{R}^I$ is well-defined. By devissage we have $\mathbf{R}^I = \mathbf{R}^T$. Recall that $\mathbf{R}^T = \sum_{\lambda \in \mathbf{X}^T} \mathbb{Z}\theta_\lambda$ is the group algebra of $\mathbf{X}^T$, see 1.5.17. We'll abbreviate $q = \theta_\delta$, $t = \theta_t$. For any commutative ring $\mathbf{A}$ we'll write

$$\mathbf{A}_t = \mathbf{A}[t, t^{-1}], \quad \mathbf{A}_q = \mathbf{A}[q, q^{-1}], \quad \mathbf{A}_{q,t} = \mathbf{A}[q, q^{-1}, t, t^{-1}].$$

Finally we define the following $\mathbb{Z}_t$-algebras

$$\mathbb{Z}_t\tilde{\mathbf{X}} = \sum_{\lambda \in \tilde{\mathbf{X}}} \mathbb{Z}_t\theta_\lambda, \quad \mathbb{Z}_t\tilde{\mathbf{Y}} = \sum_{\bar{\lambda} \in \tilde{\mathbf{Y}}} \mathbb{Z}_t\theta_{\bar{\lambda}}.$$

Note that $\mathbb{Z}_t\tilde{\mathbf{X}} = \mathbf{R}^T$.

2.2. Reminder on the affine flag manifold

2.2.1. The affine flag manifold. Let $\mathfrak{F} = \mathfrak{F}_G$ be the affine flag manifold of G. It is an ind-proper ind-scheme of ind-finite type whose set of $\mathbb{C}$-points is

$$\mathfrak{F} = G((\varpi))/I = \{ \text{ Iwahori Lie subalgebra of } \mathfrak{g}((\varpi)) \}.$$

The space $\mathfrak{F}$ can be viewed as the sheaf for the fppf topology over the flat affine site over $\mathbb{C}$, associated with the quotient pre-sheaf $\tilde{G}/\tilde{I}$. In particular there is a canonical ind-scheme homomorphism $\tilde{G} \to \mathfrak{F}$ which is an $\tilde{I}$-torsor as in 1.5.15. The set of $\mathbb{C}$-points of $\mathfrak{F}$ is simply the quotient set $\tilde{G}/\tilde{I}$. It will be convenient to regard an element of this set as an Iwahori Lie subalgebra of $\tilde{\mathfrak{g}}$. The ind-group $\tilde{G}$ acts on itself by left multiplication. This action yields a $\tilde{G}$-action on $\mathfrak{F}$. The group-scheme $\tilde{I}$ acts also on $\mathfrak{F}$, and the latter has the structure of an admissible ind-$\tilde{I}$-scheme. The $\tilde{I}$-orbits are numbered by the elements of $\tilde{W}$

$$\mathfrak{F} = \bigsqcup_{w \in \tilde{W}} \overset{\circ}{\mathfrak{F}}_w.$$

Let $\leqslant$ be the Bruhat order on $\tilde{W}$. We have

$$\mathfrak{F} = \operatorname*{colim}_w \mathfrak{F}_w, \quad \mathfrak{F}_w = \bigsqcup_{v \leqslant w} \overset{\circ}{\mathfrak{F}}_v.$$

Further $\mathfrak{F}_w$ is a projective, normal, $\tilde{I}$-scheme for every w. We have

$$\mathbf{K}(\mathfrak{F}) = \operatorname*{colim}_w \mathbf{K}(\mathfrak{F}_w), \quad \mathbf{K}(\mathfrak{F}_w) = [\mathcal{C}oh(\mathfrak{F}_w)].$$

For a future use, we'll abbreviate $\mathfrak{D} = \mathfrak{F} \times \mathfrak{F}$. For each v, w let $\mathfrak{D}_{v,w} = \mathfrak{F}_v \times \mathfrak{F}_w$.

2.2.2. The Kashiwara affine flag manifold. We'll also use the Kashiwara flag manifold $\mathfrak{X}$. See [17], [20], [21] for details. It is a coherent non quasi-compact scheme locally of countable type with a left $\tilde{I}^\circ$-action. It is covered by pro-smooth open subsets. Recall that a G-scheme X is *locally free* if any point of X has a G-stable open neighborhood which is isomorphic, as a G-scheme, to $G \times Y$ for some scheme Y. In this case the quotient X/G is representable by a scheme. The Kashiwara flag manifold is constructed as a quotient $\mathfrak{X} = \tilde{G}_\infty/\tilde{I}$, where $\tilde{G}_\infty$ is a coherent scheme with a locally free left action of $\tilde{I}^\circ$ and a locally free right action of $\tilde{I}$. In particular there is a canonical scheme homomorphism $\tilde{G}_\infty \to \mathfrak{X}$ which is an $\tilde{I}$-torsor as in 1.5.15. There is an $\tilde{I}^\circ$-orbit decomposition

$$\mathfrak{X} = \bigsqcup_{w \in \tilde{W}} \overset{\circ}{\mathfrak{X}}{}^w$$

where $\overset{\circ}{\mathfrak{X}}{}^w$ is a locally closed subscheme of codimension $l(w)$ ($=$ the length of w in $\tilde{W}$) which is isomorphic to the infinite-dimensional affine space $\mathbb{A}^{\mathbb{N}}$. The Zariski closure of $\overset{\circ}{\mathfrak{X}}{}^w$ is $\bigsqcup_{v \geqslant w} \overset{\circ}{\mathfrak{X}}{}^v$. The scheme $\mathfrak{X}$ is covered by the following open subsets

$$\mathfrak{X}^w = \bigsqcup_{v \leqslant w} \overset{\circ}{\mathfrak{X}}{}^v.$$

Note that $\mathfrak{X}^w$ is an $\tilde{I}^\circ$-stable finite union of translations of the big cell $\mathfrak{X}^e$ and that $\mathfrak{X}^e \simeq \mathbb{A}^{\mathbb{N}}$. Thus $\mathfrak{X}^w$ is quasi-compact and pro-smooth. Since $\mathfrak{X}$ is not quasi-compact, we have

$$\mathbf{K}(\mathfrak{X}) = \lim_w \mathbf{K}(\mathfrak{X}^w), \quad \mathbf{K}(\mathfrak{X}^w) = [\mathcal{C}oh(\mathfrak{X}^w)].$$

For each w there is a closed immersion $\mathfrak{F}_w \subset \mathfrak{X}^w$, see [21, Proposition 1.3.2]. Therefore the restriction of $\mathcal{O}$-modules yields a functor

$$\mathcal{Q}coh(\mathfrak{X}) \to \mathcal{Q}coh(\mathfrak{F}).$$

The tensor product of quasi-coherent $\mathcal{O}_\mathfrak{X}$-modules yields a functor

$$\otimes_\mathfrak{X}: \mathcal{C}oh(\mathfrak{F}) \times \mathcal{C}oh(\mathfrak{X}) \to \mathcal{C}oh(\mathfrak{F}).$$

Since $\mathfrak{X}^w$ is pro-smooth we have also a group homomorphism

$$\otimes_\mathfrak{X}^L: \mathbf{K}(\mathfrak{F}) \otimes \mathbf{K}(\mathfrak{X}) \to \mathbf{K}(\mathfrak{F}).$$

Finally, we have the following important property.

2.2.3. Proposition. *The $\tilde{I}^\circ$-scheme $\mathfrak{X}^w$ is admissible and it satisfies $(A_{\tilde{I}^\circ})$.*

Proof. The admissibility follows from 1.4.6. Given an integer $l \geqslant 0$ we consider the quotients

$$\mathfrak{X}^{w,l} = N_l^\circ \backslash \mathfrak{X}^w, \quad \tilde{I}^{\circ,l} = \tilde{I}^\circ/N_l^\circ.$$

Note that $\tilde{I}^{\circ,l}$ is a linear group, that $\mathfrak{X}^{w,l}$ is a smooth $\tilde{I}^{\circ,l}$-scheme and that the canonical map $\mathfrak{X}^w \to \mathfrak{X}^{w,l}$ is an $\tilde{I}^\circ$-equivariant N_l°-torsor [21, Lemma 2.2.1]. A priori $\mathfrak{X}^{w,l}$ could be not separated. See the remark after [21, Lemma 2.2.1]. The separatedness is proved in [32, Section A.6]. See also 2.2.4 below. Since the $\tilde{I}^{\circ,l}$-scheme $\mathfrak{X}^{w,l}$ is Noetherian and regular it satisfies the property $(A_{\tilde{I}^{\circ,l}})$. Then $\mathfrak{X}^w$ satisfies also the property $(A_{\tilde{I}^\circ})$ by 1.5.4. $\square$

2.2.4. *Remarks.*

(a) The scheme $\mathfrak{X}^{w,l}$ above is separated if l is large enough, even in the more general case of Kac–Moody groups considered in [21]. This follows from [31, Proposition C.7] and the fact that $\mathfrak{X}^w$ is a separated scheme.

(b) The $\tilde{T}$-fixed points subsets in $\overset{\circ}{\mathfrak{F}}_w$ and $\overset{\circ}{\mathfrak{X}}^w$ are reduced to the same single point. We'll denote it by $\mathfrak{b}_w$. Note that $\mathfrak{b}_e$ is identified with the Iwahori Lie algebra $\mathfrak{i}$ (or $\tilde{\mathfrak{i}}$) for e the unit element of $\tilde{W}$.

2.2.5. Pro-finite-dimensional vector-bundles over $\mathfrak{F}$. Consider the ind-coherent ind-scheme of ind-infinite type

$$\tilde{\mathfrak{g}} \times \mathfrak{F} = \underset{w,l}{\mathrm{colim}}(\tilde{\mathfrak{g}}_l \times \mathfrak{F}_w),$$

where $l \geqslant 0$ and $\tilde{\mathfrak{g}}_l \subset \tilde{\mathfrak{g}}$ is the product of all weight subspaces which do not belong to $\tilde{\Delta}_l^\circ$. Given a Lie subalgebra $\mathfrak{b} \subset \tilde{\mathfrak{g}}$ let $\mathfrak{b}_{\mathrm{nil}}$ denote its pro-nilpotent radical. Set

$$\dot{\mathfrak{n}} = \{\, (x, \mathfrak{b}) \in \tilde{\mathfrak{g}} \times \mathfrak{F}; x \in \mathfrak{b}_{\mathrm{nil}} \,\}.$$

It is a pro-finite-dimensional vector bundle over $\mathfrak{F}$. Thus it is an ind-coherent ind-scheme such that

$$\dot{\mathfrak{n}} = \underset{w}{\mathrm{colim}}\,\dot{\mathfrak{n}}_w, \quad \dot{\mathfrak{n}}_w = \dot{\mathfrak{n}} \cap (\tilde{\mathfrak{g}} \times \mathfrak{F}_w),$$

where $\dot{\mathfrak{n}}_w$ is a compact coherent scheme for each w. Define also

$$\mathfrak{N} = \dot{\mathfrak{n}} \cap (\mathfrak{n} \times \mathfrak{F}).$$

It is an ind-coherent admissible ind-$\tilde{I}$-scheme such that

$$\mathfrak{N} = \underset{w}{\mathrm{colim}}\,\mathfrak{N}_w, \quad \mathfrak{N}_w = \dot{\mathfrak{n}} \cap (\mathfrak{n} \times \mathfrak{F}_w).$$

Note that the $\tilde{I}$-scheme $\dot{\mathfrak{n}}_w$ satisfies the property $(A_{\tilde{I}})$ because the canonical map $\dot{\mathfrak{n}}_w \to \mathfrak{F}_w$ is $\tilde{I}$-equivariant, affine, and $\mathfrak{F}_w$ is normal and projective, see 1.5.4 and 2.2.1.

2.2.6. Group actions on flag varieties and related objects. Recall that the ind-group $\tilde{G}$ acts on the ind-scheme $\mathfrak{F}$ by left multiplication. It acts also on $\mathfrak{D} = \mathfrak{F} \times \mathfrak{F}$ diagonally, on $\tilde{\mathfrak{g}}$ by conjugation, and on $\tilde{\mathfrak{g}} \times \mathfrak{F}$ diagonally. For each $w \in \tilde{W}$ let $\mathfrak{D}_w \subset \mathfrak{D}$ be the smallest $\tilde{G}$-stable subset containing the pair $(\mathfrak{b}_e, \mathfrak{b}_v)$ for each $v \leqslant w$.

Similarly, the group G acts also on $\mathfrak{F}$, $\mathfrak{D}$, $\tilde{\mathfrak{g}}$ and $\tilde{\mathfrak{g}} \times \mathfrak{F}$. We simply require that an element $z \in \mathbb{C}_{\mathrm{qua}}^\times$ acts trivially on $\mathfrak{F}$ and that z acts by multiplication by z on $\tilde{\mathfrak{g}}$. This action preserves $\mathfrak{n} \times \mathfrak{F}$ and $\dot{\mathfrak{n}}$, and it restricts to an admissible I-action on both of them. Note that $\mathfrak{F}$, $\tilde{\mathfrak{g}} \times \mathfrak{F}$, $\mathfrak{n} \times \mathfrak{F}$ and $\dot{\mathfrak{n}}$ are admissible ind-coherent ind-I-schemes. We also equip $\mathfrak{X}$ with the canonical I°-action such that $\mathbb{C}_{\mathrm{qua}}^\times$ acts trivially.

For a future use let us introduce the following notation. Given $\lambda \in \mathbf{X}$ we can view θ_λ as a one-dimensional representation of I. Then for each I-scheme X and each I-equivariant $\mathcal{O}_X$-module $\mathcal{E}$ we'll write $\mathcal{E}\langle\lambda\rangle$ for the I-equivariant $\mathcal{O}_X$-module

$$\mathcal{E}\langle\lambda\rangle = \theta_\lambda \otimes \mathcal{E}.$$

2.3. K-theory and the affine flag manifold

2.3.1. Induction of ind-schemes. Recall that the Kashiwara flag manifold is endowed with a canonical $\tilde{I}$-torsor $\tilde{G}_\infty \to \mathfrak{X}$, where $\tilde{G}_\infty$ is a coherent scheme with an $(\tilde{I}^\circ \times \tilde{I})$-action. For any admissible ind-I-scheme Z we equip the quotient

$$Z_{\mathfrak{X}} = \tilde{G}_\infty \times_{\tilde{I}} Z$$

with the I°-action such that the subgroup $\tilde{I}^\circ$ acts by left multiplication on $\tilde{G}_\infty$ and $\mathbb{C}_{\mathrm{qua}}^\times$ through its action on Z. We can regard $Z_{\mathfrak{X}}$ as a bundle over $\mathfrak{X}$. For any subspace $X \subset \mathfrak{X}$ let Z_X be the restriction of $Z_{\mathfrak{X}}$ to X. We'll abbreviate $Z^{(w)} = Z_{\mathfrak{X}^w}$ and $Z_{(w)} = Z_{\mathfrak{F}_w}$ for each $w \in \tilde{W}$. The discussion in Section 1.5.15 yields the following.

2.3.2. Proposition. *Let Z be an ind-coherent admissible ind-I-scheme. Then $Z_{(w)}$ and $Z_{\mathfrak{F}}$ are ind-coherent admissible ind-I-schemes, and $Z^{(w)}$ is an ind-coherent admissible ind-I°-scheme.*

Note that $Z_{\mathfrak{X}}$ is only an ind$'$-scheme, because $\mathfrak{X}$ is not quasi-compact. Now, we discuss a few examples which are important for us.

2.3.3. *Examples.* (a) Set $Z = \mathfrak{F}$. Consider the natural projection $p \colon \mathfrak{F}_{\mathfrak{F}} \to \mathfrak{F}$ and the action map $a \colon \mathfrak{F}_{\mathfrak{F}} \to \mathfrak{F}$. The pair (p, a) gives an ind-I-scheme isomorphism

$$\mathfrak{F}_{\mathfrak{F}} \to \mathfrak{D} = \mathfrak{F} \times \mathfrak{F}, \quad (g, \mathfrak{b}) \bmod \tilde{I} \mapsto (g(\mathfrak{i}), g(\mathfrak{b})),$$

where $\tilde{I}$ acts diagonally on $\mathfrak{D}$. Under this isomorphism the maps p, a are identified with the projections $\mathfrak{D} = \mathfrak{F} \times \mathfrak{F} \to \mathfrak{F}$ to the first and the second factors respectively. Further, the ind-subscheme $(\mathfrak{F}_w)_{\mathfrak{F}}$ is taken to the ind-subscheme $\mathfrak{D}_w \subset \mathfrak{D}$.

(b) Taking $Z = \mathfrak{n} \times \dot{\mathfrak{n}}$ the induction yields an ind-scheme which is canonically isomorphic to $\dot{\mathfrak{n}} \times \dot{\mathfrak{n}}$.

(c) Taking $Z = \mathfrak{N}$ the induction yields the ind-scheme $\mathfrak{N}_{\mathfrak{F}}$. We'll abbreviate $\mathfrak{M} = \mathfrak{N}_{\mathfrak{F}}$. By 2.3.3 (a) we can view $\mathfrak{M}$ as the admissible I-equivariant pro-finite-dimensional vector bundle over $\mathfrak{D}$ whose total space is

$$\{ (x, \mathfrak{b}, \mathfrak{b}') \in \tilde{\mathfrak{g}} \times \mathfrak{D}; x \in \mathfrak{b}_{\mathrm{nil}} \cap \mathfrak{b}'_{\mathrm{nil}} \}.$$

The I-action is the diagonal one. In 2.2.5 we have defined $\mathfrak{N}$ as an ind-subscheme of $\dot{\mathfrak{n}}$ and of $\mathfrak{n} \times \mathfrak{F}$. We may also regard it as an ind-subscheme of $\mathfrak{n} \times \dot{\mathfrak{n}}$ by taking a pair $(x, \mathfrak{b}) \in \mathfrak{N}$ to the pair $(x, (x, \mathfrak{b})) \in \mathfrak{n} \times \dot{\mathfrak{n}}$. Hence we have an inclusion $\mathfrak{M} \subset \dot{\mathfrak{n}} \times \dot{\mathfrak{n}}$

which takes a triple $(x, \mathfrak{b}, \mathfrak{b}')$ to the pair $((x, \mathfrak{b}), (x, \mathfrak{b}'))$. Composing this inclusion with the obvious projections

$$q: \dot{\mathfrak{n}} \times \dot{\mathfrak{n}} \to \dot{\mathfrak{n}} \times \mathfrak{F}, \quad p: \dot{\mathfrak{n}} \times \dot{\mathfrak{n}} \to \mathfrak{F} \times \dot{\mathfrak{n}}$$

we can also view $\mathfrak{M}$ as a good ind-subscheme either of $\dot{\mathfrak{n}} \times \mathfrak{F}$ or of $\mathfrak{F} \times \dot{\mathfrak{n}}$. For each v, w we'll write

$$\mathfrak{M}_v = \{ (x, \mathfrak{b}, \mathfrak{b}') \in \mathfrak{M}; (\mathfrak{b}, \mathfrak{b}') \in \mathfrak{D}_v \}, \quad \mathfrak{M}_{w,u} = \{ (x, \mathfrak{b}, \mathfrak{b}') \in \mathfrak{M}; \mathfrak{b} \in \mathfrak{F}_w, \mathfrak{b}' \in \mathfrak{F}_u \}.$$

Note that $(\mathfrak{N}_v)_{\mathfrak{F}} \simeq \mathfrak{M}_v$ and that $\mathfrak{M} = \mathrm{colim}_{w,u} \, \mathfrak{M}_{w,u}$.

(d) Taking $Z = \mathfrak{n}$ the induction yields a pro-finite-dimensional vector bundle $\mathfrak{n}^{(w)}$ over $\mathfrak{X}^w$, and a pro-finite-dimensional vector bundle $\mathfrak{n}_{(w)}$ over $\mathfrak{F}_w$ for each w. Note that we have $\dot{\mathfrak{n}}_w = \mathfrak{n}_{(w)}$, see 2.2.5. For any integer $l \geqslant 0$ we'll set $\dot{\mathfrak{n}}_w^l = (\mathfrak{n}^l)_{(w)}$. The canonical projection $\mathfrak{n} \to \mathfrak{n}^l$ yields a smooth affine morphism $\dot{\mathfrak{n}}_w \to \dot{\mathfrak{n}}_w^l$. Both maps are denoted by the symbol p.

2.3.4. Induction of I-equivariant sheaves. Fix an admissible ind-coherent ind-I-scheme Z. Consider the induced ind-scheme $Z^{(w)}$ over $\mathfrak{X}^w$ for each $w \in \tilde{W}$. For any elements $v, w \in \tilde{W}$ such that $v \leqslant w$ the open embedding $\mathfrak{X}^v \subset \mathfrak{X}^w$ yields an open embedding of ind-schemes $Z^{(v)} \subset Z^{(w)}$. Fix a closed subgroup $S \subset T$. We obtain an inverse system of categories $(\boldsymbol{Qcoh}^S(Z^{(w)}))$, an inverse system of categories $(\boldsymbol{Coh}^S(Z^{(w)}))$ and an inverse system of $\mathbf{R}^S$-modules $\mathbf{K}^S(Z^{(w)})$, see 1.3.6. We define

$$\boldsymbol{Qcoh}^S(Z_{\mathfrak{X}}) = 2 \varprojlim_w \boldsymbol{Qcoh}^S(Z^{(w)}),$$

$$\boldsymbol{Coh}^S(Z_{\mathfrak{X}}) = 2 \varprojlim_w \boldsymbol{Coh}^S(Z^{(w)}),$$

$$\mathbf{K}^S(Z_{\mathfrak{X}}) = \varprojlim_w \mathbf{K}^S(Z^{(w)}).$$

The discussion in Section 1.5.15 implies the following.

- For $w \in \tilde{W}$ the induction yields exact functors $\boldsymbol{Qcoh}^I(Z) \to \boldsymbol{Qcoh}^S(Z^{(w)})$ and $\boldsymbol{Coh}^I(Z) \to \boldsymbol{Coh}^S(Z^{(w)})$ which commute with tensor products and a group homomorphism $\mathbf{K}^I(Z) \to \mathbf{K}^S(Z^{(w)})$.

- Taking the inverse limit over all w's, we arrive at functors $\boldsymbol{Qcoh}^I(Z) \to \boldsymbol{Qcoh}^S(Z_{\mathfrak{X}})$, $\boldsymbol{Coh}^I(Z) \to \boldsymbol{Coh}^S(Z_{\mathfrak{X}})$ which commute with tensor products and a group homomorphism $\mathbf{K}^I(Z) \to \mathbf{K}^S(Z_{\mathfrak{X}})$.

- For each $w \in \tilde{W}$ and each I-equivariant quasi-coherent $\mathcal{O}_Z$-module $\mathcal{E}$ the restriction of the induced $\mathcal{O}_{Z^{(w)}}$-module to the ind-scheme $Z_{(w)}$ is naturally I-equivariant. Hence the induction yields also functors $\boldsymbol{Qcoh}^I(Z) \to \boldsymbol{Qcoh}^I(Z_{(w)})$, $\boldsymbol{Coh}^I(Z) \to \boldsymbol{Coh}^I(Z_{(w)})$, as well as a group homomorphism $\mathbf{K}^I(Z) \to \mathbf{K}^I(Z_{(w)})$.

For each $\mathcal{E} \in \boldsymbol{Qcoh}^I(Z)$ we'll write $\mathcal{E}_{\mathfrak{X}}$, $\mathcal{E}_{(w)}$ and $\mathcal{E}^{(w)}$ for the induced $\mathcal{O}$-modules over $Z_{\mathfrak{X}}$, $Z_{(w)}$ and $Z^{(w)}$ respectively.

2.3.5. *Examples.* (a) For each $\lambda \in \mathbf{X}$ let $\mathcal{O}_{\mathfrak{X}}(\lambda)$ be the line bundle over $\mathfrak{X}$ induced from the character θ_λ. The local sections of $\mathcal{O}_{\mathfrak{X}}(\lambda)$ are the regular functions

$f \colon \tilde{G}_\infty \to \mathbb{C}$ such that $f(xb) = \lambda(b)f(x)$ for each $x \in \tilde{G}_\infty$ and $b \in \tilde{I}$. Note that $\mathcal{O}_{\tilde{x}}(t) = \mathcal{O}_{\tilde{x}}\langle t\rangle$, where t is as in 2.1.8.

Restricting $\mathcal{O}_{\tilde{x}}(\lambda)$ to $\mathfrak{F}$ we get also a line bundle $\mathcal{O}_{\mathfrak{F}}(\lambda)$ over the ind-scheme $\mathfrak{F}$. We'll write $\mathcal{O}_X(\lambda) = f^*\mathcal{O}_{\tilde{x}}(\lambda)$ or $f^*\mathcal{O}_{\mathfrak{F}}(\lambda)$ for any map $f \colon X \to \mathfrak{X}$ or $f \colon X \to \mathfrak{F}$. For instance we have the line bundles $\mathcal{O}_{\mathfrak{n}_{\tilde{x}}}(\lambda)$, $\mathcal{O}_{\dot{\mathfrak{n}}}(\lambda)$ and $\mathcal{O}_{\mathfrak{N}}(\lambda)$. For any $\mathcal{O}_X$-module $\mathcal{E}$ we'll abbreviate

$$\mathcal{E}(\lambda) = \mathcal{E} \otimes_X \mathcal{O}_X(\lambda).$$

(b) Given $\lambda, \mu \in \mathbf{X}$ we can consider the I-equivariant line bundle $\mathcal{O}_{\mathfrak{F}}(\mu)\langle\lambda\rangle$ over $\mathfrak{F}$. By induction and 2.3.3 (a) it yields an I-equivariant line bundle over $\mathfrak{D}$. Recall that the I-action on $\mathfrak{D}$ is the diagonal one. Restricting the induced bundle $(\mathcal{O}_{\mathfrak{F}}(\mu)\langle\lambda\rangle)_{\tilde{x}}$ over $\mathfrak{F}_{\tilde{x}}$ to $\mathfrak{F}_{\mathfrak{F}} \simeq \mathfrak{D}$ we get the line bundle

$$\mathcal{O}_{\mathfrak{D}}(\lambda, \mu) = \mathcal{O}_{\mathfrak{F}}(\lambda) \boxtimes \mathcal{O}_{\mathfrak{F}}(\mu).$$

For any map $Z \to \mathfrak{D}$ we'll write $\mathcal{O}_Z(\lambda, \mu) = f^*\mathcal{O}_{\mathfrak{D}}(\lambda, \mu)$. We write also $\mathcal{O}_{\mathfrak{N}_{\tilde{x}}}(\lambda, \mu) = (\mathcal{O}_{\mathfrak{N}}(\mu)\langle\lambda\rangle)_{\tilde{x}}$.

2.3.6. Convolution product on $\mathbf{K}^I(\mathfrak{N})$. The purpose of this section is to define an associative multiplication

$$\star \colon \mathbf{K}^I(\mathfrak{N}) \otimes \mathbf{K}^I(\mathfrak{N}) \to \mathbf{K}^I(\mathfrak{N}).$$

Fix $\mathcal{E}, \mathcal{F} \in \mathcal{C}oh^I(\mathfrak{N})$. Recall that

$$\mathcal{C}oh^I(\mathfrak{N}) = 2 \operatorname*{colim}_{w} \mathcal{C}oh^I(\mathfrak{N}_w).$$

Choose $v, w \in \tilde{W}$ such that $\mathcal{E} \in \mathcal{C}oh^I(\mathfrak{N}_w)$ and $\mathcal{F} \in \mathcal{C}oh^I(\mathfrak{N}_v)$. We can regard $\mathcal{E}$ as a coherent $\mathcal{O}_{\dot{\mathfrak{n}}_w}$-module and $\mathcal{F}$ as a quasi-coherent $\mathcal{O}_{\mathfrak{n} \times \dot{\mathfrak{n}}_v}$-module. Note that the closed embedding $\mathfrak{N}_v \subset \mathfrak{n} \times \dot{\mathfrak{n}}_v$ is not good. Fix $u \in \tilde{W}$ such that the isomorphism $(\mathfrak{n} \times \dot{\mathfrak{n}})_{\mathfrak{F}} = \dot{\mathfrak{n}} \times \dot{\mathfrak{n}}$ in 2.3.3 (b) factors to a good embedding

$$(2.3.2) \qquad \nu \colon (\mathfrak{n} \times \dot{\mathfrak{n}}_v)_{(w)} \to \dot{\mathfrak{n}}_w \times \dot{\mathfrak{n}}_u.$$

Consider the obvious projections

$$\dot{\mathfrak{n}}_w \xleftarrow{\ f_2\ } \dot{\mathfrak{n}}_w \times \dot{\mathfrak{n}}_u \xrightarrow{\ f_1\ } \dot{\mathfrak{n}}_u.$$

Then $f_2^*(\mathcal{E})$ and $\nu_*(\mathcal{F}_{(w)})$ are both I-equivariant quasi-coherent $\mathcal{O}$-modules over $\dot{\mathfrak{n}}_w \times \dot{\mathfrak{n}}_u$, and we can define the following complex in $\mathcal{D}^I(\dot{\mathfrak{n}}_w \times \dot{\mathfrak{n}}_u)_{qc}$

$$(2.3.3) \qquad \mathcal{G} = f_2^*(\mathcal{E}) \overset{L}{\otimes}_{\dot{\mathfrak{n}}_w \times \dot{\mathfrak{n}}_u} \nu_*(\mathcal{F}_{(w)}).$$

We'll view it as a complex of I-equivariant quasi-coherent $\mathcal{O}$-modules over the ind-scheme $\dot{\mathfrak{n}} \times \dot{\mathfrak{n}}$ supported on the subscheme $\dot{\mathfrak{n}}_w \times \dot{\mathfrak{n}}_u$. We want to consider its direct image by the map f_1. Since the schemes $\dot{\mathfrak{n}}_w$, $\dot{\mathfrak{n}}_u$ are not of finite type, this requires some work.

2.3.7. Proposition. *The complex of $\mathcal{O}$-modules $\mathcal{G}$ over $\dot{\mathfrak{n}} \times \dot{\mathfrak{n}}$ does not depend on the choices of u, v, w up to quasi-isomorphisms. It is cohomologically bounded. Its direct image $R(f_1)_*(\mathcal{G})$ is a cohomologically bounded pseudo-coherent complex over $\dot{\mathfrak{n}}_u$ with cohomology sheaves supported on $\mathfrak{N}_u$. The assignment $\mathcal{E} \otimes \mathcal{F} \mapsto R(f_1)_*(\mathcal{G})$ yields a group homomorphism $\star \colon \mathbf{K}^I(\mathfrak{N}) \otimes \mathbf{K}^I(\mathfrak{N}) \to \mathbf{K}^I(\mathfrak{N})$.*

Proof. We'll abbreviate

$$T = (\mathfrak{n} \times \dot{\mathfrak{n}}_v)_{(w)}, \quad Y = \dot{\mathfrak{n}}_w \times \dot{\mathfrak{n}}_u, \quad \phi_2 = f_2 \circ \nu, \quad \phi_1 = f_1 \circ \nu.$$

Thus we have the following diagram

(2.3.4)

$$
\begin{array}{ccccc}
\dot{\mathfrak{n}}_w & \xleftarrow{\; f_2 \;} & Y & \xrightarrow{\; f_1 \;} & \dot{\mathfrak{n}}_u \\
 & {}_{\phi_2}\nwarrow & \Big\uparrow{\scriptstyle \nu} & \nearrow_{\phi_1} & \\
 & & T. & &
\end{array}
$$

The map ϕ_2 is flat. We claim that the complex of I-equivariant quasi-coherent $\mathcal{O}_T$-modules

$$\mathcal{H} = \phi_2^*(\mathcal{E}) \overset{L}{\otimes}_T \mathcal{F}_{(w)}$$

is cohomologically bounded. It is enough to prove that the complex for$(\mathcal{H})$ is cohomologically bounded. Since the derived tensor product commutes with the forgetful functor we may forget the I-action everywhere. Hence we can use base change and the projection formula in full generality. To unburden the notation in the rest of the proof we'll omit the functor for.

Now, for an integer $l \geqslant 0$ we have the maps in 2.3.3 (d)

(2.3.5) $$p \colon \mathfrak{n} \to \mathfrak{n}^l, \quad p \colon \dot{\mathfrak{n}}_w \to \dot{\mathfrak{n}}_w^l.$$

Since $\mathcal{E}$ is an object of $\mathbf{Coh}(\dot{\mathfrak{n}}_w)$, by 1.2.18 (a) there is an l and an object $\mathcal{E}^l$ of $\mathbf{Coh}(\dot{\mathfrak{n}}_w^l)$ such that $\mathcal{E} = p^*(\mathcal{E}^l)$. Next, recall that $\mathcal{F}$ is an object of $\mathbf{Qcoh}(\mathfrak{n} \times \dot{\mathfrak{n}}_v)$. In the commutative diagram

$$
\begin{array}{ccccc}
\mathfrak{n} \times \dot{\mathfrak{n}}_v & \xrightarrow{\; p \times 1 \;} & \mathfrak{n}^l \times \dot{\mathfrak{n}}_v & \xrightarrow{\; 1 \times p \;} & \mathfrak{n}^l \times \dot{\mathfrak{n}}_v^l \\
 & \nwarrow & \Big\uparrow & & \\
 & & \mathfrak{N}_v & &
\end{array}
$$

the right vertical map is a good inclusion. Thus the $\mathcal{O}$-module $(p \times 1)_*(\mathcal{F})$ over $\mathfrak{n}^l \times \dot{\mathfrak{n}}_v$ is coherent. So if l is large enough there is a coherent sheaf $\mathcal{F}^l$ over $\mathfrak{n}^l \times \dot{\mathfrak{n}}_v^l$ such that

$$(p \times 1)_*(\mathcal{F}) = (1 \times p)^*(\mathcal{F}^l).$$

We'll abbreviate $T^l = (\mathfrak{n}^l \times \dot{\mathfrak{n}}^l_v)_{(w)}$. Set $\phi_{2,l} = f_{2,l} \circ \nu_l$ and $\phi_{1,l} = f_{1,l} \circ \nu_l$, where ν_l, $f_{2,l}$ and $f_{1,l}$ are the obvious inclusion and projections in the diagram

$$\dot{\mathfrak{n}}^l_w \xleftarrow{f_{2,l}} \dot{\mathfrak{n}}^l_w \times \dot{\mathfrak{n}}^l_u \xrightarrow{f_{1,l}} \dot{\mathfrak{n}}^l_u$$

with $\phi_{2,l}$, ν_l, $\phi_{1,l}$ and T^l.

Let us consider the complex $\mathcal{H}^l = \phi^*_{2,l}(\mathcal{E}^l) \overset{L}{\otimes}_{T^l} \mathcal{F}^l_{(w)}$ over T^l. The projection p in (2.3.5) gives a chain of maps

$$T \xrightarrow{q} (\mathfrak{n}^l \times \dot{\mathfrak{n}}_v)_{(w)} \xrightarrow{r} T^l .$$

Note that $Rq_* = q_*$ and $Lr^* = r^*$ because q is affine and r is flat. Thus a short computation using base change and the projection formula implies that

$$q_*(\mathcal{H}) = r^*(\mathcal{H}^l).$$

So to prove that $\mathcal{H}$ is cohomologically bounded it is enough to prove that $\mathcal{H}^l$ itself is cohomologically bounded. This can be proved using the Kashiwara affine flag manifold as follows. Write $X^l = \dot{\mathfrak{n}}^l_w$ and

$$T' = (\mathfrak{n}^l \times \dot{\mathfrak{n}}^l_v)^{(w)}, \quad X' = (\mathfrak{n}^l)^{(w)}.$$

Consider the Cartesian square

$$\begin{array}{ccc} T' & \xrightarrow{\phi'_2} & X' \\ \scriptstyle i \uparrow & & \uparrow \scriptstyle i \\ T^l & \xrightarrow{\phi_{2,l}} & X^l, \end{array}$$

where the vertical maps are the embeddings induced by the inclusion $\mathfrak{F}_w \subset \mathfrak{X}^w$. Recall that $\mathcal{E}^l$ is a coherent $\mathcal{O}_{X^l}$-module and that $\mathcal{F}^l_{(w)}$ is the restriction to T^l of the coherent $\mathcal{O}_{T'}$-module $\mathcal{F}' = (\mathcal{F}^l)^{(w)}$. Since the scheme $\mathfrak{X}^w$ satisfies the property (S), by 1.2.18 there is also a Cartesian square

$$\begin{array}{ccc} T^\alpha & \xrightarrow{\phi_{2,\alpha}} & X^\alpha \\ \scriptstyle p_\alpha \uparrow & & \uparrow \scriptstyle p_\alpha \\ T' & \xrightarrow{\phi'_2} & X' \end{array}$$

where the vertical maps are smooth and affine, X^α is smooth of finite type and the composed maps $j = p_\alpha \circ i$ are closed embeddings. Further we can assume that $\mathcal{F}' = (p_\alpha)^*(\mathcal{F}^\alpha)$ for some coherent $\mathcal{O}_{T^\alpha}$-module $\mathcal{F}^\alpha$. We have

$$i_*(\mathcal{H}^l) = (\phi'_2)^* i_*(\mathcal{E}^l) \overset{L}{\otimes}_{T'} \mathcal{F}'.$$

Thus we have also

$$j_*(\mathcal{H}^l) = (\phi_{2,\alpha})^* j_*(\mathcal{E}^l) \overset{L}{\otimes}_{T^\alpha} \mathcal{F}^\alpha.$$

Now $(\phi_{2,\alpha})^* j_*(\mathcal{E}^l)$ and $\mathcal{F}^\alpha$ are both coherent $\mathcal{O}_{T^\alpha}$-modules and $j_*(\mathcal{E}^l)$ is perfect because X^α is smooth, see 1.2.11. Hence the complex $j_*(\mathcal{H}^l)$ is pseudo-coherent and cohomologically bounded. Thus the complex $\mathcal{H}^l$ is also pseudo-coherent and cohomologically bounded, because j is a closed embedding. So $\mathcal{H}$ is also cohomologically bounded.

Now we can prove that $\mathcal{G}$ and $R(f_1)_*(\mathcal{G})$ are cohomologically bounded. Once again we can omit the I-action. Since $R\nu_* = \nu_*$, using the projection formula we get $\mathcal{G} = \nu_*(\mathcal{H})$. Thus the complex $\mathcal{G}$ is cohomologically bounded. Hence $R(f_1)_*(\mathcal{G})$ is also cohomologically bounded because the derived direct image preserves cohomologically bounded complexes.

To prove that the complex $R(f_1)_*(\mathcal{G})$ is pseudo-coherent it is enough to observe that we have $R(f_1)_*(\mathcal{G}) = p^* R(\phi_{1,l})_*(\mathcal{H}^l)$ and that $\mathcal{H}^l$ is pseudo-coherent.

The first claim of the proposition is obvious and is left to the reader. For instance, since $\mathcal{G} = \nu_*(\mathcal{H})$ the complex of $\mathcal{O}$-modules $\mathcal{G}$ over the ind-scheme $\dot{n} \times \dot{n}$ does not depend on the choice of u. The independence on v, w is proved in a similar way. $\square$

The following proposition will be proved in 2.4.9 below.

2.3.8. Proposition. *The map $\star$ equips $\mathbf{K}^I(\mathfrak{N})$ with a ring structure.*

2.3.9. *Remarks.* (a) The map $\star$ is an affine analogue of the convolution product used in [7]. It is $\mathbf{R}^I$-linear in the first variable (see part (c) below) but not in the second one. The definition of $\star$ we have given here is inspired from [4, Section 7.2]. Observe, however, that the complex $\mathcal{G}$ is not a complex of coherent sheaves over $\dot{n} \times \dot{n}$, contrarily to what is claimed in loc. cit. (in a slightly different setting).

(b) Since $\mathfrak{N}_e \subset \mathfrak{N}$ is a good subscheme, for each $\lambda \in \mathbf{X}$ we have the I-equivariant coherent $\mathcal{O}_{\mathfrak{N}}$-module $\mathcal{O}_{\mathfrak{N}_e}(\lambda) = \mathcal{O}_{\mathfrak{N}_e}\langle\lambda\rangle$. Consider the diagram

$$\dot{n}_w \xleftarrow{\;f_2\;} \dot{n}_w \times \dot{n}_w \xrightarrow{\;f_1\;} \dot{n}_w$$
$$\Big\uparrow{\scriptstyle\delta}$$
$$\dot{n}_w$$

where f_1, f_2 are the obvious projections and δ is the diagonal inclusion. Given $\lambda \in \mathbf{X}$ and an object $\mathcal{E}$ in $\mathbf{Coh}^I(\mathfrak{N}_w)$ let $\mathcal{E}(\lambda)$ be the "twisted" sheaf defined in 2.3.5 (a). We have

$$\mathcal{E} \star \mathcal{O}_{\mathfrak{N}_e}(\lambda) = R(f_1)_* \big(f_2^*(\mathcal{E}) \otimes_{\dot{n}_w \times \dot{n}_w} \delta_* \mathcal{O}_{\dot{n}_w}(\lambda) \big) = \mathcal{E}(\lambda).$$

Thus the associativity of $\star$ yields

$$\mathcal{E} \star \mathcal{F}(\lambda) = (\mathcal{E} \star \mathcal{F})(\lambda), \quad \forall \mathcal{E}, \mathcal{F} \in \mathbf{Coh}^I(\mathfrak{N}).$$

(c) Consider the diagram

$$\mathfrak{n} \xleftarrow{\;f_2\;} \mathfrak{n} \times \dot{\mathfrak{n}}_v \xrightarrow{\;f_1\;} \dot{\mathfrak{n}}_v$$

$$\uparrow{\scriptstyle\delta}$$

$$\mathfrak{N}_v$$

where f_1, f_2 are the obvious projections and δ is the diagonal inclusion. Given $\lambda \in \mathbf{X}$ and an object $\mathcal{F}$ in $\boldsymbol{Coh}^I(\mathfrak{N}_v)$ let $\mathcal{F}\langle\lambda\rangle$ be the "twisted" sheaf defined in 2.2.6. We have

$$\mathcal{O}_{\mathfrak{N}_e}\langle\lambda\rangle \star \mathcal{F} = R(f_1)_*\big(f_2^*(\mathcal{O}_\mathfrak{n}\langle\lambda\rangle) \otimes_{\mathfrak{n}\times\dot{\mathfrak{n}}_v} \delta_*(\mathcal{F})\big) = \mathcal{F}\langle\lambda\rangle.$$

Thus the associativity of $\star$ yields

$$\mathcal{E}\langle\lambda\rangle \star \mathcal{F} = (\mathcal{E} \star \mathcal{F})\langle\lambda\rangle, \quad \forall\,\mathcal{E}, \mathcal{F} \in \boldsymbol{Coh}^I(\mathfrak{N}).$$

2.4. Complements on the concentration in K-theory

2.4.1. Definition of the concentration map $\mathbf{r}_\Sigma$. Let $S \subset T$ be a closed subgroup. We'll say that S is *regular* if the schemes $\mathfrak{n}^S$, $\mathfrak{X}^S$ are both locally of finite type. Note that if S is regular then we have $\mathfrak{X}^S = \mathfrak{F}^S$ (as sets, because the left-hand side is a scheme of infinite type and locally finite type while the right-hand side is an ind-scheme of ind-finite type). Next, we'll say that a subset $\Sigma \subset S$ is *regular* if we have $\mathfrak{n}^S = \mathfrak{n}^\Sigma$ and $\mathfrak{X}^S = \mathfrak{X}^\Sigma$. In this subsection we'll assume that S and Σ are both regular. Let $\mathfrak{F}(\alpha)$, $\alpha \in A$, be the connected components of $\mathfrak{F}^S$. We have

$$\dot{\mathfrak{n}}^S = \bigsqcup_{\alpha \in A} \dot{\mathfrak{n}}(\alpha),$$

where $\dot{\mathfrak{n}}(\alpha)$ is a vector bundle over $\mathfrak{F}(\alpha)$ for each α. Since S is regular we have

$$\mathfrak{N}_\mathfrak{X}^S = \mathfrak{M}^S = \bigsqcup_{\alpha,\beta} \mathfrak{M}(\alpha,\beta), \quad \mathfrak{M}(\alpha,\beta) = \mathfrak{M} \cap (\dot{\mathfrak{n}}(\alpha) \times \dot{\mathfrak{n}}(\beta)),$$

where $\mathfrak{M}$ is as in 2.3.3. Here we have abbreviated $\mathfrak{N}_\mathfrak{X}^S = (\mathfrak{N}_\mathfrak{X})^S$. We define

$$\mathbf{K}^S(\mathfrak{N}_\mathfrak{X}^S) = \lim_w \operatorname*{colim}_v \mathbf{K}^S\big((\mathfrak{N}_v)^{(w),S}\big) = \prod_\alpha \bigoplus_\beta \mathbf{K}^S(\mathfrak{M}(\alpha,\beta)).$$

Observe that in 2.3.4 we have defined the group $\mathbf{K}^S(\mathfrak{N}_\mathfrak{X})$ in a similar way by setting

$$\mathbf{K}^S(\mathfrak{N}_\mathfrak{X}) = \lim_w \operatorname*{colim}_v \mathbf{K}^S\big((\mathfrak{N}_v)^{(w)}\big).$$

Now we can define the concentration map. Consider the closed embeddings

$$(2.4.1) \qquad (\mathfrak{n} \times \mathfrak{F})_\mathfrak{X}^S \xrightarrow{\;i\;} \mathfrak{n}_\mathfrak{X}^S \times_\mathfrak{X} \mathfrak{F}_\mathfrak{X} \xrightarrow{\;j\;} (\mathfrak{n} \times \mathfrak{F})_\mathfrak{X} = \mathfrak{n}_\mathfrak{X} \times_\mathfrak{X} \mathfrak{F}_\mathfrak{X}.$$

The scheme $\mathfrak{n}_\mathfrak{X}$ has an open cover by pro-smooth open subsets. The inclusion $\mathfrak{N} \subset \mathfrak{n} \times \mathfrak{F}$ is good. Thus 1.5.20 yields a group homomorphism

$$\boldsymbol{\gamma}_\Sigma = (i_*)^{-1} \circ Lj^* \colon \mathbf{K}^S(\mathfrak{N}_\mathfrak{X}) \to \mathbf{K}^S(\mathfrak{N}_\mathfrak{X}^S)_\Sigma.$$

Composing it with the induction $\Gamma\colon \mathcal{E} \mapsto \mathcal{E}_{\tilde{x}}$ yields a group homomorphism

(2.4.2) $\mathbf{r}_\Sigma\colon \mathbf{K}^I(\mathfrak{N}) \longrightarrow \mathbf{K}^S(\mathfrak{N}_{\tilde{x}}) \longrightarrow \mathbf{K}^S(\mathfrak{N}_{\tilde{x}}^S)_\Sigma$.

The map $\mathbf{r}_\Sigma$ is called the *concentration map*.

2.4.2. *Remark.* The map $\mathbf{r}_\Sigma$ is an affine analogue of the concentration map defined in [7, Theorem 5.11.10]. It can also be described in the following way. Let $\mathcal{E}$ be an $\boldsymbol{I}$-equivariant coherent $\mathcal{O}$-module over $\mathfrak{N}$. Fix $v \in \tilde{W}$ such that $\mathcal{E} \in \boldsymbol{Coh}^I(\mathfrak{N}_v)$. Given any $w \in \tilde{W}$ we fix $\nu,\, u$ as in (2.3.2). Under the direct image by ν we can view the S-equivariant coherent $\mathcal{O}$-module $\mathcal{E}_{(w)}$ as an S-equivariant quasi-coherent $\mathcal{O}$-module over $\dot{\mathfrak{n}}_w \times \dot{\mathfrak{n}}_u$ supported on $\mathfrak{M}_v \cap \mathfrak{M}_{w,u}$. The obvious projection

$$q\colon \dot{\mathfrak{n}}_w \times \dot{\mathfrak{n}}_u \to \dot{\mathfrak{n}}_w \times \mathfrak{F}_u$$

yields a good inclusion $\mathfrak{M}_{w,u} \subset \dot{\mathfrak{n}}_w \times \mathfrak{F}_u$. Thus, under the direct image by q we can also regard $\mathcal{E}_{(w)}$ as an S-equivariant coherent $\mathcal{O}$-module over $\dot{\mathfrak{n}}_w \times \mathfrak{F}_u$. Then we consider the following chain of inclusions

(2.4.3) $\dot{\mathfrak{n}}_w^S \times \mathfrak{F}_u^S \overset{i}{\longrightarrow} \dot{\mathfrak{n}}_w^S \times \mathfrak{F}_u \overset{j}{\longrightarrow} \dot{\mathfrak{n}}_w \times \mathfrak{F}_u.$

Since $\mathcal{E}_{(w)}$ is flat over $\mathfrak{F}_w$ and since $\dot{\mathfrak{n}}_w \to \mathfrak{F}_w$ is a pro-finite-dimensional vector bundle, we have a cohomologically bounded pseudo-coherent complex $\mathcal{E}' = Lj^*(\mathcal{E}_{(w)})$ over $\dot{\mathfrak{n}}_w^S \times \mathfrak{F}_u$. Next, the Thomason theorem yields an invertible map

$$i_*\colon \mathbf{K}^S(\dot{\mathfrak{n}}_w^S \times \mathfrak{F}_u^S)_\Sigma \to \mathbf{K}^S(\dot{\mathfrak{n}}_w^S \times \mathfrak{F}_u)_\Sigma.$$

Thus we have a well-defined element $\mathcal{E}'' = (i_*)^{-1}(\mathcal{E}')$. It can be regarded as an element of $\mathbf{K}^S(\mathfrak{M}_{w,u}^S)_\Sigma$ for a reason of supports. If $w,\, u$ are large enough then $\mathfrak{M}(\alpha,\beta)$ is a closed and open subset of $\mathfrak{M}_{w,u}^S$. The component of $\mathbf{r}_\Sigma(\mathcal{E})$ in $\mathbf{K}^S(\mathfrak{M}(\alpha,\beta))_\Sigma$ is the restriction of $\mathcal{E}''$ to $\mathfrak{M}(\alpha,\beta)$.

2.4.3. Proposition. *If $S = \Sigma = T$ then $\mathbf{r}_S$ is an injective map.*

Proof. We have $\mathfrak{X}^e = N^\circ = \tilde{I}^\circ/\tilde{T}$ as an $\tilde{I}^\circ$-scheme. Thus we have $\mathfrak{N}^{(e)} = N^\circ \times \mathfrak{N}$. The induction yields an inclusion

(2.4.4) $\mathbf{K}^I(\mathfrak{N}) \to \mathbf{K}^T(\mathfrak{N}^{(e)}),\ \mathcal{E} \mapsto \mathcal{E}^{(e)}.$

Therefore the induction map $\mathbf{K}^I(\mathfrak{N}) \to \mathbf{K}^T(\mathfrak{N}_{\tilde{x}})$ is also injective, because composing it with the canonical map

$$\mathbf{K}^T(\mathfrak{N}_{\tilde{x}}) = \lim_w \mathbf{K}^T(\mathfrak{N}^{(w)}) \to \mathbf{K}^T(\mathfrak{N}^{(e)})$$

yields (2.4.4). Thus, to prove that $\mathbf{r}_T$ is injective it is enough to check that the canonical map $\mathbf{K}^T(\mathfrak{N}^{(e)}) \to \mathbf{K}^T(\mathfrak{N}^{(e)})_T$ is injective. This is obvious because the $\mathbf{R}^T$-module $\mathbf{K}^T(\mathfrak{N}^{(e)})$ is torsion-free (use an affine cell decomposition of $\mathfrak{N}$). $\qquad\square$

2.4.4. Concentration of $\mathcal{O}$-modules supported on $\mathfrak{N}_e$. Let $\mathcal{E}$ be an I-equivariant vector bundle over $\mathfrak{N}_e$. Since the inclusion $\mathfrak{N}_e \subset \mathfrak{n} \times \mathfrak{F}$ is good we may view $\mathcal{E}$ as an object of $\mathcal{C}oh^I(\mathfrak{n} \times \mathfrak{F})$. Now we consider the diagrams (2.4.1) and (2.4.2). The induced coherent sheaf $\Gamma(\mathcal{E}) = \mathcal{E}_{\tilde{x}}$ is flat over $\mathfrak{n}_{\tilde{x}}$. Thus we have $Lj^*(\mathcal{E}_{\tilde{x}}) = j^*(\mathcal{E}_{\tilde{x}})$. Thus we obtain

$$\mathbf{r}_\Sigma(\mathcal{E}) = (i_*)^{-1} j^*(\mathcal{E}_{\tilde{x}}).$$

Next we have $j^{-1}((\mathfrak{N}_e)_{\tilde{x}}) = i((\mathfrak{N}_e)_{\tilde{x}}^S)$. This implies that

$$\mathbf{r}_\Sigma(\mathcal{E}) = j^*(\mathcal{E}_{\tilde{x}}).$$

Therefore we have proved the following.

2.4.5. Proposition. *If $\mathcal{E}$ is an I-equivariant vector bundle over $\mathfrak{N}_e$ then $\mathbf{r}_\Sigma(\mathcal{E})$ is the restriction of the coherent sheaf $\mathcal{E}_{\tilde{x}}$ over $\mathfrak{N}_{\tilde{x}}$ to the fixed-points subscheme $\mathfrak{N}_{\tilde{x}}^S$.*

2.4.6. Concentration of $\mathcal{O}$-modules supported on $\mathfrak{N}'_{s_\alpha}$. Fix a simple affine root $\alpha \in \tilde{\Pi}$. Recall that s_α is the corresponding simple reflection and that $\mathfrak{n}^\circ_{s_\alpha}$ is a 1-dimensional T-module whose class in the ring $\mathbf{R}^T$ is $\theta_{t+\alpha}$. Recall also that $\mathfrak{N} \subset \mathfrak{n} \times \mathfrak{F}$ and that $\mathfrak{n}_{s_\alpha} \subset \mathfrak{b}_{\mathrm{nil}}$ are good inclusions for each $\mathfrak{b} \in \mathfrak{F}_{s_\alpha}$. Note that the I-action on $\mathfrak{n}$ preserves $\mathfrak{n}_{s_\alpha}$. So we have a good I-equivariant subscheme $\mathfrak{N}'_{s_\alpha} \subset \mathfrak{N}$ given by

$$\mathfrak{N}'_{s_\alpha} = \mathfrak{n}_{s_\alpha} \times \mathfrak{F}_{s_\alpha}.$$

By a good subscheme we mean that $\mathfrak{N}'_{s_\alpha}$ is a good ind-subscheme, as in 1.3.7, which is a scheme. Note that $\mathfrak{N}'_{s_\alpha}$ is pro-smooth, because it is a pro-finite-dimensional vector bundle over the smooth scheme $\mathfrak{F}_{s_\alpha}$. Let $\mathcal{E}$ be an I-equivariant vector bundle over $\mathfrak{N}'_{s_\alpha}$. We'll view it as an I-equivariant coherent $\mathcal{O}$-module over $\mathfrak{N}$ or $\mathfrak{n} \times \mathfrak{F}$. The purpose of this section is to compute the element $\mathbf{r}_\Sigma(\mathcal{E})$.

First we assume that $S = T$. Consider the diagrams (2.4.1) and (2.4.2). The coherent sheaf $\Gamma(\mathcal{E}) = \mathcal{E}_{\tilde{x}}$ is flat over $(\mathfrak{N}'_{s_\alpha})_{\tilde{x}}$. So it is also flat over $(\mathfrak{n}_{s_\alpha})_{\tilde{x}}$. However it is not flat over $\mathfrak{n}_{\tilde{x}}$. To compute $Lj^*(\mathcal{E}_{\tilde{x}})$ we need a resolution of $\mathcal{E}_{\tilde{x}}$ by flat $\mathcal{O}_{\mathfrak{n}_{\tilde{x}}}$-modules. For this it is enough to construct a resolution of $\mathcal{E}$ by flat $\mathcal{O}_{\mathfrak{n}}$-modules, and to apply induction to it. We have a closed immersion

$$\mathfrak{N}'_{s_\alpha} \subset \mathfrak{N}''_{s_\alpha}, \quad \mathfrak{N}''_{s_\alpha} = \mathfrak{n} \times \mathfrak{F}_{s_\alpha}.$$

The Koszul resolution of $\mathcal{O}_{\mathfrak{N}'_{s_\alpha}}$ by locally-free $\mathcal{O}_{\mathfrak{N}''_{s_\alpha}}$-modules is the complex

$$\Lambda_{\mathfrak{N}''_{s_\alpha}}(\alpha) = \left\{ \mathcal{O}_{\mathfrak{N}''_{s_\alpha}} \langle t + \alpha \rangle \to \mathcal{O}_{\mathfrak{N}''_{s_\alpha}} \right\}$$

situated in degrees $[-1, 0]$. We may assume that

$$\mathcal{E} = \mathcal{O}_{\mathfrak{n}_{s_\alpha}} \boxtimes \mathcal{F},$$

where $\mathcal{F}$ is an I-equivariant locally free $\mathcal{O}_{\mathfrak{F}_{s_\alpha}}$-module. Set

$$\mathcal{E}' = \mathcal{O}_{\mathfrak{n}} \boxtimes \mathcal{F}.$$

It is an I-equivariant locally free $\mathcal{O}_{\mathfrak{N}''_{s_\alpha}}$-module whose restriction to $\mathfrak{N}'_{s_\alpha}$ is equal to $\mathcal{E}$. We have

$$\mathbf{r}_\Sigma(\mathcal{E}) = (i_*)^{-1} Lj^* \Gamma(\mathcal{E}' \otimes_{\mathcal{O}_{\mathfrak{N}''_{s_\alpha}}} \Lambda_{\mathfrak{N}''_{s_\alpha}}(\alpha)),$$

$$= (i_*)^{-1} j^* \Gamma(\mathcal{E}') - (i_*)^{-1} j^* \Gamma(\mathcal{E}'\langle t + \alpha\rangle).$$

Since $S = T$ we have $j^{-1}((\mathfrak{N}'_{s_\alpha})_{\mathfrak{x}}) = j^{-1}((\mathfrak{N}''_{s_\alpha})_{\mathfrak{x}})$. Thus, for each $\mathcal{O}_{\mathfrak{N}''_{s_\alpha}}$-module $\mathcal{F}$ we have $j^* \Gamma(\mathcal{F}) = j^* \Gamma(\mathcal{F}|_{\mathfrak{N}'_{s_\alpha}})$. This implies that

$$\mathbf{r}_\Sigma(\mathcal{E}) = (i_*)^{-1} j^* \Gamma(\mathcal{E}) - (i_*)^{-1} j^* \Gamma(\mathcal{E}\langle t + \alpha\rangle).$$

Next, observe that $\mathfrak{N}_{\mathfrak{x}}^S = \mathfrak{D}^S$. Thus the map

$$i \colon (\mathfrak{N}'_{s_\alpha})_{\mathfrak{x}}^S \to j^{-1}((\mathfrak{N}'_{s_\alpha})_{\mathfrak{x}})$$

is equal to the obvious inclusion

$$\mathfrak{D}_{s_\alpha}^S \subset \mathfrak{D}'_{s_\alpha}, \quad \mathfrak{D}'_{s_\alpha} = \mathfrak{D}_{s_\alpha} \cap (\mathfrak{F}^S \times \mathfrak{F}).$$

Now, we have an exact sequence of $\mathcal{O}_{\mathfrak{D}'_{s_\alpha}}$-modules

$$0 \to \mathcal{O}_{\mathfrak{D}'_{s_\alpha}}(0, -\alpha) \to \mathcal{O}_{\mathfrak{D}'_{s_\alpha}} \to \mathcal{O}_{\mathfrak{D}_{s_\alpha}^S} \to 0.$$

Therefore we have

$$\mathbf{r}_\Sigma(\mathcal{E}) = (1 - \mathcal{O}_{\mathfrak{N}_{\mathfrak{x}}^S}(\alpha + t, 0)) (1 - \mathcal{O}_{\mathfrak{N}_{\mathfrak{x}}^S}(0, -\alpha))^{-1} \mathcal{E}_{\mathfrak{x}}|_{\mathfrak{N}_{\mathfrak{x}}^S},$$

where $\mathcal{E}_{\mathfrak{x}}|_{\mathfrak{N}_{\mathfrak{x}}^S}$ is the restriction to $\mathfrak{N}_{\mathfrak{x}}^S$ of the induced sheaf $\mathcal{E}_{\mathfrak{x}}$ over $\mathfrak{N}_{\mathfrak{x}}$.

For any $S \subset T$ we obtain in the same way the following formula, compare [32, (2.4.6)].

2.4.7. Proposition. *We have*

$$\mathbf{r}_\Sigma(\mathcal{E}) = \begin{cases} (1 - \mathcal{O}_{\mathfrak{N}_{\mathfrak{x}}^S}(\alpha + t, 0)) (1 - \mathcal{O}_{\mathfrak{N}_{\mathfrak{x}}^S}(0, -\alpha))^{-1} \mathcal{E}_{\mathfrak{x}}|_{\mathfrak{N}_{\mathfrak{x}}^S} & \text{if } \theta_\alpha \neq 1, t, \\ (1 - \mathcal{O}_{\mathfrak{N}_{\mathfrak{x}}^S}(\alpha + t, 0)) \mathcal{E}_{\mathfrak{x}}|_{\mathfrak{N}_{\mathfrak{x}}^S} & \text{if } \theta_\alpha = 1 \neq t, \\ (1 - \mathcal{O}_{\mathfrak{N}_{\mathfrak{x}}^S}(0, -\alpha))^{-1} \mathcal{E}_{\mathfrak{x}}|_{\mathfrak{N}_{\mathfrak{x}}^S} & \text{if } \theta_\alpha = t \neq 1, \\ \mathcal{E}_{\mathfrak{x}}|_{\mathfrak{N}_{\mathfrak{x}}^S} & \text{if } \theta_\alpha = t = 1. \end{cases}$$

2.4.8. Multiplicativity of $\mathbf{r}_\Sigma$. Let $S \subset T$ be a regular closed subgroup. We have

$$\mathfrak{M}^S = \operatorname*{colim}_{w, u} \mathfrak{M}_{w, u}^S, \quad \mathbf{K}(\mathfrak{M}^S) = \operatorname*{colim}_{w, u} \mathbf{K}(\mathfrak{M}_{w, u}^S) = \bigoplus_{\alpha, \beta} \mathbf{K}(\mathfrak{M}(\alpha, \beta)),$$

where $\mathfrak{M}(\alpha, \beta)$ is as in 2.4.1. We have also

$$\mathbf{K}(\mathfrak{N}_{\mathfrak{x}}^S) = \prod_\alpha \bigoplus_\beta \mathbf{K}(\mathfrak{M}(\alpha, \beta)).$$

Therefore the group $\mathbf{K}(\mathfrak{N}_{\mathfrak{x}}^S)$ can be regarded as the completion of $\mathbf{K}(\mathfrak{M}^S)$. Note that $\mathfrak{M}(\alpha, \beta)$ is a closed subscheme of $\dot{\mathfrak{n}}(\alpha) \times \dot{\mathfrak{n}}(\beta)$ and the latter is smooth and of finite type because S is regular. So $\mathbf{K}(\mathfrak{M}^S)$, $\mathbf{K}(\mathfrak{N}_{\mathfrak{x}}^S)$ are both equipped with an associative convolution product. See Section 3.1 and the proof of the proposition below for details.

2.4.9. Proposition. *The map* $\star$ *yields a ring structure on* $\mathbf{K}^I(\mathfrak{N})$. *If the group* S *is regular then the map* $\mathbf{r}_\Sigma \colon \mathbf{K}^I(\mathfrak{N}) \to \mathbf{K}^S(\mathfrak{N}_{\bar{\mathfrak{x}}}^S)_\Sigma$ *is a ring homomorphism.*

Proof. Since the group S acts trivially on $\mathfrak{N}_{\bar{\mathfrak{x}}}^S$ we have

$$\mathbf{K}^S(\mathfrak{N}_{\bar{\mathfrak{x}}}^S)_\Sigma = \mathbf{K}(\mathfrak{N}_{\bar{\mathfrak{x}}}^S) \otimes \mathbf{R}_\Sigma^S.$$

The multiplication on the left-hand side is deduced by base change from the product on $\mathbf{K}(\mathfrak{N}_{\bar{\mathfrak{x}}}^S)$ mentioned above. It is enough to check that we have

$$\mathbf{r}_\Sigma(x \star y) = \mathbf{r}_\Sigma(x) \star \mathbf{r}_\Sigma(y), \quad \forall x, y.$$

Indeed, setting $S = \Sigma = T$, this relation and 2.4.3 imply that $\mathbf{K}^I(\mathfrak{N})$ is a subring of $\mathbf{K}(\mathfrak{N}_{\bar{\mathfrak{x}}}^S) \otimes \mathbf{R}_S^S$.

Fix $v, w \in \tilde{W}$. Let $u \in \tilde{W}$ be as in 2.3.6. Fix $\mathcal{E} \in \boldsymbol{Coh}^I(\mathfrak{N}_w)$ and $\mathcal{F} \in \boldsymbol{Coh}^I(\mathfrak{N}_v)$. Recall that $\mathcal{E}$, $\mathcal{F}$ denote also the corresponding classes in $\mathbf{K}^I(\mathfrak{N})$ and that $\mathcal{E}\star\mathcal{F}$ is the class of an I-equivariant cohomologically bounded pseudo-coherent complex over $\mathfrak{N}_u$. Let us recall the construction of this complex. We'll regard $\mathcal{E}$ as an I-equivariant coherent $\mathcal{O}_{\dot{\mathfrak{n}}_w}$-module and $\mathcal{F}$ as an I-equivariant quasi-coherent $\mathcal{O}_{\mathfrak{n}\times\dot{\mathfrak{n}}_v}$-module. Consider the diagram (2.3.4) that we reproduce below for the comfort of the reader

$$\dot{\mathfrak{n}}_w \xleftarrow{\;f_2\;} Y \xrightarrow{\;f_1\;} \dot{\mathfrak{n}}_u$$

$$\uparrow{\scriptstyle\nu}$$

$$T.$$

The map f_2 is flat and we have

$$\mathcal{E} \star \mathcal{F} = R(f_1)_*(\mathcal{G}), \quad \mathcal{G} = f_2^*(\mathcal{E}) \overset{L}{\otimes}_Y \nu_*(\mathcal{F}_{(w)}).$$

We want to compute $\mathbf{r}_\Sigma(\mathcal{E} \star \mathcal{F})$. Fix an element $x \in \tilde{W}$. First, we consider the induced complex $(\mathcal{E}\star\mathcal{F})_{(x)}$ over $(\dot{\mathfrak{n}}_u)_{(x)}$. Under induction the maps f_1, f_2 yield flat morphisms

$$(\dot{\mathfrak{n}}_w)_{(x)} \xleftarrow{\;f_{2,(x)}\;} Y_{(x)} \xrightarrow{\;f_{1,(x)}\;} (\dot{\mathfrak{n}}_u)_{(x)}.$$

The induction is exact and it commutes with tensor products. Thus we have

$$(\mathcal{E} \star \mathcal{F})_{(x)} = R(f_{1,(x)})_* \big(f_{2,(x)}^*(\mathcal{E}_{(x)}) \overset{L}{\otimes}_{Y_{(x)}} (\nu_*\mathcal{F}_{(w)})_{(x)} \big).$$

Fix $y, z \in \tilde{W}$ such that the canonical isomorphisms $\dot{\mathfrak{n}}_{\mathfrak{F}} = \mathfrak{F} \times \dot{\mathfrak{n}}$ and $(\mathfrak{n} \times \dot{\mathfrak{n}})_{\mathfrak{F}} = \dot{\mathfrak{n}} \times \dot{\mathfrak{n}}$ yield inclusions

$$\lambda \colon (\dot{\mathfrak{n}}_w)_{(x)} \to \mathfrak{F}_x \times \dot{\mathfrak{n}}_y, \quad \mu \colon (\dot{\mathfrak{n}}_u)_{(x)} \to \mathfrak{F}_x \times \dot{\mathfrak{n}}_z, \quad \nu \colon (\mathfrak{n} \times \dot{\mathfrak{n}}_v)_{(y)} \to \dot{\mathfrak{n}}_y \times \dot{\mathfrak{n}}_z.$$

We put

$$\mathcal{G}' = \mu_*((\mathcal{E} \star \mathcal{F})_{(x)}), \quad \mathcal{E}' = \lambda_*(\mathcal{E}_{(x)}), \quad \mathcal{F}' = \nu_*(\mathcal{F}_{(y)}).$$

We have

$$(2.4.5) \qquad \mathcal{G}' = R(\pi_2)_* \big(\pi_3^*(\mathcal{E}') \overset{L}{\otimes}_{Y'} \pi_1^*(\mathcal{F}') \big),$$

where $Y' = \mathfrak{F}_x \times \dot{\mathfrak{n}}_y \times \dot{\mathfrak{n}}_z$ and π_1, π_2, π_3 are the obvious projections

$$\dot{\mathfrak{n}}_y \times \dot{\mathfrak{n}}_z \xleftarrow{\quad\pi_1\quad} Y' \xrightarrow{\quad\pi_2\quad} \mathfrak{F}_x \times \dot{\mathfrak{n}}_z$$

$$\downarrow{\scriptstyle\pi_3}$$

$$\mathfrak{F}_x \times \dot{\mathfrak{n}}_y.$$

As explained in 2.4.2 we can regard the complex $\mathcal{G}'$, which is supported on $\mathfrak{M}_{x,z}$, as a complex over $\dot{\mathfrak{n}}_x \times \mathfrak{F}_z$. Let $\mathcal{G}''$ denote the latter. We have

$$\mathbf{r}_\Sigma(\mathcal{E} \star \mathcal{F}) = \gamma_\Sigma(\mathcal{G}'').$$

Now we compute $\mathcal{G}''$. Applying the base change formula to (2.4.5) we obtain the following equality in $\mathbf{K}^S(\dot{\mathfrak{n}}_x \times \mathfrak{F}_z)$

$$(2.4.6) \qquad \mathcal{G}'' = R(p_2)_* \big(p_3^*(\mathcal{E}'') \overset{L}{\otimes}_{Y''} p_1^*(\mathcal{F}'') \big).$$

Here $Y'' = \dot{\mathfrak{n}}_x \times \dot{\mathfrak{n}}_y \times \mathfrak{F}_z$ and p_1, p_2, p_3 are the projections

$$\dot{\mathfrak{n}}_y \times \mathfrak{F}_z \xleftarrow{\quad p_1\quad} Y'' \xrightarrow{\quad p_2\quad} \dot{\mathfrak{n}}_x \times \mathfrak{F}_z$$

$$\downarrow{\scriptstyle p_3}$$

$$\dot{\mathfrak{n}}_x \times \dot{\mathfrak{n}}_y.$$

Further $\mathcal{E}''$, $\mathcal{F}''$ are S-equivariant quasi-coherent $\mathcal{O}$-modules over $\dot{\mathfrak{n}}_x \times \dot{\mathfrak{n}}_y$, $\dot{\mathfrak{n}}_y \times \mathfrak{F}_z$ respectively which are characterized by the following properties

$$p_*(\mathcal{E}'') = \mathcal{E}', \quad \mathcal{F}'' = q_*(\mathcal{F}'), \quad \mathcal{E}'' \text{ is supported on } \mathfrak{M}_{x,y},$$

where p, q are the obvious maps

$$\mathfrak{F}_w \times \dot{\mathfrak{n}}_u \xleftarrow{\quad p\quad} \dot{\mathfrak{n}}_w \times \dot{\mathfrak{n}}_u \xrightarrow{\quad q\quad} \dot{\mathfrak{n}}_w \times \mathfrak{F}_u .$$

Now, recall that we must prove that the following formula holds in $\mathbf{K}^S(\mathfrak{N}_{\mathfrak{F}}^S)_\Sigma$

$$\mathbf{r}_\Sigma(\mathcal{E} \star \mathcal{F}) = \mathbf{r}_\Sigma(\mathcal{E}) \star \mathbf{r}_\Sigma(\mathcal{F}).$$

Let i, j be as in 2.4.2. The left-hand side is

$$(2.4.7) \qquad \mathbf{r}_\Sigma(\mathcal{E} \star \mathcal{F}) = \gamma_\Sigma(\mathcal{G}'') = (i_*)^{-1} L j^*(\mathcal{G}'').$$

Let us describe the right-hand side. We'll abbreviate

$$N = \dot{\mathfrak{n}}^S, \quad M = \mathfrak{M}^S.$$

Both are regarded as ind-schemes of ind-finite type. Thus we have

$$\mathbf{K}(N) = \bigoplus_\alpha \mathbf{K}(\dot{\mathfrak{n}}(\alpha)), \quad \mathbf{K}(M) = \bigoplus_{\alpha,\beta} \mathbf{K}(\mathfrak{M}(\alpha,\beta)).$$

Note that $\mathbf{K}(M) = \mathbf{K}(N^2 \text{ on } M)$ for the inclusion $M \subset N^2$ given by

$$(x,\mathfrak{b},\mathfrak{b}') \mapsto \big((x,\mathfrak{b}),(x,\mathfrak{b}')\big).$$

Given $a = 1, 2, 3$ let $q_a \colon N^3 \to N^2$ be the projection along the ath factor. We define the convolution product on $\mathbf{K}(M)$ by

$$(2.4.8) \qquad x \star y = R(q_2)_* \big(q_3^*(x) \overset{L}{\otimes}_{N^3} q_1^*(y) \big), \quad \forall x, y \in \mathbf{K}(M).$$

Note that N is a disjoint union of smooth schemes of finite type and that q_1, q_3 are flat maps. We'll use another expression for $\star$. For this, we write

$$F = \mathfrak{F}^S, \quad NF = N \times F, \quad N^2 F = N \times N \times F.$$

The obvious projections below are flat morphisms

$$NF \xleftarrow{\ \rho_1\ } N^2 F \xrightarrow{\ \rho_2\ } NF$$
$$\downarrow{\rho_3}$$
$$N^2.$$

We have also $\mathbf{K}(M) = \mathbf{K}(NF \text{ on } M)$ for the inclusion $M \subset NF$ given by

$$(x, \mathfrak{b}, \mathfrak{b}') \mapsto (\mathfrak{b}, (x, \mathfrak{b}')).$$

The projection formula yields

$$(2.4.9) \qquad x \star y = R(\rho_2)_* \big(\rho_3^*(x) \overset{L}{\otimes}_{N^2 F} \rho_1^*(y) \big).$$

Note that $N^2 F$ is also a disjoint union of smooth schemes of finite type. Finally we must compute $\mathbf{r}_\Sigma(\mathcal{E})$ and $\mathbf{r}_\Sigma(\mathcal{F})$. Once again, as explained in 2.4.2, we must first regard $\mathcal{E}'$, $\mathcal{F}'$ as complexes of $\mathcal{O}$-modules over $\dot{\mathfrak{n}}_x \times \mathfrak{F}_y$, $\dot{\mathfrak{n}}_y \times \mathfrak{F}_z$ respectively, and then we apply the map $(i_*)^{-1} \circ Lj^*$ to their class in K-theory.

Now, by (2.4.6), (2.4.7) and (2.4.9) we are reduced to prove the following equality

$$(i_*)^{-1} Lj^* R(p_2)_* \big(p_3^*(\mathcal{E}'') \overset{L}{\otimes}_{Y''} p_1^*(\mathcal{F}'') \big) =$$
$$= R(\rho_2)_* \big(\rho_3^*(i_*')^{-1} L(j')^*(\mathcal{E}'') \overset{L}{\otimes}_{N^2 F} \rho_1^*(i_*)^{-1} Lj^*(\mathcal{F}'') \big).$$

Here i, i', j and j' are the obvious inclusions in the following diagram

$$
\begin{array}{ccccc}
\dot{\mathfrak{n}}^S \times \mathfrak{F}^S & \xrightarrow{\ i\ } & \dot{\mathfrak{n}}^S \times \mathfrak{F} & \xrightarrow{\ j\ } & \dot{\mathfrak{n}} \times \mathfrak{F} \\
\big\uparrow{q} & & \big\uparrow{q} & & \big\uparrow{q} \\
\dot{\mathfrak{n}}^S \times \dot{\mathfrak{n}}^S & \xrightarrow{\ i'\ } & \dot{\mathfrak{n}}^S \times \dot{\mathfrak{n}} & \xrightarrow{\ j'\ } & \dot{\mathfrak{n}} \times \dot{\mathfrak{n}}.
\end{array}
$$

This is an easy consequence of the base change and of the projection formula. $\qquad \square$

2.5. Double affine Hecke algebras

2.5.1. Definitions. Recall that G is a simple, connected and simply connected linear group over $\mathbb{C}$. The double affine Hecke algebra (= DAHA) associated with G is the associative $\mathbb{Z}_{q,t}$-algebra $\mathbf{H}$ with 1 generated by the symbols T_w, X_λ with $w \in \tilde{W}$, $\lambda \in \tilde{\mathbf{X}}$ such that the T_w's satisfy the braid relations of $\tilde{W}$ and such that

$$X_\delta = q, \quad X_\mu X_\lambda = X_{\lambda+\mu}, \quad (T_{s_\alpha} - t)(T_{s_\alpha} + 1) = 0,$$

$$X_\lambda T_{s_\alpha} - T_{s_\alpha} X_{\lambda-r\alpha} = (t-1)X_\lambda(1 + X_{-\alpha} + \cdots + X_{-\alpha}^{r-1}) \quad \text{if } \langle \lambda, \check{\alpha} \rangle = r \geq 0.$$

Here α is any simple affine root. Let $\mathbf{H}^f \subset \mathbf{H}$ be the subring generated by t and the T_w's with $w \in W$. Let $\mathbf{R} \subset \mathbf{H}$ be the subring generated by $\{X_\lambda; \lambda \in \tilde{\mathbf{X}}\}$. To avoid confusions we may write $\mathbf{R}_X = \mathbf{R}$. Finally let $\mathbf{R}_Y \subset \mathbf{H}$ be the subring generated by $\{Y_{\check{\lambda}}; \check{\lambda} \in \hat{\mathbf{Y}}\}$, where $Y_{\check{\lambda}} = T_{\xi_{\check{\lambda}_1}} T_{\xi_{\check{\lambda}_2}}^{-1}$ with $\check{\lambda} = \check{\lambda}_1 - \check{\lambda}_2$ and $\check{\lambda}_1, \check{\lambda}_2$ dominant. The following fundamental result has been proved by Cherednik. We'll refer to it as the PBW theorem for $\mathbf{H}$.

2.5.2. Proposition. *The multiplication in $\mathbf{H}$ yields $\mathbb{Z}_{q,t}$-isomorphisms*

$$\mathbf{R} \otimes \mathbf{H}^f \otimes \mathbf{R}_Y \to \mathbf{H}, \quad \mathbf{R}_Y \otimes \mathbf{H}^f \otimes \mathbf{R} \to \mathbf{H}.$$

The $\mathbb{Z}_t$-algebra $\mathbf{H}^f$ is isomorphic to the Iwahori–Hecke algebra (over the commutative ring $\mathbb{Z}_t$) associated with the Weyl group W. The rings $\mathbf{R}$, $\mathbf{R}_Y$ are the group-rings associated with the lattices $\tilde{\mathbf{X}}$, $\hat{\mathbf{Y}}$ respectively.

2.5.3. *Remark.* The algebra $\mathbf{H}$ is the one considered in [33]. It is denoted by the symbol $\hat{\mathbf{H}}$ in [32]. Note that we have $X_{\omega_0} T_{s_0} X_{\omega_0}^{-1} = X_{\alpha_0} T_{s_0}^{-1}$. Thus $\mathbf{H}$ is a semidirect product $\mathbb{C}[X_{\omega_0}^{\pm 1}] \ltimes \mathcal{H}(\tilde{W}, \mathbf{X} \oplus \mathbb{Z}\delta)$ with the notation in [12, Section 5]. Changing the lattices in the definition of $\mathbf{H}$ yields different versions of the DAHA whose representation theory is closely related to the representation theory of $\mathbf{H}$. These different algebras are said to be *isogeneous*. In this paper we'll only consider the case of $\mathbf{H}$ to simplify the exposition. For more details the reader may consult [32, Section 2.5].

Let $\mathcal{O}(\mathbf{H})$ be the category of all right $\mathbb{C}\mathbf{H}$-modules which are finitely generated, locally finite over $\mathbf{R}$ (i.e., for each element m the $\mathbb{C}$-vector space $m\mathbf{R}$ is finite-dimensional), and such that q, t act by multiplication by a complex number. It is an Abelian category. Any object has a finite length. For any module M in $\mathcal{O}(\mathbf{H})$ we have

$$M = \bigoplus_{h \in \tilde{T}} M_h, \quad M_h = \bigcap_{\lambda \in \tilde{\mathbf{X}}} \bigcup_{r \geqslant 0} \{m \in M; m(X_\lambda - \lambda(h))^r = 0\}.$$

We'll call M_h the *h-weight subspace*. It is finite dimensional. Next, we set

$$\widehat{M} = \prod_h M_h.$$

The vector space $\widehat{M}$ is equipped with the product topology, the M_h's being equipped with the discrete topology. Note that $M \subset \widehat{M}$ is a dense subset. The $\mathbb{C}\mathbf{H}$-action on M extends uniquely to a continuous $\mathbb{C}\mathbf{H}$-action on $\widehat{M}$.

Fix an element $h = (s, \tau)$ of $\tilde{T}$, i.e., we let $s \in T_{\mathrm{cen}}$ and $\tau \in \mathbb{C}_{\mathrm{rot}}^{\times}$. For each $\zeta \in \mathbb{C}_{\mathrm{qua}}^{\times}$ we can form the corresponding tuple $(h, \zeta) \in \boldsymbol{T}$. Let $\mathcal{O}_{h,\zeta}(\mathbf{H})$ be the full subcategory of $\mathcal{O}(\mathbf{H})$ consisting of the modules M such that $q = \tau$, $t = \zeta$ and $M_{h'} = 0$ if h' is not in the orbit of h relatively to the $\tilde{W}$-action on $\tilde{T}$ in 2.1.10. Let $\tilde{W}$ act on $\boldsymbol{T}$ so that it acts on $\tilde{T}$ as in 2.1.10 and it acts trivially on $\mathbb{C}_{\mathrm{qua}}^{\times}$. We have

$$\mathcal{O}(\mathbf{H}) = \bigoplus_{h,\zeta} \mathcal{O}_{h,\zeta}(\mathbf{H}),$$

where (h, ζ) varies in a set of representatives of the $\tilde{W}$-orbits [32, Lemma 2.1.3, 2.1.6].

2.5.4. Geometric construction of the DAHA. We can now give a geometric construction of the $\mathbb{Z}_{q,t}$-algebra $\mathbf{H}$. First, let us introduce a few more notations. For each $\lambda \in \mathbf{X}$ we consider the following element of $\mathbf{K}^I(\mathfrak{N})$

$$x_\lambda = \mathcal{O}_{\mathfrak{N}_e}(\lambda) = \mathcal{O}_{\mathfrak{N}_e}\langle\lambda\rangle.$$

Next, given a simple affine root $\alpha \in \tilde{\Pi}$ we have the good I-equivariant subscheme $\mathfrak{N}_{s_\alpha}' \subset \mathfrak{N}$ introduced in 2.4.6. For each weights $\lambda, \mu \in \mathbf{X}$ we define the I-equivariant coherent sheaf $\mathcal{O}_{\mathfrak{N}_{s_\alpha}'}(\lambda, \mu)$ over $\mathfrak{N}$ as the direct image of the I-equivariant vector bundle $\mathcal{O}_{\mathfrak{N}_{s_\alpha}'}(\mu)\langle\lambda\rangle$ over $\mathcal{O}_{\mathfrak{N}_{s_\alpha}'}$, see 2.3.5 (b). Assume further that

$$(2.5.1) \qquad\qquad \lambda + \mu = -\alpha, \quad \langle\lambda, \check{\alpha}\rangle = \langle\mu, \check{\alpha}\rangle = -1.$$

Then we consider the following element of $\mathbf{K}^I(\mathfrak{N})$ given by

$$t_{s_\alpha} = -1 - \mathcal{O}_{\mathfrak{N}_{s_\alpha}'}(\lambda, \mu).$$

2.5.5. Lemma. *The element t_{s_α} is independent of the choice of λ, μ as above.*

Proof. It is enough to observe that if $\langle\lambda', \check{\alpha}\rangle = 0$ then the I-equivariant line bundle $\mathcal{O}_{\mathfrak{N}_{s_\alpha}'}(\lambda', -\lambda')$ is trivial. $\qquad\square$

The assignment $\theta_\lambda \mapsto X_\lambda$ identifies $\mathbf{R}^{\tilde{T}}$ with the ring $\mathbf{R} = \mathbf{R}_X$, and $\mathbf{R}_t = \mathbf{R}^T$ with the subring of $\mathbf{H}$ generated by t and $\mathbf{R}$. Now we can prove the main result of this section.

2.5.6. Theorem. *There is a unique ring isomorphism $\Phi\colon \mathbf{H} \to \mathbf{K}^I(\mathfrak{N})$ such that $T_{s_\alpha} \mapsto t_{s_\alpha}$ and $X_\lambda \mapsto x_\lambda$ for each $\alpha \in \tilde{\Pi}$, $\lambda \in \tilde{\mathbf{X}}$. Under Φ and the forgetting map $\mathbf{R}^I = \mathbf{R}^T$, the canonical (left) $\mathbf{R}^I$-action on $\mathbf{K}^I(\mathfrak{N})$ is identified with the canonical (left) $\mathbf{R}_t$-action on $\mathbf{H}$.*

Proof. First we prove that the assignment

$$T_{s_\alpha} \mapsto t_{s_\alpha}, \quad X_\lambda \mapsto x_\lambda, \quad \forall \alpha \in \tilde{\Pi}, \lambda \in \tilde{\mathbf{X}},$$

yields a $\mathbb{Z}_{q,t}$-algebra homomorphism

$$\Phi \colon \mathbf{H} \to \mathbf{K}^I(\mathfrak{N}).$$

We must check that the elements t_{s_α}, x_λ satisfy the defining relations of $\mathbf{H}$. To do so let $S = \Sigma = T$ and consider the group homomorphism

$$\mathbf{r}_S \colon \mathbf{K}^I(\mathfrak{N}) \to \mathbf{K}^S(\mathfrak{N}_{\bar{x}}^S)_S.$$

Note that $\mathfrak{N}_{\bar{x}}^S = \mathfrak{D}^S$, because $S = T$. Thus we have an $\mathbf{R}^S$-module isomorphism

$$\mathbf{K}^S(\mathfrak{N}_{\bar{x}}^S) = \lim_w \mathrm{colim}_v \mathbf{K}^S\big((\mathfrak{D}_v)_{(w)}^S\big)$$

$$= \lim_w \mathrm{colim}_u \mathbf{K}^S(\mathfrak{D}_{w,u}^S)$$

(2.5.2)

$$= \prod_w \bigoplus_u \mathbf{R}^S \mathbf{x}_{w,u}.$$

Here the symbol $\mathbf{x}_{w,u}$ stands for the fundamental class of the fixed point $(\mathfrak{b}_w, \mathfrak{b}_u)$, see 2.2.4 (b). The convolution product is $\mathbf{R}^S$-linear and is given by

$$\mathbf{x}_{v,w} \star \mathbf{x}_{y,z} = \begin{cases} \mathbf{x}_{v,z} & \text{if } w = y, \\ 0 & \text{else.} \end{cases}$$

Let λ, μ be as in (2.5.1). Under the isomorphism above we have

$$\mathcal{O}_{\mathfrak{N}'_{s_\alpha}}(\lambda, \mu)_{\bar{x}}|_{\mathfrak{D}^S} = \mathcal{O}_{\mathfrak{D}_{s_\alpha}^S}(\lambda, \mu)$$

$$= \sum_w^\infty (\theta_{w\lambda + w\mu} \mathbf{x}_{w,w} + \theta_{w\lambda + ws_\alpha\mu} \mathbf{x}_{w,ws_\alpha})$$

$$= \sum_w^\infty (\theta_{-w\alpha} \mathbf{x}_{w,w} + \mathbf{x}_{w,ws_\alpha}).$$

Here the symbol $\sum^\infty$ denotes an infinite sum. Thus 2.4.7 yields

$$\mathbf{r}_S(1 + t_{s_\alpha}) = -\mathbf{r}_S(\mathcal{O}_{\mathfrak{N}'_{s_\alpha}}(\lambda, \mu))$$

$$= -(1 - \mathcal{O}_{\mathfrak{D}_{s_\alpha}^S}(\alpha + t, 0))(1 - \mathcal{O}_{\mathfrak{D}_{s_\alpha}^S}(0, -\alpha))^{-1}\mathcal{O}_{\mathfrak{D}_{s_\alpha}^S}(\lambda, \mu)$$

(2.5.3)

$$= \sum_w^\infty \frac{1 - t\theta_{w\alpha}}{1 - \theta_{w\alpha}}(\mathbf{x}_{w,w} - \mathbf{x}_{w,ws_\alpha}).$$

By 2.4.4 we have also

(2.5.4)
$$\mathbf{r}_S(x_\lambda) = \mathbf{r}_S(\mathcal{O}_{\mathfrak{N}_e}\langle\lambda\rangle) = \sum_w^\infty \theta_{w\lambda} \mathbf{x}_{w,w}.$$

Using (2.5.3) and (2.5.4) the relations are reduced to a simple linear algebra computation which is left to the reader.

Next we prove that Φ is surjective. First note that $\mathbf{R}_t = \mathbf{R}^T = \mathbf{R}^I$. We have

$$\mathbf{K}^I(\mathfrak{N}) = \mathrm{colim}_w \mathbf{K}^I(\mathfrak{N}_w), \quad \mathfrak{N}_w = \bigsqcup_{v \leqslant w} \overset{\circ}{\mathfrak{N}}_v, \quad \overset{\circ}{\mathfrak{N}}_v = \mathfrak{N} \cap (\mathfrak{n} \times \overset{\circ}{\mathfrak{F}}_v).$$

Further, we have **T**-scheme isomorphisms

$$\overset{\circ}{\mathfrak{F}}_v = \mathfrak{n}_v^{\circ}, \quad \overset{\circ}{\mathfrak{N}}_v = \mathfrak{n}_v \times \overset{\circ}{\mathfrak{F}}_v = \mathfrak{n}.$$

In particular $\overset{\circ}{\mathfrak{N}}_v$ is an affine space. Let $\mathfrak{N}'_v$ be the Zariski closure of $\overset{\circ}{\mathfrak{N}}_v$ in $\mathfrak{N}$ and let $\mathbf{g}_v = \mathcal{O}_{\mathfrak{N}'_v}$, regarded as an element of $\mathbf{K}^I(\mathfrak{N})$. The direct image by the inclusion $\mathfrak{N}_w \subset \mathfrak{N}$ identifies the $\mathbf{R}_t$-module $\mathbf{K}^I(\mathfrak{N}_w)$ with the direct summand

$$\bigoplus_{v \leqslant w} \mathbf{R}_t \, \mathbf{g}_v \subset \mathbf{K}^I(\mathfrak{N}).$$

See [7, Section 7.6] for a similar argument for non-affine flags. On the other hand the PBW theorem for $\mathbf{H}$ implies that $\mathbf{H} = \bigoplus_{w \in \tilde{W}} \mathbf{R}_t T_w$ as a left $\mathbf{R}_t$-module. Set $\mathbf{H}_w = \bigoplus_{v \leqslant w} \mathbf{R}_t T_v$. We must prove that Φ restricts to a surjective $\mathbf{R}_t$-module homomorphism $\mathbf{H}_w \to \mathbf{K}^I(\mathfrak{N}_w)$ for each w. This is proved by induction on the length $l(w)$ of w. More precisely this is obvious if $l(w) = 1$ and we know that

$$l(vw) = l(v) + l(w) \;\Rightarrow\; \mathbf{H}_v \, \mathbf{H}_w = \mathbf{H}_{vw}, \;\; \mathbf{K}^I(\mathfrak{N}_v) \star \mathbf{K}^I(\mathfrak{N}_w) \subset \mathbf{K}^I(\mathfrak{N}_{vw}).$$

Therefore we are reduced to prove that under the previous assumption we have

$$\mathbf{g}_v \star \mathbf{g}_w \equiv a_{v,w} \, \mathbf{g}_{vw},$$

with $a_{v,w}$ a unit of $\mathbf{R}_t$. Here the symbol $\equiv$ means an equality modulo lower terms for the Bruhat order. To do that we fix $S = \mathbf{T}$ and we consider the image of $\mathbf{g}_w$ by $\mathbf{r}_\Sigma$. It is, of course, too complicated to compute the whole expression, but we only need the terms $g_{y,yz}^{(z)}$ with $l(yz) = l(y) + l(z)$ in the sum below

$$\mathbf{r}_\Sigma(\mathbf{g}_x) = \sum_{y,z}^{\infty} g_{y,z}^{(x)} \, \mathbf{x}_{y,z},$$

because the coefficient a above is given by the following relation

$$g_{v,vw}^{(w)} \, g_{e,v}^{(v)} = a \, g_{e,vw}^{(vw)}.$$

The same computation as in 2.4.6 shows that

$$g_{y,yz}^{(z)} = \prod_{\alpha \in \tilde{\Delta}_z^{\circ}} \frac{1 - t\theta_{y\alpha}}{1 - \theta_{-y\alpha}}.$$

Now, recall that

$$l(vw) = l(v) + l(w) \;\Rightarrow\; \tilde{\Delta}_{vw}^{\circ} = \tilde{\Delta}_v^{\circ} \sqcup v(\tilde{\Delta}_w^{\circ}).$$

Thus we have $a_{v,w} = 1$.

Finally, since Φ restricts to a surjective $\mathbf{R}_t$-module homomorphism $\mathbf{H}_w \to \mathbf{K}^I(\mathfrak{N}_w)$ for each w and both sides are free $\mathbf{R}_t$-modules of rank $l(w)$ necessarily Φ is injective.

The last claim of the theorem follows from 2.3.9 (c). $\square$

2.5.7. *Remark.* By 2.3.9 the convolution product

$$\star\colon \mathbf{K}^I(\mathfrak{N}) \otimes \mathbf{K}^I(\mathfrak{N}) \to \mathbf{K}^I(\mathfrak{N})$$

is $\mathbf{R}^I$-linear in the first variable. Recall that forgetting the group action yields an isomorphism $\mathbf{K}^I(\mathfrak{N}) \to \mathbf{K}^T(\mathfrak{N})$. Further, since $\mathfrak{N}$ has a partition into affine cells a standard argument implies that the forgetting map gives an isomorphism

$$\mathbf{R}^S \otimes_{\mathbf{R}^T} \mathbf{K}^T(\mathfrak{N}) = \mathbf{K}^S(\mathfrak{N})$$

for each closed subgroup $S \subset T$. Thus the map $\star$ factors to a group homomorphism

$$\star\colon \mathbf{K}^S(\mathfrak{N}) \otimes \mathbf{K}^I(\mathfrak{N}) \to \mathbf{K}^S(\mathfrak{N}).$$

The assignment $\theta_\lambda \mapsto X_\lambda$ identifies $\mathbf{R}^T$ with the subring $\mathbf{R}_t \subset \mathbf{H}$ generated by t and the X_λ's. By 2.5.6 the group homomorphism above is identified, via the map Φ, with the right multiplication of $\mathbf{H}$ on $\mathbf{R}^S \otimes_{\mathbf{R}_t} \mathbf{H}$.

3. Classification of the simple admissible modules of the double affine Hecke algebra

3.1. Constructible sheaves and convolution algebras

The purpose of this section is to revisit the sheaf-theoretic analysis of convolution algebras in [7, Section 8.6] in a more general setting including the case of schemes locally of finite type. It is an expanded version of [33, Section 6, Appendix B].

3.1.1. Convolution algebras and schemes which are locally of finite type.
Let $N = \bigsqcup_{\alpha \in A} N(\alpha)$ be a disjoint union of smooth quasi-projective connected schemes. We'll assume that the set A is countable. We'll view N as an ind-scheme, by setting

$$N = \operatorname*{colim}_{B \subset A} N(B), \quad N(B) = \bigsqcup_{\alpha \in B} N(\alpha),$$

where B is any finite subset of A. Let C be a quasi-projective scheme (possibly singular) and $\pi\colon N \to C$ be an ind-proper map. For each $\alpha, \beta \in A$ we set

$$M(\alpha, \beta) = N(\alpha) \times_C N(\beta)$$

(the reduced fiber product). It is a closed subscheme of $N(\alpha) \times N(\beta)$. Note that $N(\alpha)$, $M(\alpha, \beta)$ are complex varieties which can be equipped with their transcendental topology. The symbol $\mathbf{H}_*(., \mathbb{C})$ will denote the Borel–Moore homology with complex coefficients. We'll view $M = \bigsqcup_{\alpha, \beta} M(\alpha, \beta)$ as an ind-scheme in the obvious way. We set

$$\mathbf{H}_*(M, \mathbb{C}) = \bigoplus_{\alpha, \beta} \mathbf{H}_*(M(\alpha, \beta), \mathbb{C}), \quad \widehat{\mathbf{H}}_*(M, \mathbb{C}) = \prod_\alpha \bigoplus_\beta \mathbf{H}_*(M(\alpha, \beta), \mathbb{C}).$$

We'll view $\widehat{\mathbf{H}}_*(M, \mathbb{C})$ as a topological $\mathbb{C}$-vector space in the following way

- $\bigoplus_\beta \mathbf{H}_*(M(\alpha, \beta), \mathbb{C})$ is given the discrete topology for each α,
- $\widehat{\mathbf{H}}_*(M, \mathbb{C})$ is given the product topology.

We also equip $\mathbf{H}_*(M, \mathbb{C})$ with a convolution product $\star$ as in [7, Section 8]. The following is immediate.

3.1.2. Lemma. *The multiplication on $\mathbf{H}_*(M, \mathbb{C})$ is bicontinuous and yields the structure of a topological ring on $\widehat{\mathbf{H}}_*(M, \mathbb{C})$.*

3.1.3. *Remark.* We may also consider the K-theory rather than the Borel–Moore homology. Since M, N are ind-schemes of ind-finite type we have

$$\mathbf{K}(N) = \bigoplus_{\alpha} \mathbf{K}(N(\alpha)), \quad \mathbf{K}(M) = \bigoplus_{\alpha, \beta} \mathbf{K}(M(\alpha, \beta)).$$

We'll also set

$$\widehat{\mathbf{K}}(M) = \prod_{\alpha} \bigoplus_{\beta} \mathbf{K}(M(\alpha, \beta)).$$

Thus $\widehat{\mathbf{K}}(M)$ is again a topological ring. The multiplication in $\mathbf{K}(M)$, $\widehat{\mathbf{K}}(M)$ is the convolution product associated with the inclusion $M \subset N^2$. It is defined as in (2.4.8). By [7, Theorem 5.11.11] the bivariant Riemann–Roch map yields a topological ring homomorphism

$$RR\colon \mathbb{C}\widehat{\mathbf{K}}(M) \to \widehat{\mathbf{H}}_*(M, \mathbb{C})$$

which maps $\mathbb{C}\,\mathbf{K}(M(\alpha, \beta))$ to $\mathbf{H}_*(M(\alpha, \beta), \mathbb{C})$ for each α, β. It is invertible if all $\mathbf{H}_*(M(\alpha, \beta), \mathbb{C})$'s are spanned by algebraic cycles.

3.1.4. Admissible modules over the convolution algebra. Let $\mathcal{D}(C)^b_{\mathbb{C}\text{-c}}$ be the derived category of bounded complexes of constructible sheaves of $\mathbb{C}$-vector spaces over the quasi-projective scheme C. Given two complexes $\mathcal{L}$, $\mathcal{L}'$ in $\mathcal{D}(C)^b_{\mathbb{C}\text{-c}}$ we'll abbreviate

$$\mathrm{Ext}^n(\mathcal{L}, \mathcal{L}') = \mathrm{Hom}(\mathcal{L}, \mathcal{L}'[n]), \quad \mathrm{Ext}(\mathcal{L}, \mathcal{L}') = \bigoplus_{n \in \mathbb{Z}} \mathrm{Ext}^n(\mathcal{L}, \mathcal{L}'),$$

where the homomorphisms are computed in the category $\mathcal{D}(C)^b_{\mathbb{C}\text{-c}}$. Now, we set

$$\mathcal{C}_\alpha = \mathbb{C}_{N(\alpha)}[\dim(N(\alpha))], \quad \mathcal{L}_\alpha = \pi_*(\mathcal{C}_\alpha), \quad \forall \alpha \in A.$$

Each $\mathcal{L}_\alpha$ is a semi-simple complex by the decomposition theorem. Assume that there is a finite set $\mathcal{X}$ of irreducible perverse sheaves over C such that

$$\mathcal{L}_\alpha \simeq \bigoplus_{n \in \mathbb{Z}} \bigoplus_{\mathcal{S} \in \mathcal{X}} L_{\mathcal{S}, \alpha, n} \otimes \mathcal{S}[n],$$

where $L_{\mathcal{S}, \alpha, n}$ are finite-dimensional $\mathbb{C}$-vector spaces. We set

$$L_{\mathcal{S}, \alpha} = \bigoplus_{n \in \mathbb{Z}} L_{\mathcal{S}, \alpha, n}, \quad L_{\mathcal{S}} = \bigoplus_{\alpha \in A} L_{\mathcal{S}, \alpha}, \quad L = \bigoplus_{\mathcal{S} \in \mathcal{X}} L_{\mathcal{S}}.$$

For each complexes $\mathcal{L}, \mathcal{L}', \mathcal{L}''$ the Yoneda product is a bilinear map

$$\mathrm{Ext}(\mathcal{L}, \mathcal{L}') \times \mathrm{Ext}(\mathcal{L}', \mathcal{L}'') \to \mathrm{Ext}(\mathcal{L}, \mathcal{L}'').$$

By [7, Lemma 8.6.1, 8.9.1] we have an algebra isomorphism

$$\widehat{\mathbf{H}}_*(M, \mathbb{C}) = \prod_\alpha \bigoplus_\beta \mathrm{Ext}(\mathcal{L}_\alpha, \mathcal{L}_\beta),$$

where the right-hand side is given the Yoneda product. We have the following decomposition as $\mathbb{C}$-vector spaces $\widehat{\mathbf{H}}_*(M, \mathbb{C}) \simeq R \oplus J$, where

$$R = \bigoplus_{\mathcal{S} \in \mathcal{X}} \mathrm{End}(L_\mathcal{S}), \quad J = \bigoplus_{\mathcal{S}, \mathcal{T} \in \mathcal{X}} \bigoplus_{n > 0} \mathrm{Hom}(L_\mathcal{T}, L_\mathcal{S}) \otimes \mathrm{Ext}^n(\mathcal{T}, \mathcal{S}).$$

Further J is a nilpotent two-sided ideal of $\widehat{\mathbf{H}}_*(M, \mathbb{C})$ and the $\mathbb{C}$-algebra structure on $\widehat{\mathbf{H}}_*(M, \mathbb{C})/J$ is the obvious $\mathbb{C}$-algebra structure on R. Before we explain what is the topology on R recall the following basic fact.

3.1.5. Definitions.

(a) Let $\mathbf{A}$ be any ring and let M, N be $\mathbf{A}$-modules. The *finite topology* on $\mathrm{Hom}_\mathbf{A}(M, N)$ is the linear topology for which a basis of open neighborhoods for 0 is given by the annihilator of M', for all finite sets $M' \subset M$. This is actually the topology induced on $\mathrm{Hom}_\mathbf{A}(M, N)$ from N^M (a product of topological spaces where N has the discrete topology).

(b) If $\mathbf{A}$ is a topological ring we'll say that a right $\mathbf{A}$-module is *admissible* (or *smooth*) if for each element m the subset $\{\, x \in \mathbf{A}; mx = 0 \,\}$ is open.

Now we can formulate the following lemma.

3.1.6. Lemma. *The two-sided ideal $J \subset \widehat{\mathbf{H}}_*(M, \mathbb{C})$ is closed. The quotient topology on $\widehat{\mathbf{H}}_*(M, \mathbb{C})/J$ coincides with the finite topology on R.*

Therefore, the Jacobson density theorem implies that the set of simple admissible right representations of R is $\{\, L_\mathcal{S}; \mathcal{S} \in \mathcal{X} \,\}$, see, e.g., [33, Section B]. This yields the following.

3.1.7. Proposition. *The set of the simple admissible right $\widehat{\mathbf{H}}_*(M, \mathbb{C})$-modules is canonically identified with the set $\{\, L_\mathcal{S}; \mathcal{S} \in \mathcal{X} \,\}$.*

3.2. Simple modules in the category $\mathcal{O}$

This section reviews the classification of the simple modules in $\mathcal{O}(\mathbf{H})$ from [33]. The main arguments are the same as in loc. cit., but the use of the concentration map simplifies the exposition. Note that $\mathcal{O}(\mathbf{H})$ consists of *right* $\mathbb{C}\mathbf{H}$-modules. This specification is indeed irrelevant because the $\mathbb{Z}_{q,t}$-algebra $\mathbf{H}$ is isomorphic to its opposite algebra, see, e.g., [6, Theorem 1.4.4].

3.2.1. From $\mathcal{O}(\mathbf{H})$ to modules over the convolution algebra of $\mathfrak{M}$. In this section we apply the construction from Section 3.1 in the following setting. Fix a regular closed subgroup $S \subset T$. Following [22] we define the set of the *topologically nilpotent elements* in $\tilde{\mathfrak{g}}$ by

$$\mathfrak{Nil} = \bigcup_{\mathfrak{b} \in \mathfrak{F}} \mathfrak{b}_{\mathrm{nil}}.$$

Let $N = \dot{\mathfrak{n}}^S$, $C = \mathfrak{Nil}^S$, and let $\pi\colon N \to C$ be the obvious projection. The ind-scheme M in 3.1.1 is given by $M = \mathfrak{M}^S$. It is an ind-scheme of ind-finite type. We'll use the notation from 2.4.8. Recall that

$$\mathbf{K}(\mathfrak{M}^S) = \bigoplus_{\alpha,\beta} \mathbf{K}(\mathfrak{M}(\alpha,\beta)), \quad \widehat{\mathbf{K}}(\mathfrak{M}^S) = \prod_\alpha \bigoplus_\beta \mathbf{K}(\mathfrak{M}(\alpha,\beta)).$$

Now we fix an element $(h,\zeta) = (s,\tau,\zeta)$ in $\boldsymbol{T}$, i.e., we have $h = (s,\tau) \in \tilde{\boldsymbol{T}}$, $s \in T \times \mathbb{C}^\times_{\mathrm{cen}}$, $\tau \in \mathbb{C}^\times_{\mathrm{rot}}$ and $\zeta \in \mathbb{C}^\times_{\mathrm{qua}}$. Assume that $S = \langle (h,\zeta) \rangle$, i.e., we assume that S is the closed subgroup of $\boldsymbol{T}$ generated by the element (h,ζ). Let $\tilde{G}^h$ be the centralizer of the element h in the group $\tilde{G}$.

3.2.2. Definition. We'll say that the pair (τ,ζ) is *regular* if τ is not a root of 1 and $\tau^k \neq \zeta^m$ for each $m, k > 0$.

For each set X with a $\boldsymbol{T}$-action we'll abbreviate $X^{h,\zeta}$ for the fixed points subset $X^{(h,\zeta)}$. We have the following [33, Lemma 2.13], [32, Lemma 2.4.1-2].

3.2.3. Proposition. *Assume that the pair (τ,ζ) is regular. The group $\langle (h,\zeta) \rangle$ is regular. The group $\tilde{G}^h$ is reductive and connected. The scheme $\mathfrak{Nil}^{h,\zeta}$ is of finite type and it consists of nilpotent elements of $\tilde{\mathfrak{g}}$. Further $\mathfrak{Nil}^{h,\zeta}$ contains only a finite number of $\tilde{G}^h$-orbits.*

This proposition is essentially straightforward, except for the connexity of the reductive group $\tilde{G}^h$. This is an affine analogue of a well-known result of Steinberg which says that the centralizer of a semi-simple element in a connected reductive group with simply connected derived subgroup is again connected. The proof of the connexity relies on a theorem of Kac and Peterson [16] which says that a reductive subgroup of $\tilde{G}$ is always conjugated to a subgroup of a proper Levi subgroup of $\tilde{G}$. Since the proper Levi subgroups of $\tilde{G}$ are reductive with simply connected derived subgroup, because $\tilde{G}$ is the maximal affine Kac–Moody group, the claim is reduced to the Steinberg theorem.

Therefore, if (τ,ζ) is regular then the scheme $\mathfrak{Nil}^{h,\zeta}$ is of finite type, the scheme $\mathfrak{M}^{h,\zeta}$ is locally of finite type, the homology group $\widehat{\mathbf{H}}_*(\mathfrak{M}^{h,\zeta}, \mathbb{C})$ is a topological ring by 3.1.1, and the simple admissible right $\widehat{\mathbf{H}}_*(\mathfrak{M}^{h,\zeta}, \mathbb{C})$-modules are labeled by the set of irreducible perverse sheaves over $\mathfrak{Nil}^{h,\zeta}$ which occur as a shift of a direct summand of the complex $\pi_*(\mathbb{C}_{\dot{\mathfrak{n}}^{h,\zeta}})$.

Set $\Sigma = \{(h,\zeta)\}$ and $S = \langle (h,\zeta) \rangle$. We'll abbreviate

$$\mathbf{r}_{h,\zeta} = \mathbf{r}_\Sigma, \quad \mathbf{R}_{h,\zeta} = \mathbf{R}^S_\Sigma.$$

Composing Φ, $\mathbf{r}_{h,\zeta}$ and the tensor product by the character

$$\chi_{h,\zeta}\colon \mathbf{R}_{h,\zeta} \to \mathbb{C}, \quad f \mapsto f(h,\zeta),$$

we get a $\mathbb{C}$-algebra homomorphism

$$\Phi_{h,\zeta}\colon \mathbb{C}\mathbf{H} \to \mathbb{C}\widehat{\mathbf{K}}(\mathfrak{M}^{h,\zeta}).$$

Note that $\widehat{\mathbf{K}}(\mathfrak{M}^{h,\varsigma})$ is a topological ring by 3.1.3 and that the bivariant Riemann–Roch map yields a topological ring homomorphism

$$RR \colon \mathbb{C}\widehat{\mathbf{K}}(\mathfrak{M}^{h,\varsigma}) \to \widehat{\mathbf{H}}_*(\mathfrak{M}^{h,\varsigma}, \mathbb{C}).$$

We'll write

$$\Psi_{h,\varsigma} = RR \circ \Phi_{h,\varsigma} \colon \mathbb{C}\mathbf{H} \to \widehat{\mathbf{H}}_*(\mathfrak{M}^{h,\varsigma}, \mathbb{C}).$$

Throughout we'll use the following notation: for any ring homomorphism

$$\phi \colon \mathbf{A} \to \mathbf{B}$$

and for any (left or right) $\mathbf{B}$-module M let $\phi^\bullet(M)$ be the corresponding $\mathbf{A}$-module.

3.2.4. Proposition. *Assume that the pair (τ, ς) is regular.*

(a) *The map $\Phi_{h,\varsigma} \colon \mathbb{C}\mathbf{H} \to \mathbb{C}\widehat{\mathbf{K}}(\mathfrak{M}^{h,\varsigma})$ has a dense image.*

(b) *The map $RR \colon \mathbb{C}\widehat{\mathbf{K}}(\mathfrak{M}^{h,\varsigma}) \to \widehat{\mathbf{H}}_*(\mathfrak{M}^{h,\varsigma}, \mathbb{C})$ is an isomorphism.*

(c) *The pull-back by the composed map $\Psi_{h,\varsigma} = RR \circ \Phi_{h,\varsigma}$ gives a bijection between the set of simple right $\mathbb{C}\mathbf{H}$-modules in $\mathcal{O}_{h,\varsigma}(\mathbf{H})$ and the set of simple admissible right $\widehat{\mathbf{H}}_*(\mathfrak{M}^{h,\varsigma}, \mathbb{C})$-modules.*

The proof of 3.2.4 is given in 3.2.7 below. Before this we need more material.

3.2.5. The regular representation of H. First we define a right representation of $\widehat{\mathbf{K}}(\mathfrak{M}^{h,\varsigma})$ on $\mathbf{K}(\mathfrak{N}^{h,\varsigma})$. We'll use the same notation as in the previous subsection. In particular $S = \langle\!\langle (h, \zeta) \rangle\!\rangle$ is a regular closed subgroup of T. Recall that $\dot{\mathfrak{n}}^{h,\varsigma}$ and $\mathfrak{M}^{h,\varsigma}$ are both ind-schemes of ind-finite type, that $\dot{\mathfrak{n}}^{h,\varsigma}$ is a disjoint union of smooth quasi-projective varieties, and that $\mathfrak{M}^{h,\varsigma}$ is regarded as a closed subset of $(\dot{\mathfrak{n}}^{h,\varsigma})^2$. The convolution product on $\mathbf{K}(\mathfrak{M}^{h,\varsigma})$ is given by

$$x \star y = R(q_2)_* \big(q_3^*(x) \overset{L}{\otimes}_{(\dot{\mathfrak{n}}^{h,\varsigma})^3} q_1^*(y)\big), \quad \forall x, y \in \mathbf{K}(\mathfrak{M}^{h,\varsigma}),$$

where $q_a \colon (\dot{\mathfrak{n}}^{h,\varsigma})^3 \to (\dot{\mathfrak{n}}^{h,\varsigma})^2$ is the projection along the ath factor for $a = 1, 2, 3$. The inclusion $\mathfrak{N} \subset \dot{\mathfrak{n}}$ yields an inclusion of ind-schemes $\mathfrak{M}^{h,\varsigma} \subset \dot{\mathfrak{n}}^{h,\varsigma}$. For each $x \in \mathbf{K}(\mathfrak{M}^{h,\varsigma})$ and each $y \in \mathbf{K}(\mathfrak{N}^{h,\varsigma})$ we define the following element in $\mathbf{K}(\mathfrak{N}^{h,\varsigma})$

(3.2.1) $$x \star y = R(p_1)_* \big(p_2^*(x) \overset{L}{\otimes}_{(\dot{\mathfrak{n}}^{h,\varsigma})^2} y\big),$$

where $p_a \colon (\dot{\mathfrak{n}}^{h,\varsigma})^2 \to \dot{\mathfrak{n}}^{h,\varsigma}$ is the projection along the ath factor for $a = 1, 2$. It is well known that the map (3.2.1) defines a right representation of $\mathbf{K}(\mathfrak{M}^{h,\varsigma})$ on $\mathbf{K}(\mathfrak{N}^{h,\varsigma})$, see, e.g., [7].

3.2.6. Lemma.

(a) *The right representation of $\mathbf{K}(\mathfrak{M}^{h,\varsigma})$ on $\mathbf{K}(\mathfrak{N}^{h,\varsigma})$ extends in a unique way to an admissible right representation of $\widehat{\mathbf{K}}(\mathfrak{M}^{h,\varsigma})$ on $\mathbf{K}(\mathfrak{N}^{h,\varsigma})$.*

(b) *The right $\mathbb{C}\mathbf{H}$-module $\chi_{h,\varsigma} \otimes_{\mathbf{R}_t} \mathbf{H}$ belongs to $\mathcal{O}_{h,\varsigma}(\mathbf{H})$.*

(c) *There is an isomorphism $\chi_{h,\varsigma} \otimes_{\mathbf{R}_t} \mathbf{H} \simeq \Phi_{h,\varsigma}^\bullet(\mathbb{C}\mathbf{K}(\mathfrak{N}^{h,\varsigma}))$ of right $\mathbf{H}$-modules.*

Proof. The first claim is obvious, because we have

$$\widehat{\mathbf{K}}(\mathfrak{M}^{h,\varsigma}) = \prod_{\alpha} \bigoplus_{\beta} \mathbf{K}(\mathfrak{M}(\alpha,\beta)), \quad \mathbf{K}(\mathfrak{N}^{h,\varsigma}) = \bigoplus_{\alpha} \mathbf{K}(\mathfrak{N}(\alpha)),$$

$$\mathbf{K}(\mathfrak{N}(\alpha)) \star \mathbf{K}(\mathfrak{M}(\alpha,\beta)) \subset \mathbf{K}(\mathfrak{N}(\beta)),$$

where $\mathfrak{N}(\alpha) = \mathfrak{N} \cap \dot{\mathfrak{n}}(\alpha)$. Part (b) is a standard computation, see, e.g., [33]. Let us concentrate on part (c). Composing the map $\chi_{h,\varsigma} \colon \mathbf{R}_{h,\varsigma} \to \mathbb{C}$ with the canonical map $\mathbf{R}^T \to \mathbf{R}_{h,\varsigma}$ we may regard $\chi_{h,\varsigma}$ as the one-dimensional $\mathbf{R}^T$-module given by $f \mapsto f(h,\varsigma)$. Recall that $\mathbf{R}_t = \mathbf{R}^T$, see 2.5.7. The vector space $\chi_{h,\varsigma} \otimes_{\mathbf{R}_t} \mathbf{H}$ has an obvious structure of right $\mathbf{H}$-module. The isomorphism 2.5.6 factors to a right $\mathbf{H}$-module isomorphism

$$\chi_{h,\varsigma} \otimes_{\mathbf{R}_t} \mathbf{H} \to \chi_{h,\varsigma} \otimes_{\mathbf{R}^T} \mathbf{K}^T(\mathfrak{N}).$$

We claim that there is a right $\mathbf{H}$-module isomorphism

$$\chi_{h,\varsigma} \otimes_{\mathbf{R}^T} \mathbf{K}^T(\mathfrak{N}) \to \Phi^{\bullet}_{h,\varsigma}(\mathbb{C}\,\mathbf{K}(\mathfrak{M}^{h,\varsigma})).$$

To prove this, recall that composing the maps $\mathbf{r}_{h,\varsigma}$ and $\chi_{h,\varsigma}$ yields an algebra homomorphism

$$\mathbf{K}^T(\mathfrak{N}) \to \mathbb{C}\,\mathbf{K}(\mathfrak{N}^{h,\varsigma}_{\mathfrak{x}}) = \mathbb{C}\,\mathbf{K}(\mathfrak{M}^{h,\varsigma}).$$

Thus we must construct a map $\mathbf{r} \colon \mathbf{K}^T(\mathfrak{N}) \to \mathbb{C}\,\mathbf{K}(\mathfrak{M}^{h,\varsigma})$ which intertwines the right $\star$-product of $\mathbf{K}^T(\mathfrak{N})$ on itself, see 2.3.7, with the right $\star$-product of $\mathbb{C}\,\mathbf{K}(\mathfrak{M}^{h,\varsigma})$ on $\mathbb{C}\,\mathbf{K}(\mathfrak{M}^{h,\varsigma})$, see (3.2.1), relatively to the ring homomorphism

$$\chi_{h,\varsigma} \circ \mathbf{r}_{h,\varsigma} \colon \mathbf{K}^T(\mathfrak{N}) \to \mathbb{C}\,\mathbf{K}(\mathfrak{M}^{h,\varsigma}).$$

Further the map $\mathbf{r}$ should factor to an isomorphism

$$\chi_{h,\varsigma} \otimes_{\mathbf{R}^T} \mathbf{K}^T(\mathfrak{N}) \to \mathbb{C}\,\mathbf{K}(\mathfrak{M}^{h,\varsigma}).$$

Consider the following chain of inclusions

$$(3.2.2) \qquad \mathfrak{n}^{h,\varsigma} \times \mathfrak{F}^{h,\varsigma} \xrightarrow{i} \mathfrak{n}^{h,\varsigma} \times \mathfrak{F} \xrightarrow{j} \mathfrak{n} \times \mathfrak{F}.$$

Since $\mathfrak{n}$ is pro-smooth we can consider the map Lj^* in K-theory, see 1.5.18. Since $\mathfrak{F}$ is an ind-S-scheme of ind-finite type we can consider the map i_* in K-theory, see 1.5.19. Both maps are invertible, and the composed map is an isomorphism

$$(i_*)^{-1} \circ Lj^* \colon \mathbf{K}^S(\mathfrak{n} \times \mathfrak{F})_{\Sigma} \to \mathbf{K}^S(\mathfrak{n}^{h,\varsigma} \times \mathfrak{F}^{h,\varsigma})_{\Sigma}, \quad \Sigma = \{(h,\varsigma)\}.$$

Now, recall that we have a good embedding $\mathfrak{N} \subset \mathfrak{n} \times \mathfrak{F}$. Thus, we obtain also in this way an isomorphism $\mathbf{K}^S(\mathfrak{N})_{\Sigma} \to \mathbb{C}\,\mathbf{K}(\mathfrak{M}^{h,\varsigma})$. Composing it with the obvious map $\mathbf{K}^T(\mathfrak{N}) \to \mathbf{K}^S(\mathfrak{N})_{\Sigma}$ it yields a map

$$\mathbf{r} \colon \mathbf{K}^T(\mathfrak{N}) \to \mathbb{C}\,\mathbf{K}(\mathfrak{M}^{h,\varsigma}).$$

We must check that the map $\mathbf{r}$ is compatible with the right $\star$-product, in the above sense. This is left to the reader. The proof is the same as the proof of 2.4.9. Compare (2.3.3), (2.4.3) with (3.2.1), (3.2.2). $\qquad\square$

3.2.7. Proof of Proposition 3.2.4.

(a) The map $\Phi\colon \mathbf{H} \to \mathbf{K}^I(\mathfrak{N})$ is invertible by 2.5.6. The composed map

$$\mathbf{K}^I(\mathfrak{N}) \xrightarrow{r_{h,\varsigma}} \mathbf{K}^S(\mathfrak{M}^{h,\varsigma})_{h,\varsigma} = \mathbf{R}_{h,\varsigma} \otimes \mathbf{K}(\mathfrak{M}^{h,\varsigma}) \xrightarrow{\chi_{h,\varsigma}} \mathbb{C}\mathbf{K}(\mathfrak{M}^{h,\varsigma}) \subset \mathbb{C}\widehat{\mathbf{K}}(\mathfrak{M}^{h,\varsigma})$$

has a dense image, because the image contains $\mathbb{C}\mathbf{K}(\mathfrak{M}(\alpha,\beta))$ for each α, β. Composing both maps we get $\Phi_{h,\varsigma}$. Thus $\Phi_{h,\varsigma}$ has a dense image.

(b) It is easy to see that $\mathbf{H}_*(\mathfrak{M}(\alpha,\beta),\mathbb{C})$ is spanned by algebraic cycles for all α, β. Therefore we have $\mathbb{C}\widehat{\mathbf{K}}(\mathfrak{M}^{h,\varsigma}) \simeq \widehat{\mathbf{H}}_*(\mathfrak{M},\mathbb{C})$.

(c) By part (b) it is enough to check that the map $\Phi_{h,\varsigma}^\bullet$ yields a bijection between the set of simple objects in $\mathcal{O}_{h,\varsigma}(\mathbf{H})$ and the set of simple admissible right representations of the topological $\mathbb{C}$-algebra $\mathbb{C}\widehat{\mathbf{K}}(\mathfrak{M}^{h,\varsigma})$. Our proof uses the following lemma, which will be checked later on.

3.2.8. Lemma.

(a) For each $\lambda \in \widetilde{\mathbf{X}}$ the operator of right multiplication by $\Phi_{h,\varsigma}(X_\lambda)$ in any admissible right $\mathbb{C}\widehat{\mathbf{K}}(\mathfrak{M}^{h,\varsigma})$-module is locally finite and its spectrum belongs to the set $\{\,{}^w\lambda(h); w \in \widetilde{W}\,\}$.

(b) If the elements $h, h' \in \widetilde{T}$ are $\widetilde{W}$-conjugate then the topological rings $\mathbb{C}\widehat{\mathbf{K}}(\mathfrak{M}^{h,\varsigma})$, $\mathbb{C}\widehat{\mathbf{K}}(\mathfrak{M}^{h',\varsigma})$ and the homomorphisms $\Phi_{h,\varsigma}$, $\Phi_{h',\varsigma}$ are canonically identified.

Claim 3.2.4 (c) is a corollary of 3.2.6 and 3.2.8. First, let M be a simple admissible right $\mathbb{C}\widehat{\mathbf{K}}(\mathfrak{M}^{h,\varsigma})$-module. The right $\mathbf{H}$-module $\Phi_{h,\varsigma}^\bullet(M)$ belongs to $\mathcal{O}_{h,\varsigma}(\mathbf{H})$ by 3.2.8 (a). Further $\Phi_{h,\varsigma}^\bullet(M)$ is a simple right $\mathbf{H}$-module. Indeed, since $\Phi_{h,\varsigma}(\mathbb{C}\mathbf{H})$ is dense in $\mathbb{C}\widehat{\mathbf{K}}(\mathfrak{M}^{h,\varsigma})$ by 3.2.4 (a) and since M is admissible and simple as a right $\mathbb{C}\widehat{\mathbf{K}}(\mathfrak{M}^{h,\varsigma})$-module, we have

$$x \star \Phi_{h,\varsigma}(\mathbb{C}\mathbf{H}) = x \star \mathbb{C}\widehat{\mathbf{K}}(\mathfrak{M}^{h,\varsigma}) = M, \quad \forall 0 \neq x \in M.$$

Thus M is a simple object of $\mathcal{O}_{h,\varsigma}(\mathbf{H})$.

Next, let L be a simple object of $\mathcal{O}_{h,\varsigma}(\mathbf{H})$. We claim that there is a simple admissible right $\mathbb{C}\widehat{\mathbf{K}}(\mathfrak{M}^{h,\varsigma})$-module M such that $L \simeq \Phi_{h,\varsigma}^\bullet(M)$. Indeed, since L belongs to $\mathcal{O}_{h,\varsigma}(\mathbf{H})$ there is an element $h' \in \widetilde{W} \cdot h$ such that the h'-weight subspace $L_{h'}$ is non-zero. Since L is simple, it is therefore a quotient of the right $\mathbb{C}\mathbf{H}$-module $\chi_{h',\varsigma} \otimes_{\mathbf{R}_t} \mathbf{H}$. The latter is isomorphic to $\Phi_{h',\varsigma}^\bullet(\mathbb{C}\mathbf{K}(\mathfrak{N}^{h',\varsigma}))$ by 3.2.6. Let J be the kernel of the quotient map $\Phi_{h',\varsigma}^\bullet(\mathbb{C}\mathbf{K}(\mathfrak{N}^{h',\varsigma})) \to L$. Hence, J is a right $\Phi_{h',\varsigma}(\mathbb{C}\mathbf{H})$-submodule of $\mathbb{C}\mathbf{K}(\mathfrak{N}^{h',\varsigma})$. Hence it is also a right $\mathbb{C}\widehat{\mathbf{K}}(\mathfrak{M}^{h',\varsigma})$-module because $\Phi_{h',\varsigma}(\mathbb{C}\mathbf{H}) \subset \mathbb{C}\widehat{\mathbf{K}}(\mathfrak{M}^{h',\varsigma})$ is dense and $\mathbb{C}\mathbf{K}(\mathfrak{N}^{h',\varsigma})$ is admissible. By 3.2.8 (b) we can regard $\mathbb{C}\mathbf{K}(\mathfrak{N}^{h',\varsigma})$ as a right $\mathbb{C}\widehat{\mathbf{K}}(\mathfrak{M}^{h,\varsigma})$-module and J as a right $\mathbb{C}\widehat{\mathbf{K}}(\mathfrak{M}^{h,\varsigma})$-submodule of $\mathbb{C}\mathbf{K}(\mathfrak{N}^{h',\varsigma})$. Then the quotient $\mathbb{C}\mathbf{K}(\mathfrak{N}^{h',\varsigma})/J$ is again a right $\mathbb{C}\widehat{\mathbf{K}}(\mathfrak{M}^{h,\varsigma})$-submodule and we have

$$L \simeq \Phi_{h,\varsigma}^\bullet(\mathbb{C}\mathbf{K}(\mathfrak{N}^{h',\varsigma})/J)$$

as right **H**-modules. Further, since L is a simple right $\mathbb{C}$**H**-module the quotient $\mathbb{C}\mathbf{K}(\mathfrak{M}^{h',\varsigma})/J$ is a simple admissible right $\mathbb{C}\widehat{\mathbf{K}}(\mathfrak{M}^{h,\varsigma})$-module.

Finally if M, M' are admissible right $\mathbb{C}\widehat{\mathbf{K}}(\mathfrak{M}^{h,\varsigma})$-modules such that $\Phi^\bullet_{h,\varsigma}(M)$, $\Phi^\bullet_{h,\varsigma}(M')$ are isomorphic as right **H**-modules then M, M' are isomorphic as right $\mathbb{C}\widehat{\mathbf{K}}(\mathfrak{M}^{h,\varsigma})$-modules, because they are isomorphic as right $\Phi_{h,\varsigma}(\mathbb{C}\mathbf{H})$-modules and $\Phi_{h,\varsigma}(\mathbb{C}\mathbf{H})$ is a dense subring of the topological ring $\mathbb{C}\widehat{\mathbf{K}}(\mathfrak{M}^{h,\varsigma})$. $\qquad\square$

Proof of Lemma 3.2.8. (a) For each $\lambda \in \tilde{\mathbf{X}}$ we have

$$\Phi_{h,\varsigma}(X_\lambda) = x_\lambda = \mathcal{O}_{\mathfrak{N}_e}\langle \lambda \rangle = \mathcal{O}_{\mathfrak{N}_e}(\lambda) \in \mathbf{K}^I(\mathfrak{N}).$$

Note that the set $\mathfrak{M}_e(\alpha, \beta) = \mathfrak{M}_e \cap \mathfrak{M}(\alpha, \beta)$ is empty if $\alpha \neq \beta$ and that it is the diagonal of $\dot{\mathfrak{n}}(\alpha)$ else. Recall that $S = \langle (h, \varsigma) \rangle$. For each α let $\lambda_\alpha \colon S \to \mathbb{C}^\times$ be the character of the group S such that any element $g \in S$ acts on the equivariant line bundle $\mathcal{O}_{\mathfrak{F}(\alpha)}(\lambda)$ by fiberwise multiplication by the scalar $\lambda_\alpha(g)$. It is well known that for each α there is an element $w \in \tilde{W}$ such that $\lambda_\alpha = ({}^w\lambda)|_S$. By 2.4.5 we have

$$\Phi_{h,\varsigma}(X_\lambda) = \sum_\alpha^\infty \lambda_\alpha(h)\, \mathcal{O}_{\mathfrak{M}_e(\alpha,\alpha)}(\lambda, 0) = \sum_\alpha^\infty \lambda_\alpha(h)\, \mathcal{O}_{\mathfrak{M}_e(\alpha,\alpha)}(0, \lambda) \in \mathbb{C}\widehat{\mathbf{K}}(\mathfrak{M}^{h,\varsigma}).$$

Thus the operator of multiplication by $\Phi_{h,\varsigma}(X_\lambda)$ in any admissible $\mathbb{C}\widehat{\mathbf{K}}(\mathfrak{M}^{h,\varsigma})$-module is locally finite and its spectrum belongs to the set $\{\,{}^w\lambda(h); w \in \tilde{W}\,\}$. See [33, Lemma 4.8] for details.

(b) Since h and h' are $\tilde{W}$-conjugate they are also $\tilde{G}$-conjugate. The group $\tilde{G}$ acts on $\mathfrak{M}$. This yields an ind-scheme isomorphism $\mathfrak{M}^{h,\varsigma} \simeq \mathfrak{M}^{h',\varsigma}$. The rest of the claim is obvious. $\qquad\square$

3.2.9. The classification theorem. We can now compose 3.1.7 with 3.2.4 (c). We get the following theorem [33, Theorem 7.6], [32, Proposition 2.5.1] whose proof uses the connexity of the reductive group $\tilde{G}^h$ in 3.2.3. To state the theorem we need more material. Assume that the pair (τ, ς) is regular. As above, we'll write $S = \langle (h, \varsigma) \rangle$. Let $\mathcal{X}_{h,\varsigma}$ be the set of irreducible perverse sheaves over $\mathfrak{Nil}^{h,\varsigma}$ which are direct summands (up to some shift) of the complex

$$\bigoplus_\alpha (\pi_{h,\varsigma})_* \mathbb{C}_{\dot{\mathfrak{n}}(\alpha)}, \qquad \pi_{h,\varsigma} \colon \dot{\mathfrak{n}}^{h,\varsigma} = \bigsqcup_\alpha \dot{\mathfrak{n}}(\alpha) \to \mathfrak{Nil}^{h,\varsigma}.$$

Here the map $\pi_{h,\varsigma}$ is the obvious projection. There is a finite number of $\tilde{G}^h$-orbits in $\mathfrak{Nil}^{h,\varsigma}$. For each closed point $x \in \mathfrak{Nil}^{h,\varsigma}$ let $A(h, \varsigma, x)$ be the group of connected components of the isotropy subgroup of x in $\tilde{G}^h$. The group $A(h, \varsigma, x)$ acts in an obvious way on the homology space

$$H_*(\pi_{h,\varsigma}^{-1}(x), \mathbb{C}) = \bigoplus_\alpha H_*(\pi_{h,\varsigma}^{-1}(x) \cap \dot{\mathfrak{n}}(\alpha), \mathbb{C}).$$

Let $\mathrm{Irr}(A(h,\zeta,x))$ be the set of irreducible representations of the finite group $A(h,\zeta,x)$. Each representation in $\mathrm{Irr}(A(h,\zeta,x))$ can be regarded as a $\tilde{G}^h$-equivariant irreducible local system over the $\tilde{G}^h$-orbit O of x. Therefore we may regard $\mathcal{X}_{h,\zeta}$ as a set of pairs (x,χ) in $\bigsqcup_x \mathrm{Irr}(A(h,\zeta,x))$.

3.2.10. Theorem. *Assume that (τ,ζ) is regular.*

(a) *The set $\{\,\Psi^\bullet_{h,\zeta}(L_\mathcal{S}); \mathcal{S} \in \mathcal{X}_{h,\zeta}\,\}$ is the set of all simple objects in $\mathcal{O}_{h,\zeta}(\mathbf{H})$.*

(b) *The set $\mathcal{X}_{h,\zeta}$ is identified with the set of pairs (x,χ) such that $\chi \in \mathrm{Irr}(A(h,\zeta,x))$ is a Jordan–Hölder factor of the $A(h,\zeta,x)$-module $H_*(\pi^{-1}_{h,\zeta}(x),\mathbb{C})$.*

(c) *The simple right $\mathbf{H}$-modules $\Psi^\bullet_{h,\zeta}(L_{x,\chi})$ and $\Psi^\bullet_{h',\zeta}(L_{x',\chi'})$ are isomorphic if and only if the triplets (h,x,χ) and (h',x',χ') are $\tilde{G}$-conjugate.*

References

[EGAI] Grothendieck, A., *Éléments de géométrie algébrique. I: Le langage des schémas*, Publ. Math. Inst. Hautes Étud. Sci. **4** (1960).

[EGAIII] Grothendieck, A., *Éléments de géométrie algébrique. III: Étude cohomologique des faisceaux cohérents*, Publ. Math. Inst. Hautes Étud. Sci. **11** (1962); ibid. **17** (1963).

[EGAIV] Grothendieck, A., *Éléments de géométrie algébrique. IV: Étude locale des schémas et des morphismes de schémas*, Publ. Math. Inst. Hautes Étud. Sci. **20** (1964); ibid. **24** (1965); ibid. **28** (1966); ibid. **32** (1967).

[SGA4] *Séminaire de géométrie algébrique du Bois-Marie 1963–1964. Théorie des topos et cohomologie étale des schémas* (SGA 4), dirigé par M. Artin, A. Grothendieck, J.L. Verdier, avec la collaboration de N. Bourbaki, P. Deligne, B. Saint-Donat, LNM **269** (1972), **270** (1972), **305** (1973), Springer.

[SGA6] *Séminaire de géométrie algébrique du Bois-Marie 1966/67. Théorie des intersections et théorème de Riemann–Roch* (SGA 6), dirigé par P. Berthelot, A. Grothendieck et L. Illusie, avec la collaboration de D. Ferrand, J.P. Jouanolou, O. Jussilia, S. Kleiman, M. Raynaud et J.P. Serre, LNM **225** (1971), Springer.

[1] Alonso Tarrio, L., Jeremiaz Lopez, A., Souto Salorio, M.J., *Localization in categories of complexes and unbounded resolutions*, Canadian Math. **52** (2000), 225–247.

[2] Beilinson, A., Drinfeld, V., *Quantization of Hitchin's integrable system and Hecke eigensheaves*, Preprint.
Available at http://www.math.uchicago.edu/~mitya/langlands.html.

[3] Bernstein, J., Lunts, V., Equivariant sheaves and functors, LNM **1578** (1994), Springer.

[4] Bezrukavnikov, R., Finkelberg, M., Mirkovic, I., *Equivariant homology and K-theory of affine Grassmannians and Toda lattices*, Compositio Math. **141** (2005), 746–768.

[5] Bökstedt, M., Neeman, A., *Homotopy limits in triangulated categories*, Compositio Math. **86** (1993), 209–234.

[6] Cherednik, I., *Double affine Hecke algebras*, London Mathematical Society Lecture Note Series **319** (2005), Cambridge University Press.

[7] Chriss, N., Ginzburg, V., *Representation theory and complex geometry*, Birkhäuser, 1997.

[8] Drinfeld, V. *Infinite-dimensional vector bundles in algebraic geometry: an introduction*, The unity of Mathematics, Progress in Mathematics, vol. **244**, Birkhäuser, 2006, 263–304.

[9] Garland, H., *The arithmetic theory of loop groups*, Publ. Math. Inst. Hautes Étud. Sci. **52** (1980), 5–136.

[10] Grojnowski, I., Garland, H., *Affine Hecke algebras associated to Kac–Moody groups*, preprint, arXiv:q-alg/9508019.

[11] Grothendieck, A., Dieudonné, J., *Éléments de géométrie algébrique*, I, Springer, 1971.

[12] Haiman, M., *Cherednik algebras, Macdonald polynomials and combinatorics*, International Congress of Mathematicians, Vol. III, European Math. Soc., 2006, 843–872.

[13] Hartshorne, R., *Residues and duality*, LNM **20** (1966), Springer.

[14] Jantzen, J.C., *Representations of algebraic groups*, 2nd ed., American Mathematical Society, 2003.

[15] Kac, V., *Infinite dimensional Lie algebras*, 3rd ed., Cambridge Univ. Press, 1990.

[16] Kac, V., Peterson, H., *On geometric invariant theory for infinite-dimensional groups* in *Algebraic Groups* (Utrecht, Netherlands, 1986), LNM **1271** (1987), Springer, 109–142.

[17] Kashiwara, M. *The flag manifold of a Kac-Moody Lie algebra*, Algebraic analysis, geometry and number theory, Johns Hopkins Univ. Press, Baltimore 1990.

[18] Kashiwara, M., Schapira, P., *Sheaves on manifolds*, Springer, 1990.

[19] Kashiwara, M., Schapira, P., *Categories and sheaves*, Springer, 2006.

[20] Kashiwara, M., Tanisaki, T., *Kazhdan–Lusztig conjecture for symmetrizable Kac–Moody Lie algebras, II. Intersection cohomologies of Schubert varieties* in *Operator Algebras, Unitary Representations, Enveloping Algebras, and Invariant Theory*, Progr. Math **92** (1990), Birkhäuser, 159–195.

[21] Kashiwara, M., Tanisaki, T., *Kazhdan-Lusztig conjecture for affine Lie algebras with negative level.*, Duke Math. J. **77** (1995), 21–62.

[22] Kazhdan, D., Lusztig, G., *Fixed point varieties on affine flags manifolds*, Israel J. Math. **62** (1988), 129–168.

[23] Kumar, S., *Kac–Moody groups, their flag varieties and representation theory*, Birkhäuser, 2002.

[24] Lipman, J., *Notes on derived functors and Grothendieck duality*, LNM **1960** (2009), Springer.

[25] Lusztig, G., *Bases in equivariant K-theory*, Represent. Theory **2** (1998), 298–369.

[26] Serpé, C., *Resolution of unbounded complexes in Grothendieck categories*, J. Pure Appl. Algebra **177** (2003), 103–112.

[27] Spaltenstein, N., *Resolutions of unbounded complexes*, Compos. Math. **65** (1988), 121–154.

[28] Thomason, R.W., *Algebraic K-theory of group scheme actions* in *Algebraic Topology and Algebraic K-theory*, Ann. of Math. Studies **113** (1987), 539–563.

[29] Thomason, R.W., *Lefschetz–Riemann–Roch theorem and coherent trace formula*, Invent. Math. **85** (1986), 515–543.

[30] Thomason, R.W., *Equivariant resolution, linearization, and Hilbert's fourteenth problem over arbitrary base schemes*, Adv. Math. **65** (1987), 16–34.

[31] Thomason, R.W., Trobaugh, T., *Higher algebraic K-theory of schemes and of derived categories*, in *The Grothendieck Festschrift*, III, 1990, 247–435.

[32] Varagnolo, M., Vasserot, E., *Finite dimensional representations of DAHA and affine Springer fibers: the spherical case*, Duke Math. J. **147** (2009), 439–540.

[33] Vasserot, E., *Induced and simple modules of double affine Hecke algebras*, Duke Math. J. **126** (2005), 251–323.

[34] Waschkies, I., *The stack of microlocal perverse sheaves*, Bull. Soc. Math. France **132** (2004), 397–462.

Michela Varagnolo
Département de Mathématiques
Université de Cergy-Pontoise
2 av. A. Chauvin BP 222
F-95302 Cergy-Pontoise Cedex, France
e-mail: michela.varagnolo@math.u-cergy.fr

Eric Vasserot
Département de Mathématiques
Université Paris 7
175 rue du Chevaleret
F-75013 Paris, France
e-mail: vasserot@math.jussieu.fr